FOUNDATIONS OF MATHEMATICAL BIOLOGY

Volume II

Cellular Systems

CONTRIBUTORS

MICHAEL A. ARBIB
JAMES S. BECK
ALDO RESCIGNO
ROBERT ROSEN

FOUNDATIONS OF MATHEMATICAL BIOLOGY

Edited by Robert Rosen

Center for Theoretical Biology
State University of New York at Buffalo
Amherst, New York

Volume II

Cellular Systems

ACADEMIC PRESS New York and London   1972

ACADEMIC PRESS, INC.
111 Fifth Avenue, New York, New York 10003

United Kingdom Edition published by
ACADEMIC PRESS, INC. (LONDON) LTD.
24/28 Oval Road, London NW1

LIBRARY OF CONGRESS CATALOG CARD NUMBER: 71-159622

PRINTED IN THE UNITED STATES OF AMERICA

To the memory of Nicolas Rashevsky
1899–1972

CONTENTS

LIST OF CONTRIBUTORS

Numbers in parentheses indicate the pages on which the authors' contributions begin.

MICHAEL A. ARBIB, Department of Computer and Information Sciences, University of Massachusetts, Amherst, Masschusetts (142)

JAMES S. BECK, Division of Medical Biophysics, Faculty of Medicine, University of Calgary, Calgary, Alberta, Canada (255)

ALDO RESCIGNO, Department of Physiology, University of Minnesota, Minneapolis, Minnesota (255)

ROBERT ROSEN, Center for Theoretical Biology, State University of New York at Buffalo, Amherst, New York (1, 79, 217)

PREFACE TO VOLUME II

One of the most effective strategies for the study of complex systems, such as biological organisms, is to attempt to decompose them into simpler units. These units should be simple enough to be effectively studied in isolation, and should be so meaningfully related to the properties of the overall system that the properties of the units will allow us to understand the behavior of the larger system. The present volume is devoted to the specification of such simpler units, to the manner in which such units can be studied as systems in their own right, and to a determination of how the properties of the units are reflected in higher-level organizations constructed from them.

By far the most important kind of unit for the study of biological systems is the cell. The essential assertion of the "cell theory," attributed to Schleiden and Schwann, is precisely that all organisms allow a natural decomposition into cellular units. That is, the cell provides us with a kind of anchor in the hierarchy of biological organization; on the one hand, we can understand the properties of cells in terms of the kinds of subcellular constituents, down to the molecular level, which were discussed in Volume I, and on the other hand, we can use the properties of cells to understand the physiology of organisms constructed from cells.

Four of the five chapters in the present volume are devoted to the property of cells in themselves, and in relation to the development of multicellular organisms. We consider a variety of apparently quite different approaches to cellular structure and organization. Chapter 1 is concerned with the relationship between the properties of individual cells and the general problems of morphogenesis in developing systems. Here the emphasis shifts from the properties of individual cells to how those properties *could* manifest themselves in morphological terms. The aim here is not immediately to provide a full explanation for a specific morphogenetic process, but rather to show that gross observable developmental phenomena are understandable and explicable in terms of the properties of their underlying cells. Thus, the intention of the material presented in this chapter is to provide a framework within which individual morphogenetic systems can be explained, rather than to directly develop such explanations.This being so, the material of Chapter 1 is further removed from direct experimental verification than is that of Chapter 2, and for this reason can be considered more speculative (though no more theoretical).

Chapter 2 is concerned with the regulation of individual chemical processes, and sequences of such processes, at both the genetic and metabolic levels. The point of departure here is that of chemical kinetics, so that the material of this chapter forms a natural extension of the material presented in Chapter 2 of Volume I. The relation of this work with experiment is exceedingly close; the phenomena of inhibition and activation, and of repression and induction (derepression) are among the best-studied parts of molecular biology. Essentially, this chapter contains very little speculative material; we are concerned here with exploring the consequences of well-documented phenomena by essentially classical means.

The mathematical tools applied in Chapter 1 are those of continuous-time systems (as indeed are those of Chapter 2): systems of rate equations, or first-order differential equations. This is far from the only framework which can be used to study these phenomena. In Chapter 3, we study much the same set of phenomena that were considered in Chapter 1, but we employ instead the mathematical theory of discrete-time systems—the theory of automata. In the theory of automata we are in a finite and discrete world, and thus are debarred from the comfortable tools of analysis; continuity, differentiability, and similar concepts are unavailable. However, a new class of concepts makes its appearance: computability, effectiveness and ease of computation, and others. It is most instructive to see how this apparently quite different set of mathematical techniques bears upon a familiar set of phenomena; but more than this, the application of automata theory suggests important relationships between the developmental systems modeled in these terms, and other kinds of biological systems that admit of such a representation (specifically, the neural systems treated at great length in Chapter 3 of Volume III). Moreover, a mathematical problem of great interest is suggested by the fact that, since the same set of phenomena is being modeled by two rather different sets of mathematical tools—in this case by dynamical systems theory and automata theory—there must be some relation, or correspondence, between the tools themselves. The relation between automata and dynamical systems, or between automata and control systems, suggested by this fact is itself a most interesting area of general systems theory research.

It should be noted that Chapters 1 and 3 of this volume require not only that individual cells should, by their properties, determine the nature of developmental processes, but conversely, that individual cells are affected by the developmental processes going on in the whole system to which they belong. This should be looked on as an exemplification of the kinds of two-way control in biological hierarchies which were discussed in Chapter 1 of Volume I.

In Chapter 4 we consider still another way of modeling cells, which is still further removed from direct experiment, yet which allows us to approach cellular processes from quite a different point of view than either automata theory or dynamical systems theory. The viewpoint here is entirely functional, whereas in our previous treatments we took an essentially structural one. The work is of course related to the previous treatments, but gives quite different kinds of insights, and raises many deep system-theoretic questions bearing on the relation between structure and function in cellular systems. This chapter should be compared with Chapter 2C of Volume III, which latter is concerned with much the same functional viewpoint.

Now the cell is a structural unit for larger organisms; we can see a cell physically, and separate individual cells from one another for further study. We can imagine other modes of decomposition of a complex system into subunits, which are not structural that is, which cannot be separated from the larger organism by physical techniques (see the discussion of "nonsubstantial" systems in Chapter 3 of Volume I). An important class of such nonstructural units is described in Chapter 5, which deals with the concept of a compartment. In this sense, a compartmental decomposition is somewhat more of an abstraction than a structural decomposition into cells; it is rather closely akin to the concept of *component* which is used in Chapter 4. Nevertheless, in spirit the purpose of compartmental analysis is exactly the same as the decomposition into cells, and should be considered in that light.

The general strictures which were set down in the Introduction to these volumes (Volume I) apply in particular to the chapters that follow. We have omitted much from this volume which might have been included. We have omitted, for example, detailed discussions of energetics in cellular pathways, and the role of membranes and other kinds of cellular organelles. These are, of course, important matters which bear on mathematical biology. But as stated earlier, good treatment of such matters already exist in the textbook literature, and we refer the reader to the Bibliography to the Introduction for an access to that literature. Moreover, much of the material which we have presented can easily be extended to include the kinds of material we have omitted. For instance, problems of permeability often fall within the formalism of chemical reactions, with a simple rate-constant playing the role of a diffusion coefficient. Much of specific transport mechanisms requires the synthesis of specific carriers, which are subject to exactly the same kinds of control as those we do discuss in detail. Indeed, the reader will get the most from this volume, and from the others, if he asks himself how apparently untreated phenomena of biology are related to the matters discussed, and if he reads other text material in this light also.

CONTENTS OF OTHER VOLUMES

Chapter 1

MORPHOGENESIS

Robert Rosen

Center for Theoretical Biology
State University of New York at Buffalo
Amherst, New York

I. Introduction

Problems of pattern and form dominate biology at all levels of organization. At the molecular level, it is the capacity of important macromolecules to assume a proper "conformation" which confers upon them the remarkable specificities and catalytic properties upon which the properties of organisms depend. Such conformations, manifested either in macromolecules or in organized complexes of macromolecules, also provide the basis for the recognition and selection mechanisms essential for the maintenance of multicellularity; receptor sites on cell membranes, binding sites on cell organelles, and antibody activity are some of the many examples which could be given. At higher levels of organization, the form of an animal or plant, and of its organs, is its most immediately apprehensible characteristic, and the relation of the form of an organism to its behavior or mode of life is a dominant

1

component of physiology on the one hand and an understanding of evolutionary processes on the other. The literature dealing with biological forms, going back to Aristotle, is in its sheer volume quite overwhelming.

On the other hand, except in the vaguest and most general terms, the problems of organic form have been essentially untouched by the enormous advances in biochemistry which characterize what is called modern biology. This is perhaps not surprising, since many of the important problems of pattern and form in biology pertain to higher levels of organization than those with which molecular biology deals, and, as has been repeatedly emphasized in these pages, problems connected with drawing effective inferences across levels of organization are among the most difficult and troublesome in science. Thus, to obtain some understanding of the problems connected with form and pattern in biology, the most effective strategy has proved to be: to begin with a phenomenological approach directly at the level of organization at which a particular form of interest is manifested, and then to attempt to relate this phenomenology to deeper aspects of organization; rather than to immediately attempt to derive aspects of organic form from the lowest organizational levels.

In the present chapter, we shall consider only one aspect of the many complex problems dealing with organic form. We shall be concerned specifically with problems of *development*—the problems involving the generation of form and pattern in biological systems. The principal experimental or observational root of such problems lies, of course, in embryology, which seeks to understand how the adult form of an organism is generated out of what is essentially a clone of cells derived by division from a single initial cell, the zygote. However, as noted above, similar problems arise at other levels of biological organization, and we shall use the word *morphogenesis* quite generally, to describe any process whereby a system changes in time so as to produce or manifest a particular kind of pattern of form. To the extent that this is possible, we shall leave aside such profound questions as the relation of form to function, and the hereditary or evolutionary implications of biological form; these are treated to some extent in Chapters 2, 3, and 4 in this volume. We shall be instead primarily concerned with the kinds of dynamical processes which are involved in the generation of form and pattern, and most particularly with an assessment of the *morphogenetic capabilities* of such processes; that is, with a determination of the kinds of patterns which a particular kind of dynamical process can generate. Such a procedure will, among other things, serve to sharply circumscribe the kinds of dynamical processes which could be responsible for the generation of a particular kind of biological form.

It was recognized relatively early that there were at least three kinds of processes which contribute, or can contribute, to any particular morphogenetic process. These processes have been given various names, but we can refer to them most succinctly as (1) morphogenetic movements, (2) differentiation, and (3) growth. For our purposes, we shall define them as follows:

1. *Morphogenetic movements* shall refer to patterns or forms generated by rearrangements or reshufflings of an array of units, without change either in the properties of the individual units themselves, or in their number.

2. *Differentiation* shall refer to patterns which can be generated by differential changes occurring within the units in an array, without any change in their number or arrangement.

3. *Growth* shall refer to differential increases (or decreases) in the number of units in an array, without any change in the properties of the units.

Clearly, real morphogenetic processes represent combinations of these elementary kinds of mechanisms. And equally clearly, these three elementary kinds of mechanisms cannot in reality be regarded as entirely independent of one another. For instance, as we shall see, changes in the properties of individual units (differentiation) may cause a change in the morphogenetic movements which the units are undergoing. Likewise, differential increases or decreases in the numbers of units of particular types can be regarded on the one hand as a kind of differentiation; on the other hand, it is clear that changes in numbers of units can affect both morphogenetic movements and differentiation. However, the strategy we are employing for the study of morphogenetic mechanisms involves, as stated above, the assessment of the capability of *each* of these elementary mechanisms, by itself, to generate a particular kind of pattern or form. As we shall see, this can readily be done in terms of mathematical models, or in terms of real experimental systems in which one or another of the elementary mechanisms predominates. In the course of this chapter and the next, we shall see many examples of real systems for which this is possible.

One of our main concerns will be not only with the generation of a particular kind of pattern, but also with its *maintenance* in the face of external perturbations. That is, we shall be concerned to a large extent with questions of *stability*. As we shall see abundantly, questions of generation of pattern and form are inextricably tied up with questions of maintenance or stability, and therefore our work in morphogenesis will necessarily have a variety of physiological connotations, most particularly to problems of *regeneration*. It was recognized rather early, in fact, that problems of embryology and problems of regeneration possessed many points in common; this will emerge most clearly from many of the mathematical models of morphogenetic processes to be discussed.

II. Morphogenetic Movements

A. Subunits and Subunit Assemblies

All biological structures are notable for the fact that they can be regarded as being constructed out of (and hence analyzable into) subunits or sub-assemblies of particular kinds. On the one hand, this assertion is quite trivial, since we may say that the subunits involved are simply the atoms of the chemical elements of which the organism is composed. The remarkable thing is that this idea of assembly of organic structure out of subunits persists at higher levels of organization. For example, all of the biologically important macromolecules are in fact polymers. In particular, the proteins are polymers formed from a set of approximately twenty different kinds of monomeric molecules, the amino acids; the nucleic acids are polymers formed from a set of four or five different kinds of monomeric molecules involving purine and pyrimidine bases; biological membranes may be regarded as a polymerized form of lipid, while many structural materials take the form of polysaccharides. Many biological organelles can be regarded as being polymers built out of subunits, each of which consists of a whole molecule which is itself of polymeric structure, while the whole of a multicellular organism can be regarded as decomposable into cellular subunits (this indeed is the major importance of the cell theory). The suggestion occurs most strongly that *the properties of the subunits determine the properties of the structures built out of these subunits*, and hence, since it is easier to study the subunits than the finished structure, an efficient strategy for studying any kind of biological organization is to find an appropriate family of techniques for decomposing it into a correspondingly appropriate class of subunits. Indeed, this is the basic approach common to all "reductionistic" approaches to the organism.

The construction of biological organization out of subunits that are themselves assemblies of still smaller subunits, on down to the atomic level, is one of the things which gives biological structures their pronounced hierarchical characteristics. Furthermore, the universality of this mode of organization in biology suggests strongly that there is something particularly "efficient" about such a hierarchical assembly procedure. In the course of this section we shall consider certain generalities about this circle of ideas, to set the stage for later applications.

Let us consider first how one goes about generating a large and complex structure, or polymer, out of a family of elementary units (monomers). There are basically only two ways in which this can be done, depending upon whether the monomers themselves are *passive* (that is, having no particular affinity for each other, and hence no capability for spontaneously generating more complex structures) or *active* (that is, having definite intrinsic affinities for each other, and hence a definite tendency for the formation of larger structures). In the passive case, if we are to succeed at all in assembling our monomers into polymers, it must be through the intervention of accessory structures. Such structures typically take the form of *templates*. Such templates may be regarded as containing additional "information" regarding the finished polymeric structure to be formed, typically through a specific binding capacity in its various regions for specific monomeric units. Thus, although the monomers have no particular affinity for each other, their specific affinities for different regions of the template confer a spatial order into the set of monomers. Once this spatial order has been manifested, there may be either a spontaneous formation of association between the ordered subunits, or else this association may be generated through the activity of specific catalysts, resulting in the formation of a polymer that mirrors, in a sense, the structure of the template. This, it will be recognized, is the qualitative picture which has emerged of the process of protein synthesis from "information" contained in a template of ribonucleic acid, and of genetic replication, where the template consists of deoxyribonucleic acid. In all such cases, it is the template, together with associated catalysts and activators, that provides the "information" required to organize the inert monomers into polymers with a specific structure and organization. Thus, in these cases, the structure to be formed is determined independently of the structure of the passive monomers themselves, by means of external sources of "information."

The other situation, that of active monomers, is in many ways more interesting. Here it is assumed at the outset that the monomers possess intrinsic binding affinities for each other, and hence that organized polymers will be formed spontaneously within the system. Furthermore, it is clear that the binding properties of the subunits will exclusively determine the kinds of polymeric structures which will be so generated. Thus, if we look at such a system of active monomers from the outside, it looks as if it is ordering itself and generating successively more and more complex structures. Since the generation of this order does occur spontaneously, without the intervention of external templates, catalysts, or other sources of information arising outside of the system, the behavior of such a system is often called *self-organization* or *self-assembly*. It is primarily with such systems with which we shall be concerned in the present chapter.

Let us consider some simple examples of systems composed of active subunits, which have the capacity to spontaneously generate particular kinds of organizations.

Case 1: Consider a well-shaken mixture of immiscible liquids, such as mineral oil, water, and mercury. The subunits here can be regarded as the individual liquid molecules, or as the small droplets of these liquids that are randomly intermixed. On allowing the system to stand, a definite organization will appear; the liquids will settle out of the system into three distinct layers. If we repeat such an experiment many times, we find that the same organization is always generated: the heaviest liquid will be on the bottom and the lightest on top, no matter now the system is initially shaken. This phenomenon of *phase separation*, in simple physical systems, is thus a good example of the spontaneous generation of a pattern or organization without the intervention of outside sources of information, such that the properties of the final pattern depend only on the intrinsic properties of the units of which the system is composed.

Case 2: Consider a newly formed polypeptide as it comes off a ribosome. This polypeptide must acquire a particular geometrical organization ("tertiary structure") before it can become active. It is typically supposed [Anfinsen, 1970] that this geometrical organization, or active conformation, can be generated without the intervention of accessory information, that is, without anything like a template to "tell" the polypeptide how to fold up. If this is so, then the acquisition of an appropriate active conformation by a newly formed, randomly coiled polypeptide is again an instance of what we have called self-organization.

Case 3: Finally, consider the generation of infectious particles of simple spherical or helical viruses, such as (in the best-studied case) tobacco mosaic virus. This virus consists of a helical coat of protein surrounding a core of infectious viral nucleic acid. It has been determined [for review see Caspar, 1963] that the protein coat consists of an assembly of identical protein subunits, packed or assembled in a particular way. It is these protein subunits which are produced through the activity of the viral genome. Moreover, the virus particle itself can be decomposed artificially, by a variety of means, *in vitro*, into the coat protein subunits and nucleic acid. Under restoration of physiological conditions, it is observed that these separated units reaggregate back into infectious viral particles, morphologically and functionally indistinguishable from native viral particles. Thus, here too we have a clear example of self-organization.

In all of these examples, drawn from different levels of organization (and other examples, developed below) we find a recurring theme. In all these cases, a system consisting of active subunits, with specific affinities for each other, will spontaneously assemble to generate an organization of a particular type, without the intervention of outside information. Thus, there is a sense in which it is correct to say that the "information" required to assemble the finished structure is contained in the affinities of the subunits for each other.

In intuitive terms, the active character of the subunits confers on structures built by a mechanism of self-assembly a remarkable stability, since disruption of all or a part of the structure will typically be followed by a repair or reconstitution of the structure. Moreover, the fact that this reconstitution or repair takes place without added information (since the information is intrinsic to the structure of the subunit) confers an added degree of simplicity and interest to such systems. In biological terms, there is also a considerable saving of "genetic information," since a separate activity of the genome is not required for the generation or regeneration of the structure involved.

B. Prerequisites for Self-Assembly

Having a variety of examples in hand, we can now inquire what it is, in each specific case, that causes the final pattern or organization to be generated. From such specific information, we may hope to generalize, to obtain a few basic principles or concepts which govern all cases of self-assembly, even those pertaining to quite different levels of organization.

All of the cases we have described above may be regarded as falling within the conventional scope of the physics and physical chemistry of the systems considered. In each case, without going into specifics, we find that the specific affinities of the subunits is of a conventional electrostatic or covalent type. In each case, with every state of the system (that is, with every geometric arrangement of the subunits in space) there is a definite number that can be associated with the system, representing a *free energy* of the system. The physical character of this free energy differs from system to system, but in each case the free energy may be regarded as the energy available for binding between the subunits which is not so employed when the subunits are in the geometrical arrangement which determines the state. It is this free energy which, in a certain sense, drives the system; if the units are free to move, they will modify their geometric positions (that is, the system will change state) until a state of *minimum free energy* is thereby produced. Once the system has arrived at this state of minimal free energy, there will be no further change of state; therefore, this minimal free energy state will characterize the final structure or organization which the system can produe. Thus, in Case 1 of the immiscible liquids, it follows from the theory of phase separation in systems of this kind that the final separated state is in fact the state of minimal free energy. In Case 2 of folding of a polypeptide chain, although the definition of free energy is different in this case, the final active conformation is again a state of minimal free energy. Likewise in Case 3, although the subunits involved and the bases of their affinity are likewise different from Case 1 or 2, the final virus particle is again a state of minimal free energy for the system [Caspar, 1963].

There are thus three aspects which are involved in all of the examples of self-organization that we have considered, and which we may suppose, in a preliminary way, will enter into any discussion of a self-organizing system in which a pattern is generated from a population of active subunits:

1. The subunits must certainly possess a particular kind of differential affinity for each other.

2. Each distribution of the subunits in space (that is, every state of the system) defines a number, which plays the role of a *free energy*, and which measures the energy available for subunit binding which is not so employed in the geometry in question.

3. The subunits must be free to move about, so that the system as a whole has the capacity of exploring a variety of "neighboring" states.

If these conditions are satisfied, whatever the nature of the subunits involved, or of their affinity for each other, the system as a whole will automatically change state until a state of minimal free energy is arrived at. As noted above, this change of state is in a sense *driven* by the excess free energy. Since it involves changes in the relative geometric positions of the subunits with respect to each other (that is, it involves the *motility* of the subunits in an essential way) there is a sense in which it is correct to say that the final pattern is generated through a process of *morphogenetic movements*.

C. The Efficiency of Subassembly Processes

Before returning to the specifics of the systems in which we shall be interested, there are a few more generalities regarding the assembly of systems out of subunits which are of interest. We have seen above that the specificity of association between active subunits is one of the characteristics which determines the kinds of patterns or organizations which can be generated from a system composed of those subunits. Biological specificities, while quite remarkable, are not perfect; therefore, it is of interest to inquire into the efficiency of a self-assembly process, when there is a finite probability of error introduced into the association of the subunits. It turns out that this discussion will give another insight into the efficiency of self-assembly processes, which may point up why these processes are so universal in biology.

The prototype for this discussion is the argument of Crane [1950] regarding all kinds of hierarchical assembly processes (whether active or passive, in fact) in which the structures appearing as assemblies from one production stage are then treated as subunits for the next assembly stage. In this context it is best to think of Case 3 above, of the assembly of viral coat protein, as a specific prototype; this involves at least two assembly stages (the first of which is passive, or template-mediated, resulting in the assembly of coat protein subunits, and the second of which is active, involving the self-assembly of the coat protein itself).

Let us begin by recapitulating the arguments of Crane [1950] that are relevant to this aspect. He assumes the following:

1. We are given a pool or population of identical fundamental units (say 10^6, to take his figure) out of which our final finished structures are to be built; these units may be regarded as the output of the 0th subassembly stage.

2. At the first assembly stage these units are going to be joined together in groups of 10, forming a subassembly 10 elementary units long (which we shall refer to as a decamer). At the second assembly stage, 10 of these decamers are assembled into a structure (a decadecamer) 10^2 elementary units long. At the third and final assembly stage, 10 decadecamers are assembled into the finished structure, 10^3 units long.

3. At every stage, there is a probability that a unit from the previous stage will be added wrongly into the subassembly being constructed at that stage. We shall assume that an assembly in which such a mistake has been made is incapable of being incorporated into the next stage and, to make the discussion precise, that all the units (in this case 10 at each stage) which should have been incorporated at the stage in question are wasted. It is further assumed that the same error probability occurs at each subassembly stage.

Strategy A. To fix ideas, let us consider Crane's example in detail. We have already described the subassembly processes. We now specify that the probability of making an error in putting two units together at any stage is

$$1 - p = 0.01.$$

Thus, the probability p that two units will be put together correctly is 0.99, and the probability that an error will occur in putting together a structure 10 units long is

$$p' = 1 - (0.99)^{10} = 0.0955.$$

At the first assembly stage, we have 10^6 units, to be assembled into 10^5 decamers. The number of defective decamers is thus

$$100,000p' = 9550$$

so that the total number of correct decamers, which can enter into the next subassembly stage, is $100,000 - 9550 = 90,450$.

At the next stage, we wish to assemble these 90,450 decamers into 9045 decadecamers. Since p' is the same at this stage as at the first stage, the number of incorrect decadecamers is thus

$$9045p' = 864$$

(to the nearest integer), and the number of correct decadecamers actually formed is thus $8181 = 9045 - 864$. At the final stage, the reader may verify that the total number of correct finished units of length 1000 is 740 out of a possible 1000.

Crane used this argument to prove that the strategy of subassembly is far more efficient than trying to synthesize our 1000-unit long structure directly by joining 1000 of our elementary units together. Under our assumptions, the probability that an error will be made in assembling such a structure is

$$1 - (0.99)^{1000} = 0.99996$$

from which it follows that, instead of getting 740 finished perfect 1000-mers, we will be doing well to get a single one.

However, the question now arises as to how efficient is it possible to make this subassembly strategy; that is, how can we choose the number of subassembly stages and the number of units to be put together at each stage, in such a way as to maximize the yield of the last stage.

Strategy B. For instance, to build the same 1000-mer considered above, we might consider the following procedure:

1. At the first stage, assemble 5 basic units into a pentamer.
2. At the second stage, assemble 10 pentamers into a decapentamer.
3. At the third stage, assemble 10 decapentamers into a decadecapentamer.
4. At the fourth and last stage, assemble two decadecapentamers into a finished 1000-mer.

This procedure requires one more assembly stage than the previous one, but by repeating the above calculation the reader may verify (using the same probability p as before) that the number of correct 1000-mers produced in this case will be 762. Hence, this procedure is more efficient than the preceding, if efficiency is measured solely by the number of correct finished particles produced.

Let us see if we can now provide a general procedure for determining how to construct the most efficient subassembly process, and the extent to which this procedure can be generalized to more realistic situations. Let us therefore suppose the following:

    i. The number of subassembly stages is to be N.
    ii. The number of units produced at the $(i - 1)$st stage which are incorporated into a single unit produced at the ith stage is r_i.
    iii. The length of the finished particle, the output of the Nth stage, in terms of the elementary units, is L; that is, $L = \prod_{i=1}^{N} r_i$.
    iv. The initial pool of fundamental units has size M.
    v. All of Crane's assumptions are satisfied; the probability of putting two units together without error is always the same and is denoted by p.

Under these assumptions, the number of correct units produced at the ith subassembly stage is

$$v_i = p^{r_1 + \cdots + r_i}(M/r_1 r_2 \cdots r_i) \tag{1}$$

and the total number of finished correct particles is just

$$v_N = {}^{r_1 + \cdots + r_N}(M/L).$$ (2)

We want to choose the $N + 1$ numbers $N, r_1, \ldots, r_N$ such that (for fixed p, L, and M) the expression (2) is maximal. Looking at this expression, we see that this amounts simply to maximizing the term

$$p^{r_1 + r_2 + \cdots + r_N}$$

or, since $p < 1$, to minimizing the exponent $r_1 + \cdots + r_N$.

The problem is now one in elementary number theory. For the numbers $r_1, \ldots, r_N$ are by definition a factorization of the fixed number L, and N is the number of terms in the factorization. So what we want to do is to minimize the sum of the factors of a fixed number. By the unique factorization theorem, we can write

$$L = \pi_1^{\alpha_1} \pi_2^{\alpha_2} \cdots \pi_k^{\alpha_k},$$ (3)

where $\pi_1, \pi_2, \ldots, \pi_k$ are the prime factors of L and the corresponding exponents are the number of times each prime factor occurs.

Theorem 1. Except for one trivial case in which equality occurs, the following relation is always satisfied: If $L = r_1 r_2 \cdots r_N$ is *any* factorization of L, then

$$\alpha_1 \pi_1 + \cdots + \alpha_k \pi_k < r_1 + \cdots + r_N;$$

that is, the decomposition of L into powers of primes is the factorization with the property that the sum of the factors is minimal.

PROOF: Let $L = \pi_1^{\alpha_1} \pi_2^{\alpha_2} \cdots \pi_k^{\alpha_k}$. By unique factorization, any other factorization can be written

$$(\pi_1^{\alpha_1 - u_1} \pi_2^{\alpha_2 - u_2} \cdots \pi_k^{\alpha_k - u_k})(\pi_1^{u_1} \pi_2^{u_2} \cdots \pi_k^{u_k}).$$

Let $P = \pi_1^{u_1} \pi_2^{u_2} \cdots \pi_k^{u_k}$. For the first factorization, the sum of factors is

$$\pi_1 \alpha_1 + \cdots + \alpha_k \pi_k.$$

For the second, the sum of factors is

$$(\alpha_1 - u_1)\pi_1 + \cdots + (\alpha_k - u_k)\pi_k + P$$
$$= (\alpha_1 \pi_1 + \cdots + \alpha_k \pi_k) - (u_1 \pi_1 + \cdots + u_k \pi_k) + P.$$

So all that is needed is to show

$$\pi_1^{u_1} \cdots \pi_k^{u_k} - (u_1 \pi_1 + \cdots + u_k \pi_k) \geqslant 0,$$

which is evident; the $u_i \geqslant 0$ and the $\pi_i \geqslant 2$ are integers. The only trouble occurs when $k = 2$, $\pi_1 = \pi_2 = 2$, $u_1 = u_2 = 1$; then

$$2 \cdot 2 - 2 + 2 = 0.$$

It is clear from the proof how equality can arise if $4 = 2^2$ is a factor of L.

From this, we can extract all the information we need to solve our problem. For example, in the case of the 1000-mer for which two different subassembly strategies were considered in detail, we find that for the first strategy we have used the factorization

$$1000 = 10 \cdot 10 \cdot 10$$

so that, according to our formula (2), the total number of correct particles is

$$(0.99)^{30}(1,000,000/1000) = 740,$$

while according to the second factorization

$$1000 = 5 \cdot 10 \cdot 10 \cdot 2,$$

the number of correct particles, from (2) is

$$(0.99)^{27}(1,000,000/1000) = 762.$$

On the other hand, our theorem tells us that the best procedure is to use the prime factor decomposition

$$1000 = 2 \cdot 2 \cdot 2 \cdot 5 \cdot 5 \cdot 5$$

which gives six subassembly stages, and yields as the number of correct finished particles

$$(0.99)^{21}(1,000,000/1000) = 810.$$

Note that it makes no difference how we distribute the r_i among the subassembly stages in any subassembly process. Thus, in the case considered above, the strategies in Table I are equally efficient in the construction of a 1000-mer.

TABLE I

Stage	Strategy A	Strategy B
1	Dimerize elementary units	Pentamerize elementary units
2	Dimerize stage 1 dimer	Pentamerize stage 1 pentamer
3	Dimerize stage 2 tetramers	Pentamerize stage 2 25-mers
4	Pentamerize stage 3 octamers	Dimerize stage 3 125-mers
5	Pentamerize stage 4 40-mers	Dimerize stage 4 250-mers
6	Pentamerize stage 5 200-mers	Dimerize stage 5 500-mers

In other words: On the basis simply of the assumptions made so far, we are able to characterize the number N of optimal subassembly stages, but although we can determine the numbers r_i, we are not able to determine the order in which the r_i occur. Hence, we cannot uniquely characterize the optimal subunit sizes; note that Strategies A and B above do not produce any subassemblies of the same size. Thus, some other constraint is required in order to determine the absolute size of the subassembly produced at each stage.

Note also the curious fact that if L is already prime, then there are no subassembly processes (in the sense defined) to construct it efficiently.

Another curious fact worth noting is that the sum of the factors of a number is not a monotonic function. For instance, the sum of the factors of 26 $= 2 \cdot 13$ is 15, while the sum of the factors of 27 $= 3 \cdot 3 \cdot 3$ is 9. Thus, we can assemble 27-mers far more efficiently than we can assemble 26-mers, even though the 27-mers are longer; at least in terms of the strategies developed so far. We shall return to this point shortly, however.

There are two ways in which the simple considerations we have provided must be generalized.

1. We must introduce different probabilities of error (that is, different specificities) at each subassembly stage, incorporating at least the physical reality that subassemblies evolve new binding patterns different from the elements out of which they are composed.

2. We must allow subassemblies to be constructed out of nonidentical units. All the above considerations involve identical families of units produced at the previous subassembly stage. It is this restriction that makes it impossible to assemble a finished structure with a prime number of elementary units.

Also of general interest is the observation that the strategy of using subassembly procedures in the construction of large particles, in order to overcome the lack of perfect specificity in the combination of the subunits, ultimately relies on *redundancy*, and may be formally regarded as analogous to the coding used in information theory, to transmit messages through noisy channels. There does not seem to be any obvious way to make this analogy precise, but if it could be done, then the tools of coding theory could be brought directly to bear on the problems of the synthesis of biological structures.

As regards the first generalization, it is quite easy to write down the expression analogous to (2) above for the case in which different subassemblies can be combined with different specificities (that is, different probabilities of error in incorporation). Quite generally, let us write $p_{i,j}$ for the probability that j units of length i (in terms of the original fundamental units with which we started) are assembled without error. It is reasonable to assume that

$$p_{i,j} = (p_{i,1})^j \tag{4}$$

and that $p_{1,1}$ is what we called p before. Thus, if the $p_{i,1}$ are equal for each i, we are back in the previous case.

In general, if we are trying to assemble a unit of total length L and if $L = r_1 r_2 \cdots r_N$, the required probabilities are

$$p_{1,r_1}, p_{r_1,r_2}, \ldots, p_{r_{N-1},r_N}.$$

It is important to note that these will be different, depending on the order in which we take the factors. Then using Eq. (3), the formula for the number of correct units of length L that can be formed from a pool of elementary units of size M is just

$$(p_{1,1})^{r_1}(p_{r_1,1})^{r_2} \cdots (p_{r_{N-1},1})^{r_N}(M/L), \qquad (5)$$

which is to be maximized over all possible subassembly schemes. If the probabilities $p_{i,j}$ are all independent, this formula is not easy to apply, and we cannot in general obtain information about preferred schemes. All we can say in general is that we will prefer schemes with the greatest specificities (that is, probabilities $p_{i,j}$) raised to the smallest exponents, and that it is the independence of these probabilities which will enable us to state, for instance, that Strategy A is to be preferred over Strategy B in the assembly of 1000-mers discussed above.

Let us now turn to the second generalization, which involves strategies incorporating units of different sizes at a particular subassembly stage. For instance, instead of constructing our 1000-mer by a linear procedure, as in our previous considerations, we might imagine the 1000-mer being formed by assembling five 100-mers and four 125-mers at the final stage. In this case, the assembly process will look like a branched graph, as in the diagram below:

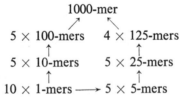

Since all our previous considerations apply to the linear parts of such an inverted tree, a moment's thought will convince the reader that there is no advantage to be gained by using such a branched strategy in the assembly of a structure like our 1000-mer. However, there are situations in which a branched strategy is the most effective. For instance, if we should want to assemble a structure of length L where L is a prime number, the most effective procedure is to assemble the structure of length $L - 1$ (which is never a prime if L is prime) and simply add a single unit. Moreover, such branched assembly strategies provide an effective means of overcoming the peculiar difficulty mentioned earlier, that it is often far more efficient to construct a longer structure (the sum of whose prime factors is small) than a shorter structure (the sum of whose prime factors is large). The nonmonotonicity of the sum of the factors arises from the fact that we can have large prime factors in the smaller number, and the only way to assemble subunits of this length is by accretion. But as we have just seen, if we allow branched strategies, we can efficiently construct units of prime length by such a strategy (a combination of subassembly and accretion). We thereby regain our intuitive feeling that the shorter a structure is, the more efficiently we should be able to assemble it.

Finally, it should be noted that the introduction of independent specificities for subassembly stages may in general give us a reason for preferring branched strategies over linear ones. For the sake of illustration, if for some specific situation we can form 130-mers and 45-mers by linear strategies far more efficiently than we can form any other substructures, it may easily turn out that the best way to form 1000-mers is to incorporate seven 130-mers and two 45-mers at the final stage of the assembly process. Thus, the study of the actual strategy employed in the synthesis of a structure out of its elementary units should give some insight into the relative specificities of the subassembly stages involved over possible competing strategies. Note, however, that we have ignored such things as competition of pathways for the basic subunits, the existence of degradative processes which may selectively attack particular kinds of structures, or which may be of a general character and so favor strategies with small numbers of subassembly stages, and many other factors which enter into real (particularly biological) assembly processes. Nevertheless, it is felt that the above considerations are of interest in setting general guidelines for the study of efficient assembly in biological systems.

III. Kinetic and Geometric Aspects of Self-Assembly

A. Spherical and Helical Viruses

As we have already noted, many biological structures may be regarded as being built from sets of identical monomeric units by a process of self-organization or self-assembly. Perhaps the best-studied among these are the spherical and cylindrical viruses, or their coats, especially tobacco mosaic virus [Caspar, 1963; Caspar and Klug, 1960, 1963]. The same principles seem to be involved, however, in the formation of other functionally active aggregates, such as hemoglobin [Perutz et al., 1965] or any of the many allosteric enzymes which have been studied [Monod et al., 1965]. In the present section, we shall study some overall properties of this phenomenon, with a particular view towards the self-assembly of viruses.

Let us first begin with some very general considerations. It was recognized very early that processes of self-assembly very much resemble crystallization.

Indeed, crystallization is an example of the class of phenomena generally regarded as *phase transitions*, which we have already noted above as exhibiting many of the properties of self-organizing systems. Therefore, it is well at the outset to consider both similarities and differences between ordinary crystallization and the formation of polymeric structures such as viral coats.

As with all self-organizing systems, the crystal organization represents a state of minimal free energy for its constituent subunits. Indeed, it is from this alone that the periodic structures so characteristic of crystals immediately follows. For if the geometric arrangement of the subunits (in other words, the manner in which the subunits are associated or bonded with each other) varied from one part of the crystal to another, there would be some regions of the crystal having lower free energy, and therefore the crystal as a whole would not be in a minimal free-energy state. From this it follows that, in the true state of minimal free energy (under a given set of external conditions, of course), *each unit of the crystal will be in an equivalent environment with every other unit.* That is, if an observer is placed at any particular point within one unit in the crystal, and is then moved to the corresponding point in another unit, he could detect no difference whatsoever in the local environments observed from the two points. This is the origin of the concept of the *unit cell* of a particular crystalline structure; a volume whose contents are periodically repeated by translations in all spatial directions. It should be observed, however, that the unit cell need bear no particular relation to the subunit of which the crystal is physically composed. This is shown in Fig. 1: the repeating structure is composed of circular units, while the unit cell may be taken as the trapezoidal region connecting the centers of four contiguous circles. In the structure shown in Fig. 1, the *crystal lattice* is obtained by translations of this unit cell in all directions, and it is easily verified that an observer placed at equivalent points in any two cells of the lattice will observe identical local environments; the lattice itself is shown in Fig. 2.

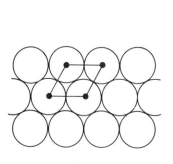

Fig. 1

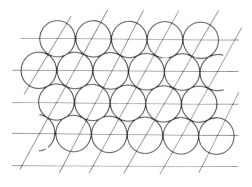

Fig. 2

The typical crystal is characterized by the fact that the bonding properties of the subunits are *isotropic* in space, so that new subunits can be added on to the structure in all directions. Thus it is that crystals are *indefinite in extent*; they have no fixed size, but are capable of growth to an arbitrary size. This is clearly not the case with the kinds of structures we are considering, which are of biological origin; they grow to a fixed size, past which no further subunits may be added to the structure. This points up one essential distinction between crystallization and the self-assembly of biological structures out of subunits, and as we have just noted, it may be regarded as consisting of an *anisotropy* in the way in which the units may combine with each other in space.

It is clear that there is a close correspondence between the particular geometrical arrangement of the subunits, mirrored in the unit cell and the lattice derived therefrom, and the specific bonding properties of the subunits. Thus, a great deal about specific bonding properties can be inferred from a study of the geometric constraints which arise from particular kinds of arrangements; indeed, much of mathematical crystallography, including the specification of the 230 different kinds of symmetrical structure which delimit crystalline organization, were well known long before any physical insight was available into the molecular forces responsible for generating this organization. Thus, we can learn a great deal simply from geometrical considerations, and then proceed to sharpen our insight into the way in which structures are assembled by means of specific energetic or bonding arguments. This is the procedure which we shall adopt.

Let us begin by seeing how it might be possible to generate closed structures, instead of the indefinitely extensible structures characteristic of ordinary crystals. Let us suppose that we have a subunit of the type shown in Fig. 3 and that such a unit is capable of polymerization through the formation of A–B bonds as shown in Fig. 4. If we think of this bond as nondeformable, only indefinitely long one-dimensional chains can be generated. Let us suppose, however, that this bond can be deformed, and that a deformation of the bond through an angle θ requires a certain amount of energy of deformation $E(\theta)$. Clearly the energy of deformation $E(\theta)$ will be a monotonic increasing function of θ; the exact form of $E(\theta)$ will depend on the physical nature of the bond, and hence should be in principle calculable through quantum mechanics.

Now let us try to fold up a linear polymer of N monomers into an N-gon as shown in Fig. 5. For a regular N-gon, the angle θ of deformation of a single bond, as shown in the figure, is of the order of $2\pi/N$. Therefore, the total energy of deformation for the configuration of Fig. 5 is $NE(2\pi/N)$. This number represents, in a sense, the "energetic cost" which must be expended in going from the open-chain configuration of Fig. 4 to a closed-chain configuration as in Fig. 5.

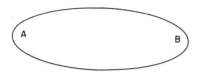

Fig. 3

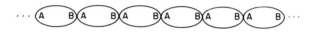

Fig. 4

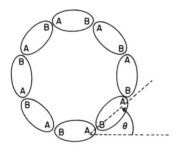

Fig. 5

If the energy of formation of a single A–B bond is E', then so long as

$$E' < NE(2\pi/N)$$

it would clearly "cost" more, in energetic terms, to close up the figure by forming an A–B bond between the two terminal subunits than would be gained by diminishing the free energy of the configuration by the formation of the bond. If, however, it should happen that the function $E(2\pi/N)$ goes to zero (as $N \longrightarrow \infty$) faster then $1/N$, then there will certainly be a value N_0 such that, for this N_0 (and all larger N), we will have

$$E' > N_0\, E(2\pi/N_0).$$

Under these conditions, the free energy decrease resulting from closing up the structure by forming an A–B bond between the two end subunits is greater than the energy which must be expended to deform the bonds of the structure. In this case, then, it is clear that the growing polymers will tend to close up into structures of a definite size, to which no new monomeric units can be added. This situation is characterized by the fact that each monomeric subunit accepts an environment (that is, a bonding pattern) which is not locally optimal, in order that the entire system may thereby reach a state of lower free energy than would be possible with locally optimal bonding. This is a first indication of the curious "local versus global" tradeoffs which are so characteristic of phase transitions, and cooperativity in general.

The closed polygonal configuration, in which there are no more bonds available in the system, and hence to which no new monomeric units may be added, is often called *saturated*. This state differs from normal crystalline growth in that the latter always has bonds available to be formed with new monomeric units; that is, it is always *unsaturated*. Ultimately, as we have seen, the distinction between unsaturated growth and the possibility of saturation lies in the function $E(\theta)$, which measures the energetic "cost" which must be paid for the deformation of the bonds of the structure, and particularly the limiting properties of this function as θ becomes small. One further property of the saturated state may be noted; in common with crystalline arrangements, each of the subunits in the saturated structure lies in an equivalent environment with every other.

Let us now consider a rather more general kind of subunit. Consider the structure shown in Fig. 6. Suppose that this unit is capable of forming bonds of A–B type, and of C–D type, so that we can now form structures in two dimensions as shown in Fig. 7. If the A–B bonds are deformable as before, but the C–D bonds are not, then the preceding arguments show that, instead of the indefinitely extendable plane structures of Fig. 7, we will instead obtain structures of the kind shown in Fig. 8, that is, cylindrical structures which may be regarded as stacks of the saturated structures shown in Fig. 5. For this reason, structures of the kind shown in Fig. 8 will be called *stacked-disk structures*. These structures are saturated around the cylinder, but unsaturated in the direction of the long axis of the cylinder. Thus the radius of the cylinder, but not its length, will be fixed by the bonding properties of the subunits.

If the C–D bonds are slightly deformable, then the plane structure of Fig. 7 can be modified (at the cost of an energy expenditure) to yield plane configurations like those shown in Fig. 9. These can generate still other cylindrical structures, different from the stacked-disk arrangement of Fig. 8. These other structures will no longer possess cylindrical symmetry, but rather helical symmetry. They can be generated from the plane lattice shown in Fig. 9 by making any of the various bonds A_1–B_2, A_1–B_3, A_1–B_4, and so forth in the figure, instead of the bond A_1–B_1 leading to the stacked-disk configuration.

Clearly the energy of the various helical configurations will differ from the stacked-disk configuration by varying amounts, but the configuration obtained, say, from the bonding A_1–B_2 will be close to that of the stacked disk. It should be noted explicitly, though, that all helical arrangements also give equivalently related subunits.

In the situation we have chosen, the symmetry of the unit is responsible for the fact that the stacked-disk conformation is that of lowest free energy.

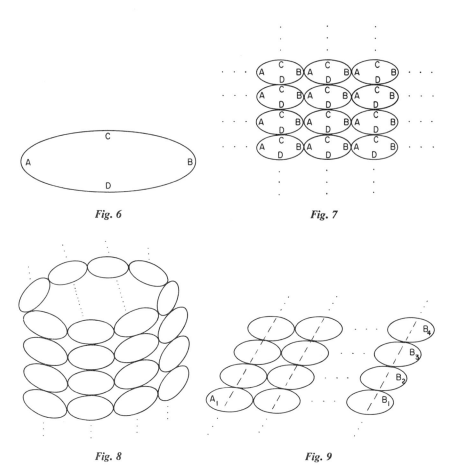

Fig. 6

Fig. 7

Fig. 8

Fig. 9

If we had chosen appropriately assymmetrical (enantiomorphic) subunits, we could equally well make one or another of the helical structures be the one which minimizes the total free energy of the system. For instance, we might choose a subunit shaped like that shown in Fig. 10. In this case it is clear that one of the helical structures will be preferred; which one will depend entirely on the local geometry (that is, the local bonding properties) of the subunit.

In any event, it is clear that in order to form cylindrical structures with cylindrical or helical symmetry, it is necessary that the subunits be capable of forming at least two different kinds of bonds (the reader may experiment for himself in trying to generate such structures out of subunits which can only form one kind of bond, say A–B bonds, with each other). In the examples we have used, two different kinds of bonds were employed, and each

Fig. 10

Fig. 11

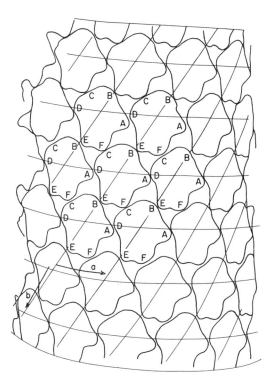

Fig. 12

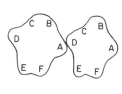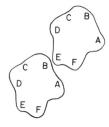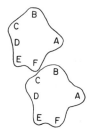

Fig. 13

subunit in the cylinder will have four nearest neighbors to which it is bonded. This is the simplest situation, but is by no means the only one. For instance, with the unit shown in Fig. 11 we can get the helical array shown in Fig. 12. In this arrangement three different kinds of bonds can be formed, and each subunit will have six neighbors.

The same considerations, of course, apply to three-dimensional units, which can by virtue of their bonding structure pack into three-dimensional cylindrical or helical shells. A striking illustration is shown in Fig. 13, taken from Caspar [1963], of the structure of the common strain of tobacco mosaic virus. Here we have a helical array in which each subunit is surrounded by six nearest neighbors, and indeed a plane cylindrical section around this figure would yield a two-dimensional structure very similar to that shown in Fig. 12.

Let us now, following Caspar, briefly consider the energetics in the aggregation of units like that shown in Fig. 11. As noted, these units can form three kinds of bonds with each other; namely A–D, B–E, and C–F bonds. Let us denote by P_n^b the aggregate of n subunits held together by b bonds (where each bond is, of course, of one of the three admissible bond types). Thus, there are three possible dimers, each containing only one bond, (that is, of type P_2^1), as shown in Fig. 13. It is readily verified that there are nine possible trimers, held together by two bonds (that is, of type P_2^2), and two possible trimers of type P_2^3, held together by three bonds; these possibilities are shown in Fig. 14a and b, respectively. Clearly the number of possible configurations of n-mers goes up very rapidly with n, while the number of bonds in an n-mer ranges from a minimum of $n - 1$ (in linear arrays) to an absolute maximum of $3n$ (in a cylindrical or helical array whose ends are bonded to form a torus).

In intuitive terms, the stability of each n-mer will depend on how many bonds are holding it together, and on the relative strengths of the bonds. We shall now consider the various possible equilibria configurations of subunits and n-mers, utilizing the simplest ideas of free energy changes for the various reactions involved, making the simplifying assumption that the energies of the three bond types involved are the same. Now the free energy change ΔF_n^b for forming an n-mer with b bonds can be written as

$$\Delta F_n^b = b\,\Delta f - n\,\Delta D_n^b - 2.3RT \log G_n^b,$$

where Δf is the negative free energy change in forming any bond, ΔD_n^b represents the distortion energy of the n-mer involved, and G_n^b is the number of geometrically distinct n-mers of type P_n^b (that is, in physical terms, the degeneracy of the corresponding energy levels of the n-mer). Now the equilibrium constant for the reaction

$$n\mathrm{P}_1 \rightleftharpoons \mathrm{P}_n^b$$

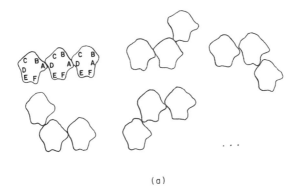

(a)

(b)

Fig. 14

is just

$$K_n^b = [P_n^b]/[P_1]^n$$

or

$$\log K_n^b = \log[P_n^b] - n \log[P_1] = -\Delta F_n^b/2.3RT.$$

It thus follows that

$$\log K_n^b = -(b\,\Delta f)/2.3RT - (n\,\Delta D_n^b)/2.3RT + \log G_n^b$$

and hence, if we put $[P_1] = 10^{-a}$, that

$$\log[P_n^b] = n[(b\,\Delta f)/2.3nRT - \Delta D_n^b/2.3RT - a] + \log G_n^b.$$

EXERCISE: Derive the corresponding expression for the reaction

$$P_n^b \rightleftharpoons P_n^{b+1}.$$

In particular, the change in free energy for the reaction $P_2^2 \rightleftharpoons P_2^3$ can be written as

$$\Delta F = \Delta f + RT \log \tfrac{9}{2},$$

where the factor $\tfrac{9}{2}$ represents the ratio of distinct noncyclic to cyclic trimer configurations. Thus

$$\log[P_3^3]/[P_3^2] = -\Delta f/2.3RT - 0.65,$$

and hence, if

$$\Delta f/2.3RT > 1.65$$

(that is, if $\Delta f < -2.3$ kcal mole^{-1} at $T = 300°$K), then the ratio of concentrations of cyclic to noncyclic trimer at equilibrium will be greater than 10. For tobacco mosaic protein subunits Caspar states that $\Delta f < -3$ kcal mole^{-1}, so that cyclic trimer will overwhelmingly dominate noncyclic trimer at equilibrium. Likewise, if we consider the reactions

$$P_2{}^1 + P_1 \rightleftharpoons P_3{}^3, \qquad 2P_1 \rightleftharpoons P_2{}^1,$$

then

$$\log([P_3{}^3]/[P_2{}^1]) = \log([P_2{}^1]/[P_1]) - \Delta f/2.3RT - \log \tfrac{9}{2}.$$

where $[P_1]$, $[P_2{}^1]$, $[P_3{}^3]$ are the concentrations of substances P_1, $P_2{}^1$, $P_3{}^2$, respectively. From this it follows that if the dimer concentration exceeds 1 % of the monomer concentration, then at equilibrium the concentration of cyclic trimer will exceed that of the dimer (using $\Delta f \sim -3.5$ kcal mole^{-1}). In other words, the negative free energy in forming cyclic trimer from monomer and dimer is greater than that involved in forming dimer from two monomers, and hence, under any conditions favoring aggregation, the cyclic trimer will predominate overwhelmingly at equilibrium over the dimers or linear trimers.

Let us now consider the relative stabilities of larger aggregates in a qualitative way (the quantitation of these arguments in terms of free energy changes is left as an exercise for the reader). First, it is clear that we need consider only those aggregates containing the maximal number of bonds. For the tetramers, therefore, we need consider only the cyclic forms $P_4{}^5$.

However, if we try to form such a cyclic tetramer from cyclic trimer and monomer, we do gain two new bonds, but we also increase the distortion energy. Therefore, the negative free energy change for the reaction

$$P_2{}^1 + P_1 \rightleftharpoons P_3{}^3, \qquad 2P_1 \rightleftharpoons P_2{}^1$$

will generally be less than for the reaction

$$P_2{}^1 + P_1 \rightleftharpoons P_3{}^3,$$

and hence the cyclic trimer will be preferred to (that is, be more stable than) the cyclic tetramer. Similar arguments may be applied to the cyclic pentamers and cyclic hexamers. Thus, under conditions favoring limited aggregation, the cyclic trimer will be the overwhelmingly predominating polymer at equilibrium.

The next larger aggregate which might be expected to have exceptional stability is the heptamer $P_7{}^{12}$, shown in Fig. 15. Repeating the same kinds of arguments we have made for the cyclic trimer, it follows that under condi-

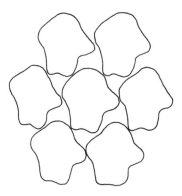

Fig. 15

tions which favor the formation of aggregates larger than the trimer, the cyclic heptamer of Fig. 15 will be favored over any smaller aggregate. And if we pursue these arguments to still larger aggregates, the next favorable aggregates are either stacked disks or helical.

Thus, we see a striking *quantization* in the kinds of aggregates formed from subunits under the various conditions favoring aggregation. This is a typical characteristic of subunits capable of self-assembly at all. However, it must be carefully noted that each favored form, such as the cyclic trimer or the heptamer of Fig. 15, represents a form of minimal free energy under a particular set of aggregation conditions; it must be carefully noted that these small stable aggregates, therefore, do not in general represent subassemblies used in the construction of the next larger units. Nevertheless, they arise under appropriate conditions because of the specific bonding properties of the subunits, and thus the kinetics of their formation must provide information related to the mechanics of formation of the ultimate structure of interest (in our case, the virus particle).

Let us now turn our attention to the possibility of formation of other kinds of structures by means of actively assembling subunits; in particular, the so-called *spherical viruses*, or "icosahedral viruses." The problem here, analogous to the one we treated for the cylindrical viruses, is how to pack (plane or solid) subunits so that they regularly cover the surface of a sphere, or form a spherical shell. The considerations of the ways in which this can be done rest ultimately on the general theory of *symmetry*. This theory is not intrinsically difficult, but it is complicated; to cover even that part of the theory which is applicable to the problem of constructing spherical structures out of subunits would require a proportionately far longer exposition than is possible in a volume of this kind. Fortunately, a variety of excellent texts are available, either of a pure mathematical character [for example,

Weyl, 1952; Toth, 1964] or crystallographic in emphasis. Nevertheless, to motivate the discussion which follows, we must say a few general words about symmetry.

In mathematical terms, symmetry is the study of those transformations of an object which bring it into coincidence with itself. There are three basic kinds of symmetry operations, from which all others can be constructed, namely *rotations, translations,* and *reflections.* An object possessing translational symmetry can be brought into coincidence with itself by means of a displacement in a particular direction and of a particular magnitude; such objects are of indefinite extent, such as an infinitely long ladder, or the kinds of crystal lattices shown in Fig. 2 (page 16). Clearly an object like a spherical virus, which is bounded in all directions, will not possess translational symmetries, so that we can exclude them, and all symmetry operations containing them (such as screw dislocations) from our considerations.

Symmetry operations involving reflections are those which interchange the "hand" of a particular configuration; a right-hand glove and a left-hand glove are related by a reflection. Now the amino acids of which protein subunits are built do possess a definite "handedness," owing to the presence of asymmetric carbon atoms, and shown by their optical activity with respect to polarized light. Because of this intrinsic "handedness" to protein molecules, it is possible to argue (as did Crick and Watson [1956], to which all of these approaches to small virus structure are essentially due) that structures which are symmetric via reflections could not be involved as possible viral structures, and hence that such symmetry elements likewise could be excluded.

This leaves rotations of two- or three-dimensional objects as the only symmetry operations which need to be considered for our purpose. If we consider an object like a wheel with n spokes, it is seen that any rotation about the hub of the wheel which is of magnitude $2\pi/n$ will bring the wheel into coincidence with itself. Thus an imaginary line through the hub and normal to the plane of the wheel, about which these rotations take place, and which itself is left invariant by the rotation, is called a *rotation axis;* because there are n distinct rotations about this axis that will carry the entire wheel into coincidence with itself, this axis is specified as an *n-fold rotation axis.*

A three-dimensional figure may, of course, have a number of distinct rotational axes. If we consider a cube, for example, it clearly possesses three four-fold axes, which are the lines joining the midpoints of opposite pairs of faces. It also possesses four threefold axes, which represent the diagonals of the cube (that is, the lines joining opposite corners; these indeed are considered characteristic of *cubic symmetry*) and in addition six twofold axes (lines joining the midpoints of opposite sides).

Let us suppose that we wish to cover the surface of a cube regularly with some arbitrary kind of unit, which may itself have no symmetry at all, and that we place one such unit down arbitrarily anywhere on one of the faces of the cube. Since the center of each face is a fourfold axis, this means that we must place down on that face three other units, so arranged that the first unit is carried identically onto each of the others by the successive rotations of the face itself, as shown in Fig. 16. Likewise, since there are six

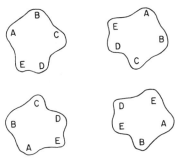

Fig. 16

faces, we must repeat this pattern symmetrically on each face of the cube. There must therefore be 24 copies of the initial asymmetric subunit, arranged in the manner indicated. Thus, the building of a cube out of asymmetric units can be accomplished; but it requires that we use exactly 24 of the subunits arranged in such a way that the symmetry elements of the cube (that is, the various two-, three-, and fourfold axes) carry these subunits onto each other.

The same thing can be done for a regular tetrahedron, which likewise is considered to have cubic symmetry, since it possesses four threefold axes (in addition to three twofold axes). Again, if we wish to cover the tetrahedron with identical asymmetric subunits, and place a subunit down anywhere on one of the faces, the highest symmetry elements (the threefold axes) require that another two subunits be placed on the same face in such a way that the first is carried onto each of the others by the rotations which carry the entire face onto itself. Since there are in this case four faces, we require exactly twelve subunits, of arbitrary symmetry themselves, but arranged in the symmetrical manner required to preserve the symmetry operations of the overall structure. In the case of a dodecahedron, in which the highest symmetry elements are six fivefold rotation axes, we require five units per face; since there are 12 faces, the total number of subunits required is 60.

Let us notice that, in all the cases discussed above, each of the subunits is equivalently related to all the others, in the sense of crystallography; that is, an observer placed at a particular point of one of the subunits, and then moved to the equivalent point in any other subunit, would be able to detect no difference in the environments seen from the two points. We can illustrate this, in the case of icosahedral symmetry, with the diagram shown in Fig. 17, showing 60 identical asymmetric polygons equivalently related, forming a quasi-spherical surface. The points labeled A, B, C in the figure locate five-fold, threefold, and fourfold rotation axes, respectively.

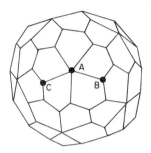

Fig. 17

Thus, we can see again a peculiar kind of quantization emerging in the construction of closed figures (in this case with cubic symmetry) out of asymmetric subunits. Regardless of the nature of the subunit, we cannot build, say, an icosahedral shell out of an arbitrary number of units; to preserve the symmetry properties we are restricted to a consideration of fixed number (in this case 60) of arbitrary units arranged in a particular way.

Although these crude arguments derived entirely from symmetry are in many ways satisfying, they raise one difficult problem. Namely, if we wish to get a larger icosahedral shell, we either need to employ subunits of a different size (but with the same number of subunits), or else give up the idea that the subunits are equivalently related. The biological evidence shows that there is a wide variation in the size of spherical viruses, but that the subunit size is in all cases comparable; hence, we must employ different numbers of subunits. It is possible to imagine that each of the subunits is itself built out of, say, n subunits; in this case we can predict that the number of subsubunits in an icosahedral shell, say, would have to be $60n$. Caspar and Klug [1962], on the other hand, have proposed that we obtain a far greater flexibility in considering the formation of closed symmetric shells from subunits if we weaken the restriction that the subunits be equivalently related. Instead of exact identity between corresponding subunit environments, they require only that, in some sense, the "average environment" seen from corresponding points on the subunits be the same. They call this notion "quasi equivalence," and have shown that, making plausible assumptions on the degree of quasi equivalence required, that there is essentially only one way in which the surface of a sphere may be covered with asymmetric identical subunits, and that this leads necessarily to icosahedral symmetry. This leads to very efficient methods of construction, essentially (as Caspar and Klug point out) the same as is used in the construction of "geodesic domes." Again, the arguments are rather too involved to be presented in detail here, but they may be illustrated in the following way.

It is well known, for example, that the surface of a sphere cannot be enclosed with hexagons, even when the hexagons are not all identical [see Thompson, 1961]. However, hexagons do cover a plane surface in a regular way. A typical hexagonal "paving" of the plane, in which all corresponding points are equivalently related, is shown in Fig. 18. Let us now distort the paving a bit, as shown in Fig. 19. It will be noticed that in each case the plane is

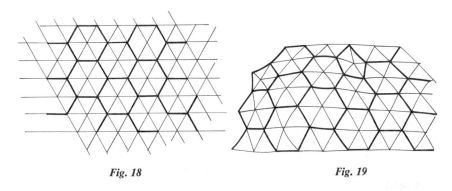

Fig. 18   *Fig. 19*

covered with triangles. In Fig. 18 these triangles are all congruent, and can be regarded as grouped equivalently into hexamers which cover the plane in a regular way. In Fig. 19, these triangles no longer are grouped exclusively into hexamers; we also find pentamers of triangles. Naturally the triangles are no longer exactly identical, nor exactly equivalently related, but the distortion involved in any one triangle from the corresponding situation in Fig. 18 is small. However, a paving configuration like that shown in Fig. 19 has the property that it may now be folded into a spherical (icosahedral) surface. This "closure" of the surface may be highly energetically favorable, for the same reasons we described in our discussion of cylindrical shells above. Thus, as before, a departure from strict local adherence to energetic requirements may result in a highly more favorable energetic situation or the configurations as a whole; this is another illustration of the "local-versus-global" tradeoffs that, as noted above, characterize such discussions. However, whereas in the cylindrical case we did not have to depart from strict equivalence of local encironments, the idea of quasi-equivalence allows the economical construction of closed shells of various sizes out of subunits of approximately the same size.

B. "Sorting Out" in Mixed Cell Populations

In this section, we turn to a class of self-organizational phenomena at still a higher level of biological organization, which plays a direct role in the morphogenetic processes occurring during development of multicellular organisms. Thus, we shall consider these systems in some detail. As will be seen, the formal treatment of these systems parallels in every way those we have already given; the only difference is in the nature of the units involved and the physical basis of the forces or interactions between them.

Let us begin by placing these phenomena into a historical perspective. The first relevant observations appear to be those of Wilson [1907], who worked with sponges, the simplest metazoan organisms. Wilson discovered that sponge tissues could be dissociated into their constituent cells by pressing the tissues through fine silk. If the dissociated sponge cells then are placed under physiological conditions, it is observed that within a day or so the cells have formed aggregates or clusters; after five or six days these initially random aggregates of dissociated sponge cells have reorganized themselves so as to generate the histological organization characteristic of the functional sponge from which the cells were initially derived. Thus, the Wilson experiments seem to have all the characteristics of a true self-organization: a system composed of subunits (cells) with the property that, when the organization of the subunits is artificially disrupted, the "information" required for the reconstitution of the structure inheres in the properties of the subunits themselves.

Naturally, these experiments of Wilson were quite striking, and this system was carefully studied by quite a number of other authors [for a review see Berrill, 1961]. Particular attention was paid to the behavior of interspecific mixtures of cells, taken from disaggregated sponges of different species. However, it turned out that the situation was not as simple as it first appeared. For one thing, the sponge organization, though comparatively of a simple kind, is in any absolute sense rather complex; many different cell types are involved (pinacocytes, archocytes, amebocytes, collencytes, collar cells, and so forth). Further, it was not clear whether the reconstituted structure was generated entirely through a simple reshuffling of cells of constant properties, or whether there was an associated dedifferentiation and redifferentiation of the cells themselves. For these reasons the sponge experiments, while suggestive, are not easy to interpret entirely in terms of morphogenetic movements leading to a reconstitution of the original morphology.

A much simpler and more unequivocal system was discovered by Holtfreter [Townes and Holtfreter, 1944], in the course of his work on "tissue affinities" in amphibian embryos. He discovered that a fragment of embryonic gastrula would, when placed in an alkaline medium, dissociate into its individual cells. On returning these dissociated cells to physiological pH, these dissociated cells would reaggregate into small spherical masses, in which the various cell types derived from the tissue fragment were intermixed at random. Thus, the behavior of the dissociated tissue fragment mirrored that observed in the behavior of dissociated sponge. However, in the case of embryonic cells the situation was considerably simpler. In the first place, there were only three cell types involved: ectoderm, endoderm, and mesoderm. Second, during the time scales involved, there was no question of dedifferentiation and redifferentiation in the individual cells. Finally, the different cell types in the embryonic tissue were clearly distinguishable from each other: the ectodermal cells are large and deeply pigmented; the mesodermal cells are smaller and lightly pigmented; the endodermal cells are smallest and unpigmented. The system thus comes equipped with natural visual markers with which the temporal course of the reaggregation process can be directly followed. Holtfreter thus could directly observe the reorganization of the initially random mass of cells into an ordered aggregate, in which all the mesodermal cells were located together in the interior of the mass, and the ectodermal and endodermal cells on the outside. This, of course, is precisely the geometrical arrangement of the same cells found in the normal gastrula.

Experiments of this type were repeated and extended by a variety of workers, with similar and perhaps even more startling results. Moscona [1965] found that the tissues of avian embryos could be dissociated into their constituent cells through the employment of trypsin; in a long series of experiments, he and his co-workers showed that the histology of complex organs, such as lung and kidney, could be strikingly approximated through the "sorting out," as it came to be called, of tissue masses formed by the initially random reaggregation of dissociated embryonic tissues. A variety of other experiments involving interspecific mixtures of cells, or systems in which the proportions of cells of various types were modified from the normal. were also carried out. An excellent review of all of these experiments, and the role of cell movements in normal morphogenesis, can be found in the monograph by Trinkaus [1969].

Naturally, it was a matter of great interest to elucidate the mechanism by which the original morphology was stubbornly reestablished in such cellular aggregates. For while the original morphogenesis of this morphology does not occur through a process of sorting out of a randomly intermixed mass of cells, it is evident that the forces responsible for the sorting out are also those which stabilize the morphology. Furthermore, it is at least curious that the initial morphogenetic mechanisms responsible for the original pattern, and those which govern sorting out (if they are different), seem always to determine the same kind of final structure. It is thus clear that a study of the response of cell masses to an artificial disruption can throw a great light on at least some of the forces responsible for real morphogenesis. Moreover, as is apparent from the above-mentioned monograph by Trinkaus, there are many morphological processes, in real morphogenesis, which do primarily involve the same kind of relative cell movements that must occur in the sorting-out process in artificially disrupted tissues.

The major advance towards an understanding of the behavior of intermixed populations of embryonic cells was made by Steinberg [1963]. This advance had two characteristic sides, one experimental and the other theoretical. On the experimental side, he recognized the artificial character of the experimental techniques by which sorting-out phenomena had been demonstrated up to that time, and had the insight to push this artificiality one step further. Specifically, instead of simply dissociating preformed embryonic organs and allowing the cells to sort out, he began to experiment with mixtures of cell types which never come in contact with each other during the normal course of morphogenesis. Such cells, he discovered, also sort out from one another, thereby generating what may be regarded as completely *artificial organs*. The use of these artificial organs, the composition of which can be carefully controlled, throws a great deal more light on the underlying mechanisms of sorting out than can be obtained from the original reaggregation experiments.

On the theoretical side, based on his experimental results with artifically generated mixtures of cells, Steinberg recognized the phenomena of sorting out as a kind of self-organization, behaving in accord with one's expectations for any population of active (that is, motile and differentially interacting) subunits. On this basis he was already able to give a good phenomenological accounting for his experimental work, which we now may summarize briefly. Let us suppose that we have a population of two cell types, which we may denote by a and b, respectively. Following Steinberg, let W_a, W_b, W_{ab} be the work done when a cell of type a combines with a cell of type a, when a cell of type b combines with a cell of type b, and when a cell of type a combines with a cell of type b, respectively. These "works of adhesion" thus are inversely related to the free energy of the system; the minimization of the free energy is correspondingly the maximization of the total "work of adhesion" in the system. There are now several possibilities.

1. If

$$W_{ab} > \tfrac{1}{2}(W_a + W_b),$$

that is, if the a–b adhesions exceed the average strength of a–a and b–b adhesions, then clearly the total work of adhesion of the system is maximized when the a and b units are alternately arranged in a so-called "checkerboard" pattern.

2. If

$$W_{ab} < \tfrac{1}{2}(W_a + W_b),$$

that is, if the a–b adhesions are weaker than either the a–a or b–b adhesions, then there will be a maximum for the total work of adhesion when the two populations are segregated from each other.

3. If in the preceding case we should have

$$W_b < W_{ab} < W_a,$$

then clearly the state of maximal total work of adhesion will be given when the morphology of the population consists of a sphere containing all the cells of type a (the more cohesive phase, since $W_a > W_b$) totally surrounded by the cells of type b (the less cohesive phase). If the works of adhesion were reversed, that is, if

$$W_a < W_{ab} < W_b,$$

then the morphology would be the same, except that the inner sphere would consist of the cells of the b population, surrounded by cells of the a population.

4. As in Case 2, if we have

$$W_{ab} < W_b < W_a,$$

then it is clear that the morphology which maximizes the total work of adhesion is given by a sphere of the more cohesive phase (the a cells, in this case), *partially* enveloped by the population of b cells. The degree of envelopment can be measured by the ratio W_{ab}/W_b; when this ratio is zero, the two cell populations fall apart into disconnected masses; when it is unity, we are at the limit of the case of complete envelopment. This situation is shown graphically in Fig. 20, taken from Steinberg [1963].

On the basis of this very simple phenomenological analysis, a number of predictions for the behavior of cell sorting systems can be made, of which one may be immediately mentioned. It follows from what we have said that the morphology of the states which maximize the work of adhesion behave like those of "thermodynamical equilibrium," which means that the same morphology will be produced by *any* such system, regardless of its initial configuration. This was strikingly confirmed with populations of two cell types (a and b) satisfying

$$W_b < W_{ab} < W_a.$$

The two initial starting configurations are shown in Fig. 21; both led as indicated to the predicted end configuration. This points to a kind of

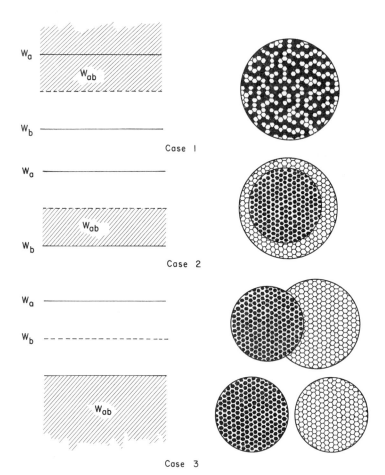

Fig. 20

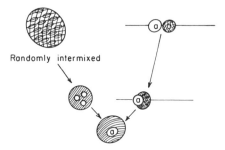

Fig. 21

global stability to the final morphology, which is indeed what one must expect of a sorting-out process, or indeed of any process of self-organization.

We are now going to embody the phenomenological ideas implicit in the experimental and phenomenological analysis given above in a concrete mathematical model, geared specifically to investigate the morphogenetic capacities of sorting out in the generation and maintenance of patterns. Although this particular model is geared to the study of sorting out, it should be borne in mind (a) that the basic method of approach can be readily modified to deal with other kinds of self-organizational problems, and (b) that there is a close relationship between these specific models and the notions of stability in arbitrary dynamical (for example, physicochemical) systems. The models are given in the simplest and most basic form; various generalizations and refinements are mentioned in the course of the development but will not be explored here; appropriate references are provided for these generalizations whenever possible.

For simplicity the models we will study initially shall consist of *two-dimensional* cells. Let us imagine a region of the plane (possibly infinite) tessellated or subdivided into equal squares. We can index these squares by means of an appropriate coordinate system in the plane. Let us further suppose that we are given fixed numbers $N_1, N_2, \ldots, N_r$ of cells of each of r types, and that each cell is assigned to a square in our grid. The assignment of cells to squares produces a configuration which we shall call a *pattern*. Any square in the grid that does not contain a cell will be assumed to contain medium, which if convenient can be treated as another cell type, designated by $r + 1$.

To each edge of contact between two cells, we assign an "affinity" λ which is a measure of the strength of binding between the cell types. Thus, λ_{aa}, λ_{ab} are the affinities between cells of type a and a, and between a and b, respectively, across the common edge separating them. It is assumed that the affinities between cells are independent of the orientation of the cells. Thus, this model is a two-dimensional analog of a system of isotropic cells. We also define an "E function," which is the sum of the number of edges multiplied by their λ values in a given pattern. This E can be regarded as the negative or reciprocal of the surface free energy, or directly as a work of adhesion in the sense of Steinberg. Thus, a maximum E will correspond to a minimum surface free energy.

We ultimately desire to study the way in which these patterns change with time. Such changes may be regarded as an interchange of cells between squares in the tessellation according to rules which define the nature of cell mobility. It is important to stress again that the model is purely phenomenological; no specific *physical* assumptions need be made either as to the physical nature of the affinities, or as to the mechanisms of cell motility.

Once the model is defined, it may be used to investigate various properties of cell sorting systems. Here we shall study three.

(1) ABSOLUTELY STABLE (OR MAXIMAL) PATTERNS: Given a finite number of cells, the number of possible patterns is finite. Given a certain choice of values of λ's, one or more of them will be configurations with maximum E. One problem, then, is to search for and define such patterns. These patterns, of course, are independent of any rules of cell motility. If all absolutely stable patterns and corresponding constraints on the values of λ's are identified, any other pattern will not be an absolutely stable one.

(2) LOCALLY STABLE PATTERNS: Realistically, a cell would not be expected to move arbitrarily from any position to any other. It would move initially to a neighboring site, constrained by the presence of other nearby cells. But, given certain motility rules, one configuration could be transformed into another through a series of such elementary cell moves. A configuration reachable from another by a single elementary move will be called a "neighbor" of the one from which it was derived. Given the concept of neighboring configurations, it is possible to visualize the existence of local maxima of E, or values of E of a configuration that are higher than all values of E of neighboring configurations. Such local maxima make a configuration locally stable, since a move away from it entails the passage through configurations of lower E. We discuss here the nature of locally stable patterns.

(3) THE PATH PROBLEM: It is intuitively obvious that, given some definite motility rules, not all configurations need be reachable from any arbitrary configuration, since the path from one to another might entail passing through a locally stable configuration. Here we discuss the motility rules which would allow a random pattern of cells to reach an absolutely stable configuration.

1. Absolutely Stable Patterns

We shall now determine absolutely stable patterns in terms of the corresponding constraints on λ's. For simplicity and to introduce the method and notation, let us first consider the case of two types of cells (which includes the case of one cell type and medium).

a. Patterns Involving Two Types of Cells. Let us denote the types of cells by b and w and let λ_{bb}, λ_{bw}, and λ_{ww} denote the λ's for bb, bw (wb), and ww edges, respectively. Let N_b and N_w denote the number of cells of types b and w, respectively.

To determine the absolutely stable patterns, we could proceed by enumeration: that is, by generating all the patterns with the same number of cells of the various (here two) types, counting the number of edges of various (in this case, three) types, finding the E function for each pattern, and thus determining which one is maximally stable for a given set of values of λ's. We find that this rather laborious procedure can be elegantly simplified by making use of the fact that since, in our simple model, the cells are nondeformable (or rigid), the number of edges of cells of each type are conserved. In other words, for any configuration

$$2N_{bb} + N_{bw} = 4N_b, \qquad (6)$$

$$2N_{ww} + N_{bw} = 4N_w. \qquad (7)$$

By definition the E function for a pattern is

$$E = \lambda_{bb}N_{bb} + \lambda_{ww}N_{ww} + \lambda_{bw}N_{bw}. \qquad (8)$$

Using Eqs. (6) and (7), Eq. (8) can be written in either of three equivalent forms:

$$E = N_{bb}\mu_1 + C_1 \qquad (9a)$$

$$= -\tfrac{1}{2}N_{bw}\mu_1 + C_2 \qquad (9b)$$

$$= N_{ww}\mu_1 + C_3, \qquad (9c)$$

where

$$\mu_1 = \lambda_{bb} + \lambda_{ww} - 2\lambda_{bw}, \qquad (10)$$

$$C_1 = 2\lambda_{ww}(N_w - N_b) + 4\lambda_{bw}N_b, \qquad (11a)$$

$$C_2 = 2\lambda_{bb}N_b + 2\lambda_{ww}N_w, \qquad (11b)$$

$$C_3 = 2\lambda_{bb}(N_b - N_w) + 4\lambda_{bw}N_w. \qquad (11c)$$

In Eqs. (9), C_1, C_2, C_3 are constants dependent on λ's and the number of cells, but independent of pattern. Therefore, for a given set of three λ's, that is, given μ_1, Eqs. (9) imply that it is enough to count the number of edges of only one type (say bb) for the purpose of determining the stability of a particular pattern. In writing the conservation relations, Eqs. (6) and (7), we need not consider the boundary of the tessellation, for the cells just external to the boundary *must* remain constant (otherwise no solution is possible). We set the boundary sufficiently distant to include all movable cells. Further, we consider the boundary to be made entirely of the predominant cell type since this leads to the most general mathematical expressions. Thus, the system of r cell types may be viewed as cells of $(r - 1)$ cell types embedded in an infinitely large mass of the predominant cell type.

If one is not aware of the simplification made by the conservation relations, Eqs. (6) and (7), one may say that, in general, one can have many absolutely stable patterns corresponding to the values we give to three λ's. But from Eqs. (9), we see that we can have at most three geometrically (or topologically) different absolutely stable patterns depending upon whether $\mu_1 \gtrless 0$. Thus, effectively, there is a reduction of three λ's (or energy parameters) to one "effective" λ (here μ_1). In general, for r types of cells, (including medium), there are $\frac{1}{2}r(r + 1)$ different kinds of edges (hence λ's) and there are r conservation relations. Thus the number of effective λ's, that is,

$$\text{number of } \mu\text{'s} = \tfrac{1}{2}r(r + 1) - r = \tfrac{1}{2}r(r - 1). \tag{12}$$

We may now list the possible patterns of absolute stability involving two cell types.

(1) "ONION" PATTERN: For $\mu_1 > 0$, from Eqs. (9) the absolutely stable pattern is the one for which N_{bb} and N_{ww} are maximum and N_{bw} is minimum, that is, the patterns in which one type of cells are closely packed and surrounded by cells of other types. Thus, for $\mu_1 > 0$, the absolutely stable structure is the "onion" structure previously described. If w are the predominant cells, b cells will be closely packed, and if b are predominant, w cells will be closely packed. It may be noted that in either of the two cases, the minimum value of N_{bw} is not zero.

(2) NO PREFERRED PATTERN: For $\mu_1 = 0$, E is constant and, therefore, all structures are equally stable. Note that this result is more general than the trivial one in which all the λ's are equal to each other.

(3) CHECKERBOARD PATTERN: For $\mu_1 < 0$, from Eqs. (9), the absolutely stable pattern is the one for which N_{bb} and N_{ww} are minimum and N_{bw} is maximum. Such a pattern involves the minority cell dispersed in the predominant cell type; in the case where the minority cells are as close together as possible, the pattern resembles a checkerboard. The three possible patterns are illustrated in Fig. 22.

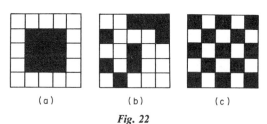

(a)   (b)   (c)

Fig. 22

Thus, there are only two absolutely stable patterns for two types of cells: the onion type and the checkerboard type, with the condition that the cells at the boundary are all of one type.

We emphasize that our procedure is valid for nondeformable cells of any shape and, also, in any dimension. To generalize the procedure, the only change we have to make is that in the conservation Eqs. (6) and (7), the coefficient 4 on the right-hand side has to be replaced by the number of edges through which a cell can make contact with another cell.

b. Patterns Involving Three Types of Cells. We now derive the absolutely stable patterns for a system with three types of cells (or two types of cell and medium). Let us denote the three types by b, w, and e and the number of cells of these types by N_b, N_w, and N_e, respectively. If N_{bb}, N_{bw}, N_{be}, N_{we}, N_{ww}, and N_{ee}, respectively, denote the number of bb, bw, be, we, ww, and ee edges for a pattern, then the conservation of edges implies

$$2N_{bb} + N_{bw} + N_{be} = 4N_b, \tag{13}$$

$$2N_{ww} + N_{bw} + N_{we} = 4N_w, \tag{14}$$

$$2N_{ee} + N_{be} + N_{we} = 4N_e. \tag{15}$$

Here, as for two types of cells, we need not consider edge effects if one cell type (here taken to be e) is predominant. The E function for a pattern is

$$E = \lambda_{bb}N_{bb} + \lambda_{aw}N_{bw} + \lambda_{ww}N_{ww} + \lambda_{be}N_{be} + \lambda_{we}N_{we} + \lambda_{ee}N_{ee}, \tag{16}$$

where λ_{bw} denotes the λ function for contact between b and w cell types, and so forth. Using Eqs. (13)–(15), the right-hand side of Eq. (16) can be expressed in terms of only three types of edges.

There are $6!/3!3! = 20$ possible distinct groups of three edges. Out of these 20 groups there are three groups, namely, those involving

$$N_{bb}, N_{bw}, \text{ and } N_{be};$$

$$N_{ww}, N_{bw}, \text{ and } N_{we};$$

$$N_{ee}, N_{be}, \text{ and } N_{we},$$

in terms of which E cannot be expressed [as can be seen from Eqs. (13)–(15)]. Thus, there are 17 possible expressions for E, each involving three types of edges. For the purpose of illustration, let us write E in terms of N_{bb}, N_{bw}, and N_{ww}. From Eqs. (13) and (14),

$$N_{be} = 4N_b - 2N_{bb} - N_{bw}, \tag{17a}$$

$$N_{we} = 4N_w - 2N_{ww} - N_{bw}. \tag{17b}$$

Substituting Eqs. (17a) and (17b) into Eq. (15), we get

$$N_{ee} = 2(N_e - N_b - N_w) + N_{bw} + N_{bb} + N_{ww}. \tag{17c}$$

Substituting Eqs. (17) into Eq. (16), we get

$$\begin{aligned} E = \; &N_{bb}(\lambda_{bb} - 2\lambda_{be} + \lambda_{ee}) + N_{bw}(\lambda_{bw} - \lambda_{be} - \lambda_{we} + \lambda_{ee}) \\ &+ N_{ww}(\lambda_{ww} - 2\lambda_{we} + \lambda_{ee}) \\ &+ 2[N_b(2\lambda_{be} - \lambda_{ee}) + N_w(2\lambda_{we} - \lambda_{ee}) + N_e\lambda_{ee}]. \end{aligned} \tag{18}$$

The term in the square brackets in this equation is a constant, (that is, independent of the pattern).

In a similar fashion we can write the other 16 equivalent forms of E. It may be noted that because of the symmetry of the function E for interchange of subscripts b, w, and e, some of the 17 forms can be obtained from other forms by simple interchange of b $\rightleftharpoons$ w or b $\rightleftharpoons$ e or w $\rightleftharpoons$ e. If

$$\mu_1 = \lambda_{bb} - 2\lambda_{bw} + \lambda_{ww}, \tag{19a}$$

$$\mu_2 = \lambda_{bb} - 2\lambda_{be} + \lambda_{ee}, \tag{19b}$$

$$\mu_3 = \lambda_{ww} - 2\lambda_{we} + \lambda_{ee}, \tag{19c}$$

so that

$$\tfrac{1}{2}(\mu_1 + \mu_2 + \mu_3) = \lambda_{bb} + \lambda_{ww} + \lambda_{ee} - \lambda_{bw} - \lambda_{be} - \lambda_{we}, \tag{19d}$$

then all the coefficients of N_{xy} ($x, y = $ b, w, e) in the seventeen equivalent forms of E can be expressed as a linear combination of these three μ_i ($i = 1, 2, 3$). This proves the statement, made for the case of two types of cells that, effectively, the number of λ's is decreased by the number of types of cells in the system.

We now consider the absolutely stable patterns for various values of λ's. If

$$\mu_1 = \mu_2 = \mu_3 = 0,$$

then, as can be seen by using any of the seventeen equivalent forms, all possible patterns are equally stable.

On the other hand, if one μ_i is zero and the other two are equal,

$$\mu_1 = \mu_2 > 0, \qquad \mu_3 = 0,$$

then, by using any of the forms, one finds that for these μ_i, the absolutely stable pattern is the one with b cells clustered together and w cells distributed outside the cluster in any form. Thus, there are an infinite number of absolutely stable patterns, all having b cells clustered together, If

$$\mu_1 = \mu_3 > 0, \qquad \mu_2 = 0,$$

then any pattern with w cells clustered together is absolutely stable. For the general case, the systematic approach to identify all absolutely stable patterns is to consider each of the seventeen forms of E separately, and to find for what values of μ_i there exists one or more absolutely stable patterns. To determine all the absolutely stable patterns from one form, we proceed as follows: In each form, E is expressed in terms of the number of edges of three types. Since E is maximum for an absolutely stable pattern, such a pattern must have the number of edges of these three types, either maximum or minimum. If we denote the maximum by 1 and the minimum by 0, we can have eight possibilities for the number of edges of the three types, namely 000, 001, 010, 100, 110, 101, 011, 111. We investigate each of these eight cases and determine whether we can have a pattern with the number of edges of three types belonging to this case. If so, this will be one of the absolutely stable patterns for suitable values of the μ_i implied by the form under consideration. In this way we exhaust all eight cases and find all the absolutely stable patterns implied by this form. We repeat this procedure for other forms.

To determine the values of the μ_i for a particular absolutely stable pattern, we set the coefficient of the number of edges of a particular type less than or equal to 0 if this number is minimum and greater than 0 if it is maximum. This procedure will ensure maximum E and hence the absolute stability of the pattern under consideration.

As an illustration of our procedure, let us consider E expressed in the form

$$E = N_{bb}\mu_1 + \tfrac{1}{2}N_{be}(\mu_1 + \mu_3 - \mu_2) + N_{ee}\mu_3 + \text{const.} \tag{20}$$

Suppose we require all N_{bb}, N_{be} to be maximum. N_{ee} is maximum when all the b cells are clustered together, but this will not allow N_{be} to be maximum because N_{be} is maximum when all the cells are intermixed with e-type cells. Likewise maximum N_{ee} is incompatible with maximum N_{be}. Therefore, N_{bb}, N_{be}, and N_{ee} cannot be maximum simultaneously and, hence, we will not have an absolutely stable pattern if the coefficients of N_{bb}, N_{be}, and N_{ee} in Eq. (20) are positive. On the other hand, if we require both N_{bb} and N_{ee} to be maximal and N_{be} to be minimal, we obtain an onion pattern with the b cells forming the inner layer, the w cells surrounding them, and these in turn being surrounded by the e cells. This pattern will be absolutely stable if

$$M_1 > 0, \qquad M_1 + M_3 \leqslant M_2, \qquad M_3 > 0.$$

We can repeat this procedure for all 17 forms, thereby determining all the absolutely stable patterns with three types of cells. Similar considerations can be applied to systems consisting of four or more cell types. However, it is clear that the number of groups of edge types goes up very rapidly with the number of cell types, and the problem of computing all of the absolutely stable patterns becomes correspondingly more complicated.

Let us conclude this discussion of absolute stability with some comments on the concept of "transitivity" as introduced by Steinberg. Steinberg [1964] states, as a summary of his experimental results, that, "If tissue a segregates internally to tissue b, and tissue b segregates internally to tissue c, then tissue a will be found to segregate internally to tissue c" [p. 344]. This is apparently also intended to be a statement of the results to be expected if the affinities are transitive, but in fact it is not a critical test of transitivity since the affinities between a and b do not imply anything as to the affinities between a and c. This can be seen as follows.

Let the absolutely stable (maximum E) pattern of b and w cells be b enclosed by w, and similarly that of w and x cells be a pattern of w enclosed by x. Let e denote the medium common to both the systems. The question we then ask is whether this implies anything about the maximum E pattern of b and x. The first two observations and the previous discussion imply

$$\lambda_{bb} - 2\lambda_{bw} + \lambda_{ww} > 0, \tag{21a}$$

$$\lambda_{we} + \lambda_{bw} \geqslant \lambda_{be} + \lambda_{ww}, \tag{21b}$$

$$\lambda_{ee} - 2\lambda_{we} + \lambda_{ww} > 0, \tag{21c}$$

$$\lambda_{ww} - 2\lambda_{wx} + \lambda_{xx} > 0, \tag{22a}$$

$$\lambda_{xe} + \lambda_{wx} \geqslant \lambda_{we} + \lambda_{xx}, \tag{22b}$$

$$\lambda_{ee} - 2\lambda_{xe} + \lambda_{xx} > 0. \tag{22c}$$

Adding Eqs. (21a) and (22a), we get

$$\lambda_{bb} + \lambda_{xx} + 2\lambda_{ww} > 2_{bw} + 2\lambda_{wx}. \tag{23}$$

Multiplying Eqs. (21b) and (22b) by 2 and adding both the resulting equations to Eq. (23), we get

$$\lambda_{bb} - \lambda_{xx} > 2(\lambda_{be} - \lambda_{xe}). \tag{24}$$

Adding Eq. (22c) and Eq. (24), we finally have

$$\lambda_{bb} + \lambda_{ee} > 2\lambda_{be}. \tag{25}$$

From Eqs. (22)–(25) we conclude that the above-mentioned observations tell us nothing about λ_{bx} and hence do not imply anything as to the absolutely stable pattern between b and x, although in the special case of $\lambda_{be} = \lambda_{xe} = \lambda_{ee} = 0$, Eq. (24) does imply that $\lambda_{bb} > \lambda_{xx}$. Therefore, the enclosure of x cells by b cells is excluded. If we further assume that $\lambda_{bx} = 0$, Eqs. (22c) and (25) and the discussion earlier in this section imply that b cells and x cells will be completely segregated. On the other hand, instead of assuming $\lambda_{bx} = 0$, we assume that $\lambda_{xx} \leqslant \lambda_{bx} < \frac{1}{2}(\lambda_{bb} + \lambda_{xx})$, then b cells will be aggregated internally to x cells. Thus, the transitive and nontransitive relations are both compatible with the model. It should be explicitly noted that whether the λ's are transitive or not, the number of possible absolutely stable patterns will not change. This is because our analysis of stable patterns has covered all of μ space (which is a complete projection of λ space).

Experimentally most results of cell-sorting experiments, including those of Steinberg, are compatible with transitive relations. But there are a few exceptions, also. For example, in the experiments of Townes and Holtfreter [1955] on the ectoderm (a), mesoderm (m), and endoderm (b) from an amphibian neurula, both ectoderm and endoderm enclose the mesoderm when these tissues are placed in contact with each other, but ectoderm and endoderm segregate from each other. Thus it may be concluded that $\lambda_{mm} > \lambda_{aa}$, $\lambda_{mm} > \lambda_{bb}$ and $\lambda_{ab} = 0$.

We may also add that our model does *not* imply that if tissue a segregates internally to tissue b, tissue b segregates internally to tissue c, and tissue a segregates internally to tissue c. Then, if all three tissues a, b, and c are allowed to aggregate, tissue a will be segregated internally to tissue c. This follows from the analysis, similar to that presented above, based on the conditions for the absolute stability of onions for two-cell types and medium and three-cell types and medium.

2. *Locally Stable Patterns*

The patterns discussed thus far have been absolutely stable, that is, they have the maximum E value for any possible pattern formed by cells having the assigned λ values. We call a pattern locally stable if its E value, even though not maximal, is greater than the E value of any neighboring pattern. A neighboring pattern is defined as one which can be reached from the given pattern, according to some cell motility rule. Since the real motility rules of cells are not definitely known, some plausible rules will be assumed and their consequences compared.

Given a set of motility rules, the number of possible patterns which are locally stable is large and thus a general study of locally stable patterns is difficult and will not be attempted here. Attention will be concentrated on the conditions for local stability of a few structures of biological interest, leaving aside the question of how such locally stable patterns were formed in the first place. It is thus obvious that, whereas in spite of the simplifications of the model, the study of absolutely stable patterns has a fairly direct bearing on the possible formation of real biological structures through the operation of cell motility and cell adhesion, a similar study of locally stable patterns is less likely to represent biological reality until the actual rules of cell motility become better known. Nevertheless, model studies of locally stable patterns are worthwhile since they throw some light on the nature of the problem.

From these remarks it is clear that the local stability of a pattern depends strongly on the motility rules, since such rules determine what is and what is not a neighboring configuration.

It is realistic to assume that a cell is restricted in its movements; the restriction will be assumed that a cell can only move to a neighboring square of the tessellation. Let us say that two patterns P_1 and P_2 differ by an elementary permutation if they are identical except that the contents of two laterally or diagonally adjacent cells are interchanged. Two elementary permutations are called compatible if they involve no overlapping of squares of the tessellation. We shall say that two patterns are neighboring if they differ by a number of compatible elementary permutations. Thus any individual pattern P_0 determines a set of other patterns which are its neighbors. It is useful to think of time as moving along a discrete set of instants so that a cell is either in one square of a tessellation or in another. Within this unit of time, the value

E of a pattern does not change; in the next unit of time, it may be the same or different.

Unfortunately the choice of rules which govern cell movement is very wide. In attempting to put some restrictions on the choice, one may suggest that movement will be such that the E value of the subsequent pattern will always increase or, at least, remain the same. In some sense this requirement is implied by the experimental results: configurations of maximum E do not change into submaximal configurations with time. As far as elementary permutations are concerned, this restriction may be interpreted to mean that the requirement is global: that is, the E of the overall pattern must increase even if some moves are locally energetically unfavorable. Since this implies that individual cells have some way of detecting and responding to information as to the global pattern of which they are a part, this does not seem very plausible. The alternative is to assume that cells will only move from an environment which is locally of lower E to an environment which is locally of higher E. Somewhat less restrictive would be the assumption that moves to equal E are permitted, or even moves to lower E with reduced probability.

We discuss the nature of the real motility rules later. For the purpose of studying local stability we shall assume merely that a cell can undergo a single flip between adjacent diagonal or nondiagonal squares of the tesselation, and individual flips will occur only when the individual flip causes an increase in E of the total pattern.

The structures whose local stability we investigate, using the above motility rules, are those which are histologically significant, that is, single layers, either as sheets (epithelia), or tubules and vesicles. In two dimensions, they may be represented by layers and rings (or rectangles) of cells. Because of the nature of our grid, tubules, and vesicles are represented by rectangular structures; this introduces some artifacts because of the presence of corners that are not present in the structures they represent. A more serious criticism is that they are ambiguous, since the figures we study are cross sections of three-dimensional figures. A ring may represent the cross section of a sphere, a tubule, and so on. These considerations may be regarded as making the similarity between our models and actual three-dimensional figures quite remote. This is indeed true but we present our results because they are an entering wedge to more realistic representations. Let us discuss separately the local stability of patterns involving two and three types of cells.

a. Patterns Involving Two Types of Cells. Let us call the patterns whose local stability we wish to study P, and denote the neighboring patterns by P'. From the remarks made earlier in connection with the absolute stability of patterns with two types of cells, it is clear that for the study of local stability in a two-cell system, it is enough to know the difference between the number of any one of three types of edges in P and P'. Let us choose this type to be bb. Also let us assume $N_w \gg N_b$. Set

$$N_{bb}(P) - N_{bb}(P') = M_1. \tag{26}$$

Therefore, from Eq. (9a),

$$E(P) - E(P') = M_1 \mu_1. \tag{27}$$

Since for any P of biological interest we can always find a P' for which $M_1 > 0$ (namely, by interchanging a b cell with a w cell), for P to be locally stable we must have

$$\mu_1 = \lambda_{bb} + \lambda_{ww} - 2\lambda_{bw} > 0. \tag{28}$$

This is the first necessary condition for the local stability of P. The other necessary condition is that there be no P' for which $M_1 \leqslant 0$. Together, these two conditions are necessary and suffucient for the local stability of P. If there exists a P' for which $M_1 = 0$ and no P' for which $M_1 < 0$, P is a meta-stable pattern. We now discuss the local stability of rings and sheets of cells.

(1) RINGS OF CELLS: For convenience we divide these rings into three categories and discuss them separately.

(a) *Single-Layered Ring:* The simplest pattern in this category is one with eight b cells, $N_b = 8$, with one w cell trapped inside (Fig. 23a). This pattern is unstable because for the pattern P' obtained from this pattern P by inter-changing the trapped w with any b, $M_1 < 0$.

The next simplest pattern is the pattern with more w cells inside but arrang-ed in the form of a layer, one cell thick (Fig. 23b). This pattern is metastable if we allow only nondiagonal flips because, for P' obtained from P by inter-changing an end internal w cell sith any neighboring b cell, $M_1 = 0$. If we also allow diagonal flips, it becomes unstable because, for the pattern P' obtained from P by interchanging an end internal w cell with a corner b cell, $M_1 < 0$. This illustrates the sensitivity of local stability to the motility rules.

The next pattern which we consider is the one in which the w-cell layer is two cells thick. The simplest such pattern is $N_b = 12$, and 4 w cells trapped inside (Fig. 23c). This pattern is stable if we allow only a single nondiagonal flip because there is no P' for which $M_1 < 0$. But, if we allow a diagonal flip, this pattern becomes metastable because, for example, for the pattern P' obtained by interchanging a corner b cell with a diggonally opposite w cell $M_1 = 0$. On the other hand, if we allow simultaneous compatible flips, then the same pattern becomes unstable (for P' obtained by interchanging the two w cells by their diagonally opposite b cells, $M_1 < 0$). Even if $N_w > 4$, as long as w cells are arranged in layers two cells thick, the results for the case $N_w = 4$ hold.

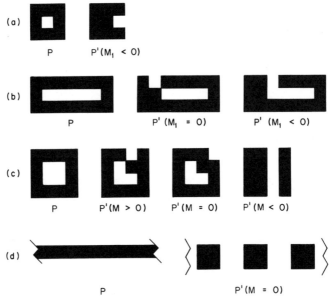

Fig. 23

We now consider single b-layered patterns with inner w cells arranged in at least three cell layers. These patterns are stable if only a single nondiagonal flip is allowed. They are metastable if a diagonal flip is also allowed because, if we take one corner w cell inside P_1 and interchange with the diagonal b corner cell, the resulting pattern P' has $M_1 = 0$. Similar results hold for larger single-layered patterns.

From the above discussion it is clear that to obtain a pattern P' such that $M_1 < 0$, we must be able to interchange b and w cells in such a fashion that the b cells on the opposite edges of P come in contact with each other. This is not possible with our motility rules with only a single flip allowed for single b-layered patterns (with the exception of those in which inner w cells are arranged in the form of layers one cell thick). However, if we allow more than one flip until E changes, then all the single b-layered patterns become unstable because, for P' obtained from P by interchanging w cells on the edges of the inside layer by diagonally opposite b cells, $M_1 < 0$.

(b) *Two-Layered Rings:* The simplest pattern in this category will have one w cell inside two layers of b cells. This pattern is locally stable with our neighbor rules because, even bringing b cells on the opposite sides of a w cell in contact, there is no increase in N_{bb}. This one w cell pattern is a metastable pattern because P' obtained by interchanging any of the inner w cells by any of its neighboring b cells has $M_1 = 0$. However, a pattern with inner w cells arranged in the form of a layer one cell thick, with two b layers, is unstable if we allow more than one consecutive flip.

The next pattern in this category with $N_w = 4$ is locally stable because any interchange of w with b will always decrease N_{bb}. In fact, if the inner w cell layer is at least two cells thick, the two b-layered rings will always be locally stable. We may note that the local stability of two b-layered rings does not change whether we do or do not allow diagonal flips.

(*c*) *Three- or More-Layered Rings:* We leave as a trivial exercise for the readers to show that whatever we have said for two-layered rings is completely valid for this case.

(2) SHEETS OF CELLS: A sheet is represented as a chain of b cells in contact with w cells on both sides (Fig. 23d). If we allow only one flip, this chain P is stable because, for any P' obtained by any interchange, $M_1 > 0$. However, if we allow simultaneous compatible flips, then if the chain is finite it will be unstable, tending to break up into the pattern,

If the chain is infinitely long, then rearrangement into the broken form is a metastable transition ($M_1 = 0$) (Fig. 23d). This is in accordance with the observed fact that, if an epithelium layer is stretched beyond a certain point, the layer disappears.

If the sheet is two cells thick, then any flip will decrease bb contacts and hence is locally stable.

b. Patterns Involving Three Types of Cells. Analysis of a system of three types of cells requires three μ's and 17 different forms of E. This makes the general analysis of local stability rather complicated. Finding the set of values of μ_i that will make a particular pattern locally stable for a given set of motility rules involves the calculation of $E(P) - E(P')$, for all P' in 17 different forms, and then finding for which values of μ's, $E(P) - E(P') > 0$ for all P'. We will not do this analysis in this paper; instead, we will find the set of μ's for which the locally stable patterns for cells of two types will remain stable, and the unstable or metastable ones either remain so or become more stable. We discuss the rings and sheets of cells.

(1) RINGS OF CELLS: Let us denote $N_{bb}(P) - N_{bb}(P')$, $N_{ww}(P) - N_{ww}(P')$, and $N_{ee}(P) - N_{ee}(P')$, by M_1, M_2, and M_3, respectively. For a ring pattern, for any flip, $M_3 \geqslant 0$. For a "capillary" ring pattern, that is, patterns in which w cells are arranged in the form of a layer one cell thick, any flip between a pair of neighboring b and w cells which are not diagonal to each other will give P' with M_1, $M_2 > 0$ and $M_3 = 0$ and the flip between diagonal cells will give P' with $M_1 = M_3 = 0$, $M_2 > 0$. For noncapillary rings, independent

of the thickness of b layers. for all P', $M_2 > 0$, and $M_1 \geqslant 0$. We can write

$$E(P) - E(P') = M_1(\mu_1 + \mu_2 - \mu_3) + M_2(\mu_1 + \mu_3 - \mu_2)$$
$$+ M_3(\mu_2 + \mu_3 - \mu_1). \tag{29}$$

Since for ring patterns our motility rules allow only $M_1 \geqslant 0$, $M_2 > 0$, $M_3 \geqslant 0$, if

$$\mu_1 + \mu_2 > \mu_3, \qquad \text{that is,} \qquad \mu_{bb} + \mu_{we} > \mu_{be} + \mu_{bw}, \tag{30a}$$
$$\mu_1 + \mu_3 \geqslant \mu_2, \qquad \text{that is,} \qquad \mu_{ww} + \mu_{be} \geqslant \mu_{bw} + \mu_{we}, \tag{30b}$$
$$\mu_2 + \mu_3 \geqslant \mu_1, \qquad \text{that is,} \qquad \mu_{bw} + \mu_{ee} \geqslant \mu_{we} + \mu_{be}, \tag{30c}$$

all these ring patterns that were stable for the two-cell types, will remain stable and those which were unstable or metastable will either remain so or become more stable.

Thus, by bringing another type of cells for suitable μ's, we have made single b-layered rings also stable. This is not very surprising because by introducing another type of cell we have artificially introduced anisotropy in the cell surface, namely, two edges of b cells are in contact with b cells, one with a w cell and one with an e cell. We had another example of stability due to anisotropy in the system with two types of cells where the double b-layered ring became stable. As shown elsewhere [Goel and Leith, 1970], the anisotropy can make epithelial, vesicular, and tubular structures absolutely stable.

We conclude the discussion on the local stability of patterns involving three types of cells by showing that we can make single b-layered rings more stable than complete and partial multi-b-layered rings for the same number of w cells by choosing appropriate values of the λ's. To find these values we note that the maximum number of b cells that can be accommodated around w cells in a single-layer ring is

$$N_b = 2N_w + 6. \tag{31}$$

Therefore, for the purpose of the present discussion, the relevant number of b cells is $N_b \leqslant 2N_w + 6$. For single or multilayer structure, $N_{we} = 0$. Let us choose the form in Eq. (7) of the E function, that is,

$$E = N_{bb}(\lambda_{bb} - 2\lambda_{be} + \lambda_{ee}) + N_{ww}(\lambda_{ww} - 2\lambda_{bw} + 2\lambda_{be} - \lambda_{ee}). \tag{32}$$

Since for a multi-b-layered ring, both N_{bb} and N_{ww} are greater than those for a single b-layered ring, the sufficient conditions for a single b-layered ring to be most stable among b-ring structures, with w cells inside, are

$$\lambda_{bb} < (2\lambda_{be} - \lambda_{ee}),$$
$$\lambda_{ww} + (2\lambda_{be} - \lambda_{ee}) < 2\lambda_{bw}.$$

It may be noted that these conditions imply

$$\lambda_{bb} + \lambda_{ww} < 2\lambda_{bw}.$$

We may add that our one-layer pattern is metastable: that is, there are more than one b-layered pattern (though topologically identical) that have the same energies. One such pattern is (for $N_w = 4$) as shown,

bbbbb,   bwwwb,   bbbwb,   bbb.

Although these cases of local stability are presented primarily for illustration only, it is worthwhile to compare the conclusions with histological observations. We consider two cases, the occurrence of single and double sheets of cells, and the reaggregations of embryonic lung tissue.

(2) SHEETS OF CELLS: Given any of a number of cell motility rules, we find that single layers of cells which have the same environment on both sides are locally unstable. Two or more layers, however, are locally stable. It is therefore of interest that, in general, when sheets of cells occur, they do not have the same environment on both sides of the sheet. This is the case of all epithelia. When there is a sheet of cells with the same environment on both sides, the sheet is double.

It may be noted that this generalization and, in fact, much of the discussion on cell sorting, has little relevance to plant histology. In plants the cells are held rigidly by cell walls with respect to each other, and much morphogenesis is a matter of the pattern of cell division.

3. Reaggregation of Embryonic Lung Tissue

Most experimental work on vertebrate reaggregation has dealt with the formation of "onions." Studies on the appearance of more complex patterns in reaggregates [Moscona, 1965] have not been detailed. However, Grover [1961, 1962, 1963] has described in considerable detail the reaggregation of dissociated embryonic chick lung tissue, which it is of interest to consider from the point of view of our model. Dissociated embryonic lung contains mainly two cell types, epithelial and mesenchymal. (We will here not discuss the vascular tissue although this may affect reaggregation pattern [Grover, 1962].) The course of reaggregation in a hanging drop depends on the relative proportions of the two types of cells.

When the proportions are optimal, reaggregation produces masses of mesenchyme which are coated with a single layer of epithelial cells. Within the mesenchyme are islands of epithelial cells which eventually form vesicles and tubules lined with epithelium.† This produces a rough outline of the

---

†Tubular structures are rare in reaggregates of 7-day lung, but develop well from cells of lung 11–12 days old.

Robert Rosen

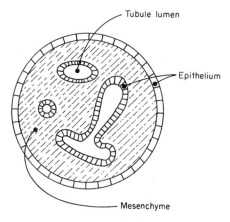

Tubule lumen

Epithelium

Mesenchyme

Fig. 24

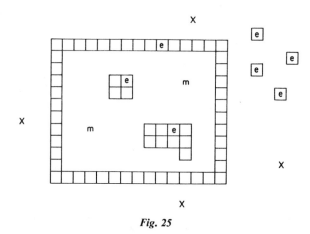

Fig. 25

embryonic lung structure, shown schematically in Fig. 24. In terms of our model, this can be represented as in Fig. 25.

To analyze this behavior in terms of our model, we can consider as a standard the major pattern observed by Grover [1962], a mesenchymal mass covered with epithelium and containing an epithelium-lined tubular system. We then consider the energy changes, or the ΔE vector, which is associated with various transformations away from this pattern.

It is useful to express the energy E of a pattern containing three cell types: m, mesenchymal; e, epithelial; x, medium as

$$2E = -\mu_1 N_{me} - \mu_2 N_{ex} - \mu_3 N_{xm}, \tag{33}$$

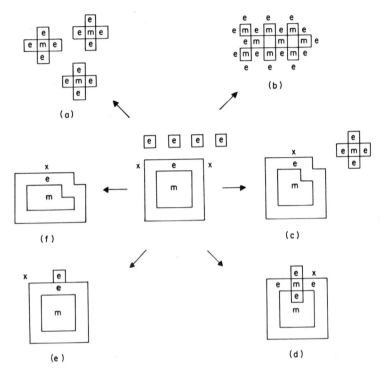

Fig. 26

where

$$\mu_1 = \lambda_{mm} + \lambda_{ee} - 2\lambda_{me}, \tag{34a}$$

$$\mu_2 = \lambda_{ee} + \lambda_{xx} - 2\lambda_{ex}, \tag{34b}$$

$$\mu_3 = \lambda_{xx} + \lambda_{mm} - 2\lambda_{xm}. \tag{34c}$$

Let us first consider the stability of the major pattern observed by Grover, namely, a mesenchymal mass coated with additional epithelian cells in suspension and see if we can choose a set of μ's such that this pattern (P) is more stable than other patterns (P'). Let

$$2E = 2(E(P) - E(P')) = \mu_1 n_1 + \mu_2 n_2 + \mu_3 n_3, \tag{35}$$

where

$$n_1 = N_{me}(P') - N_{me}(P), \tag{36a}$$

$$n_2 \equiv N_{ex}(P') - N_{ex}(P), \tag{36b}$$

$$n_3 \equiv N_{xm}(P') - N_{xm}(P). \tag{36c}$$

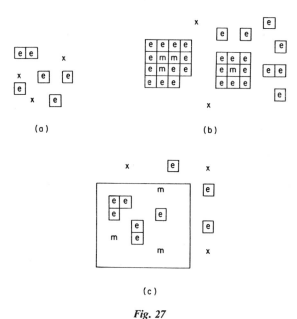

(a)

(b)

(c)

Fig. 27

We require $\Delta E > 0$ for all P'. For our pattern, $N_{xm}(P) = 0$. Let us first consider those P' for which $N_{xm}(P') = 0$, so that $n_3 = 0$. Some of the typical P' are as shown in Fig. 26a–f. Pattern (a) is formed when additional e cells are recruited from the medium and m–e interaction is increased by the formation of a checkerboard between m and e. Pattern (b) is formed when sufficient e cells will be recruited from the medium to completely surround each m cell with clusters each of one m cell and four e cells left suspended in the medium. For both P and P', n_1, $n_2 > 0$ but n_2 is not too much greater than zero. Pattern (c) is also formed by a similar process. Patterns (d), (e), and (f) involve changes of one or two cell positions. Pattern (d) has a preliminary checkerboard configuration; (e) involves a suspended cell "precipitating" onto the structure to make the multilayered epithelium, and (f) involves an increase of epithelial area through recruitment of cells from suspension with consequent distortion of the mass. For the patterns (c), (d), (e), and (f), $n_1 > 0$ and $n_2 < 0$. If we choose $\mu_1 > 0$ and $\mu_2 \simeq 0$, then $\Delta E > 0$ for all P' shown in Fig. 26.

   We now consider those P' for which $N_{xm} \neq 0$. For such P' (Fig. 26) at least a single e cell should leave the continuous epithelium, exposing one M cell to the medium. For the pattern (a) of Fig. 26, $n_1 = -1$, $n_2 = 5$, $n_3 = 1$. We have to choose μ_i such that $-\mu_1 + 5\mu_2 + \mu_3 > 0$. This choice has to be consistent with the condition derived above, namely, $\mu_2 > 0$, $\mu_2 \simeq 0$. Such a choice could be $\mu_1 = 5$, $\mu_2 = 1$, $\mu_3 = 1$. It may be noted that this choice will energetically allow the dissolution of a discontinuous epithelium, as apparently occurs. These dissolutions are shown in Fig. 27, the transformation leading to (b) corresponds to $n_1 = 1$, $n_2 = -3$, $n_3 = -1$, and the one leading to (c) corresponds to $n_1 = 1$, $n_2 = -1$, $n_3 = -1$.

The choice of μ's in this fashion explains almost all of Grover's observed reaggregation patterns as patterns of local stability. It stabilizes the general pattern of an epithelium coating a mesenchymal mass. It provides for excess epithelial cells remaining in the medium in a loosely associated fashion. These epithelial cells will not intercalate into the epithelium to distort the masses from a near-spherical shape. However, these cells would add on to existing epithelia with slight increase in energy; it is necessary to assume that, for some reason, suspended cells are unable to add to the mass. Thus, the assumptions that cell properties do not change with time, and that the cells are isotropic so far as their adhesive properties are concerned, seem to suffice to explain most but not all the phenomena of reaggregating lung.

In this and in other cases, it is likely that sorting out is the major process in early pattern formation, but it is apparent that at some point simple reaggregation processes end and tissues begin to differentiate by virtue of changes in the cell properties. Presumptive tubule cells secrete fluid, for example, and in this way change in configuration from a clump to a shell. This process involves cells secreting in a polarized fashion and it also may involve the formation of "tight junctions" [Wood, 1965; Wartiovaara, 1966]. Clearly the events beyond the initial stages must be considered in terms of modified cell properties. Nevertheless, it is clear that cell sorting mechanisms may serve to generate a morphology which may then be stabilized through other means (for example, tight junctions) so that the cells are now free to undergo specific further differentiations which would otherwise change their surface properties and hence thereby disrupt the morphology.

4. The Concept of an Epithelium

The histological structures whose local stability has been considered are all essentially single sheets of cells, sometimes closed into tubules or vesicles. They are thus examples of epithelial structures. We wish to stress at this point that we define an epithelium as something more than merely a single layer of cells spread over an underlying mass of cells of another type. By an epithelium we mean a single cell layer which has an intrinsic tendency to remain as such when more cells are added to it. The additional cells will not add to the original single layer, but rather will intercalate into the original layer. To accommodate them, the epithelial cells may be distorted, or the underlying cell mass may be distorted to increase its surface-to-volume ratio.

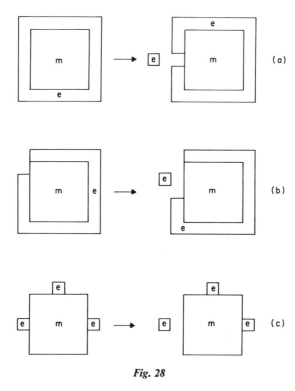

Fig. 28

The difference between a true epithelium and a fortuitous single-cell layer, resulting from a restriction on the number of cells present, is illustrated in Fig. 28.

Now as we have shown, an epithelium is not an absolutely stable structure if the cells are isotropic, although it may be a locally stable configuration. Some further investigations by Goel and Leith (1970) have shown, however, that if cells are nonisotropic, having μ's which vary with direction, then an epithelium is one of the absolutely stable structures. It is very likely that, in fact, epithelial cells are nonisotropic as is suggested, for example, by the polarized formations of basal lamellae or adhesive junctional complexes. This would probably make it possible to greatly improve the concordance between the model and such observations as the reaggregation of lung. It is, nevertheless, of interest that even local stability goes some way to explain the persistence of certain histological structures.

5. The Path Problem and Cell Mobility

We now turn to the path problem and the role of cell motility in determining the path to stable structure. We begin by recalling that Steinberg's hypothesis provides a method by which cells can produce a maximum E configuration from other configurations, including random ones. Nevertheless, *a priori*, it is by no means obvious that such patterns can be reached from any configurations. Indeed, the experimental results show that under certain conditions patterns are reached which are locally, but not absolutely, stable. One therefore has to investigate what are the requirements which have to be met when pattern P_2 has to be reached from pattern P_1. This is the general path problem. Here we consider a restricted part of this problem: which motility rules make it possible to reach a configuration of concentric spheres, starting with a random distribution of cells? Another part of the problem is as follows: Given a set of motility rules, in general, cells can follow more than one path when pattern P_2 is to be reached from pattern P_1. Some of these paths are "shorter" than others; that is, some require smaller numbers of cell movements than others. Further, in general, in different paths the value of the E function will change differently with every cell movement. The problem is: Which of the paths are the cells going to follow in a real cell sorting? We will not discuss this part of the problem here. A discussion may be found in Goel and Leith [1970].

There is one final remark which may be made regarding this entire phenomenological approach to cell sorting. This concerns the curious interplay between the *local* behavior of the individual cells and the *global* behavior of the pattern as a whole. The function E which defines the global change of pattern is a summation of all the local E changes arising from local cell motility, and all that is required is that a given pattern make the transition to that neighboring pattern whose E value is maximal. This global constraint, however, would say nothing about the kinds of patterns which are being generated in smaller subregions of the pattern. Thus, in rough intuitive terms, we can generate what look locally like energetically unfavorable patterns in parts of the overall system, provided we pay for them by constructing highly favorable patterns in other parts of the system. This may perhaps provide an insight into why there is so much "parallel processing" in development, and points up that we must pay attention to the system as a whole in attempting to obtain an overall understanding of developmental processes.

One final point is relevant here also. Computer simulations [Goel *et al.*, 1970] show that local motility rules requiring only locally favorable pattern changes tend to trap the system in patterns which are locally stable but far removed from the global E maximum. This is in distinction to the experimental results on cell sorting, and suggests that, in the real situation, a cell can, as it were, "see" several moves ahead, and make an energetically unfavorable move if by doing so it arrives at a position from which it can make an exceedingly favorable one in a subsequent transition. This is a temporal analog of the spatial local-versus-global problem mentioned in the preceding paragraph. This would mean that the cells of a population would have a means for obtaining information about circumstances relatively remote (for example, two or three cell diameters) from their present position; computer simulations [see Goel and Leith, 1970] indicate that this is sufficient to avoid pattern trapping in locally stable configurations.

IV. Differentiation

A. Asymmetries and Their Generation

In preceding sections, we have discussed at length the manner in which biological patterns and forms can be constructed through the morphogenetic movements of an appropriate set of subunits. As we saw abundantly there, the ultimate patterns formed in such systems tended to display a high degree of regularity; indeed, in the case of the spherical and cylindrical virus coats, which are constructed out of only a single kind of subunit, conditions of symmetry played a major role in understanding how the properties of the subunits determined both the assembly process and the form of the final structure.

However, there are many striking morphogenetic processes in biology that lead rather to the generation of asymmetric structures from symmetric precursors. It turns out that such phenomena are not convenient to describe in terms of the assembly of unchanging subunits in the sense in which we have described them, and that it is more useful to think of such morphogenetic phenomena in terms of actual quantitative and qualitative changes in the properties of the units themselves. This leads us at once into the realm of *differentiation*. In the present section, we shall explore some generalities about differentiation processes, which will be made more explicit in the specific examples which will follow in subsequent sections.

Just as with morphogenesis by means of relative movements of subunits, which could either be of a passive or an active kind, so morphogenesis by differentiation mechanisms can be either passive or active. By *passive* we shall understand in this context a process which is driven by the response of the differentiating unit (for example, a cell) to an external environment. In dynamical terms, such behavior is customarily called a *forced response* of the system to an input or forcing which arises externally to the differentiating system itself. An *actively* differentiating system shall refer to a system the internal structure and properties of which are changing with time as a consequence of its own intrinsic, autonomous dynamical structure (although it may be capable of accepting "signals" or "triggers" from the environment, which we shall regard essentially as a resetting of initial conditions or parameters in such systems). This terminology accords well with the use of the corresponding terms in self-assembly, and indeed arises from deeper dynamical homologies between differentiation and assembly processes. As with the self-assembly case, we shall be most interested in the active situation.

Many of the most important questions of morphogenesis find their prototype in the following simple question: How can an initially symmetrical, homogeneous structure spontaneously generate an asymmetry or an inhomogeneity? This problem arises at many morphological levels. We cite three examples.

1. How can a single, initially symmetric cell spontaneously establish a *polarity*, in which one end of the cell becomes structurally and functionally different from the other end?

2. How can a symmetric aggregate of initially identical cells (which may be regarded as an idealized blastula) spontaneously generate the regionalized and sequential cellular modifications which are characteristic of development in higher organisms?

3. How can an initially uniform distribution of organisms spontaneously pass to a highly nonuniform distribution, as in the aggregation of initially free-living myxamoebas to form the highly aggregated slugs characteristic of one part of the life cycle of the cellular slime molds?

Questions regarding such spontaneous generation of states of asymmetry and inhomogeneity from an initially symmetric and homogeneous state, which are among the most characteristic observations of biology, have always been difficult to understand for those familiar only with conventional classical physics. In physics one typically identifies asymmetries and inhomogeneities in an assembly of units or particles with the imposition of an *order* on the system, or in thermodynamic terms, with a state of low entropy. Left to itself, any such ordering in an isolated system will tend to disappear with time; this is the essential content of the second law of thermodynamics, which states that the entropy of such a system must increase (or, in other words, that such a system must ultimately pass from an inhomogeneous to a homogeneous state, the exact opposite of the developmental phenomena we are considering). Indeed, for a long time such developmental phenomena were often exhibited as examples of nonphysical processes in biology, processes governed by vitalistic principles which could never be explained in conventional physical terms.

If we try to explain the generation of inhomogeneities in terms of ordinary thermodynamic ideas, we can argue qualitatively that (a) the energy required to maintain the ordered structure against the disordering effect of the environment must be generated from within the system, and that one of the main functions of cell metabolism is to supply this energy, or (b) that the entropy decrease that seems to be characteristic of such developmental processes must be compensated for by "ingesting" order (or "negative entropy") from the environment in the form of foodstuffs themselves derived from ordered biological structures [Schroedinger, 1944]. Such explanations enable us to rationalize, if not to understand, developmental phenomena in purely physical terms, by plausibly establishing a consistency between our customary views of the natural evolution of a system from an ordered to a disordered state, and the behavior of a system partially open to appropriate physicochemical structures.

The examples of the subsequent sections, however, will suggest a rather different viewpoint for the relation between entropic and developmental processes. It must be remembered that the classic example of a thermodynamic system is a gas, composed of a large number of molecules. This system has the property that, whatever the initial distribution of molecules within a vessel, it will spontaneously and autonomously proceed to a state in which the molecules are uniformly distributed with constant density throughout the vessel. Any state of the gas for which the molecules are not uniformly distributed is considered an ordered state; it requires energy to maintain the gas in such a state. The uniform or homogeneous state is thus the state of maximal disorder (or maximal entropy) and any other state must therefore be of lower entropy. Entropy itself is *entirely a function of state*; in its definition no mention is made of the dynamical processes which govern the way in which the state of the gas changes in time. This dynamical element enters only implicitly, in the statement of the second law that the system dynamics is such that entropy always increases and tends to a maximum.

This example suggests that there are two intrinsically different ways employed in this simple example to measure the state of "disorder" of a thermodynamic system. On the one hand, the "maximally disordered" state is the one (or ones) to which the system will *autonomously* tend. In general, the properties of such a state cannot be posited in advance from simply knowing the set of states accessible to the system, but must be inferred from the system dynamics. On the other hand, in a gas, the dynamics of the system are such that the gas will in fact autonomously tend to a uniform or homogeneous state. Ordinary physical entropy, which is *only* a function of state, and not of system dynamics, is so constructed as to be maximized in the homogeneous state; but it is clear that it will exhibit its normal properties *only* for systems whose autonomous dynamics are like those of the gas. In any other kind of system, it will not locate the states of "maximal disorder"; that is, those states to which the system will autonomously tend, by where it assumes its maxima. For such systems, the ordinary physical "entropy," which simply measures the deviation of a state from the homogeneous state, must be supplemented by some explicit consideration of the overall intrinsic system dynamics. It seems that this restriction of entropy as simply a function of state, initially derived from a simple class of systems whose autonomous behavior does yield a homogeneous end-state, and which does not directly generalize to other situations, is at the root of much of the difficulty encountered in relating thermodynamic ideas to developmental phenomena (which, from a simple thermodynamic viewpoint, generate "order" instead of "disorder").

Indeed, all of the examples we shall discuss in this section are based on the following idea: The homogeneous state of the system is *unstable*, and the intrinsic dynamics of the system are such that the system has other stable steady states available to it, and to which it will thus naturally tend. These other steady states are necessarily inhomogeneous (since the homogeneous state is unstable) and therefore appear "ordered" from the simple standpoint of thermodynamic entropy. However, in purely dynamical terms, they are indeed "states of maximal disorder" in the sense that they are the states to which the system will autonomously tend, and in that it requires energy to maintain the system in a different state.

B. Rashevsky's Model for Cell Polarities

As we have seen abundantly in these pages, biology confronts us with many novel and complex situations, to which our relatively simpler intuitions derived from nonbiological systems are not initially suited. As a first step in understanding how biological processes actually do work, it is often most important to get some feeling for how they *could* work. That is, it is often most helpful to approach such difficult conceptual problems by initially constructing a class of "model systems," which are more or less plausible in themselves (that is, they can be regarded as constructible from biologically realistic elements) and which exhibit, at least qualitatively, the *kinds* of phenomena under discussion. Such a preliminary investigation of the *capability* of a set of models to exhibit a behavior of interest is often of the greatest importance; it is psychologically important to the investigator to be able to identify his phenomenon in a class of tangible and plausible systems, and further it gives him the opportunity to form insights and hypotheses unobtainable from other sources, which may lead to a clarification or solution of the problems of ultimate interest. In previous chapters we have seen many examples of such "capability" arguments, such as those involving the capability of open chemical systems to exhibit thresholds and hence complex switching behavior in response to environmental signals. To condemn such explorations of capability out of hand as "unbiological" is simply to miss their essential point.

The examples to be described in the next few sections are all, in this sense, arguments of capability; they will exhibit the *capacity* of certain kinds of systems, involving only biologically meaningful processes like chemical reactions and diffusion, to manifest the kinds of morphological changes which we have collectively called differentiation; or more particularly, they will exhibit the crucial property of spontaneously and autonomously passing from an initially homogeneous state to a nonhomogeneous state. Indeed, in this case such arguments of capability become most significant because the generation of inhomogeneity seems to contradict ordinary physical ideas.

One of those most adept at constructing arguments of "capability," in this sense, was Rashevsky, an early and effective champion of the view that even apparently hopelessly complex biological phenomena could be understood in effective formal terms through the use of appropriate modeling techniques. In the years 1929–1940, he developed a class of "capability" models for understanding cell division in terms of systems in which chemical reactions and diffusion (both biologically plausible processes) were taking place. The basic point of departure for these models was the fact that a gradient of chemical concentration could exert a mechanical force on the membrane bounding a region in which such a gradient existed, and that under appropriate (and reasonable) conditions, a membrane-bound droplet in which chemical sources existed could be rendered unstable by the mechanical forces so generated, and spontaneously split into two smaller droplets. This class of models for cell division, or cytokinesis, was a first natural attempt at an understanding of the forces responsible for a ubiquitous and mysterious biological phenomenon. And although its biological plausibility may be smaller now than when it was first proposed, it was certainly not "unbiological," nor should its psychological and methodological role be underestimated.

Rashevsky noted that the same mechanical forces that appear on a cell membrane would also be exerted on particles suspended in a medium in which a chemical gradient is being maintained. He utilized this fact to provide the first class of models in which asymmetries could be spontaneously generated, and thus the beginnings of a plausible physical basis by which the generation of inhomogeneities and polarities in biological systems could be understood on physical grounds.

Qualitatively, the argument is simple. Consider a spherical cell which contains sources for some chemical substance distributed uniformly through the cell. Suppose further that the cell contains particles which inhibit these sources. From the general theory of the mechanical forces acting on suspended particles in a concentration gradient, it follows that these particles will tend to accumulate in regions of low concentration of the substance; this would be true for any kind of suspended particle. But in particular this specific particle also inhibits the sources of the substance producing the gradient; therefore the effect of the diffusion forces is to *enhance* the gradient. The effect is essentially that of a positive feedback loop; the initial gradient causes the inhibitory particles to tend toward regions of low concentration, which further enhances the gradient, and so forth.

It can readily be seen that if the initial distribution of sources and inhibitory particles were spherically symmetrical, then the system would remain spherically symmetrical through the gradient enhancement. But this spherically symmetrical situation is *unstable*; any departure from spherical symmetry in the initial distribution of sources would be magnified by exactly the same positive feedback loop we have discussed. It follows from Rashevsky's treatment, which may be found in his book [Rashevsky, 1961], that all of the substance will ultimately reside in the hemisphere which initially showed the larger concentration, while all of the inhibitory particles will ultimately reside in the hemisphere which initially showed the smaller concentration. This, as Rashevsky points out, is a system which shows true self-regulation, in the sense that no matter how the cell is stirred, or how it is divided, the same final pattern will be reestablished.

Thus, we have here a true example of a system which will automatically generate an inhomogeneity or a polarity. The qualitative resemblance of this system to the one involving the separation of immiscible liquids (phase separation) is quite striking, but it will be evident that the dynamical bases of the two situations are quite different.

C. The Turing Models

A further class of models, employing the same basic strategy as that of Rashevsky which we have just considered, but of a rather simpler mathematical form, was proposed by the English mathematician Turing [1952]. In order to make the basic idea as clear as possible, we shall begin by considering a still simpler and less realistic example than that originally proposed by Turing, but one that retains the same basic properties.

Let us suppose that the *state* of an abstract cell can be specified at an instant of time t_0 by a single number, $x(t_0)$ which represents the concentration of a single substance x. Following Turing, we shall call such a substance a *morphogen*, and suppose that this concentration plays a role in determining the later development or form of the cell. For simplicity we suppose that this morphogen is coming into the cell, from an outside source, at a constant rate S, and is being destroyed in the cell at a rate proportional to its concentration. Then the concentration $x(t)$ of morphogen x satisfies the simple differential equation

$$dx/dt = -ax + S. \tag{37}$$

The solutions of this equation are plotted in Fig. 29. Note that the system has a single steady state available to it, namely, $x^* = S/a$, and that this steady

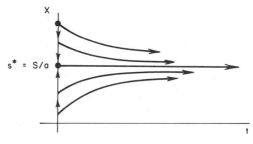

Fig. 29

state is stable; it is approached asymptotically by every solution regardless of initial conditions. If we look at the projections of the solution curves on the X axis, we can see explicitly the way in which the state of the cell changes in time; these projections represent the *system trajectories*, and all approach the point $x^* = S/a$ asymptotically. Thus, this simple cellular system exhibits an entirely standard behavior.

Now let us consider a system consisting of two identical cells of the type we have just considered. The state of each of them at an instant of time t_0 is specified by the morphogen concentrations in each of them at time t_0; we shall denote these morphogen concentrations by $x_1(t_0)$, $x_2(t_0)$, respectively. If we assume that the cells are not interacting in any way, then the two-cell system is governed by the pair of independent equations

$$dx_1/dt = -ax_1 + S, \qquad dx_2/dt = -ax_2 + S. \tag{38}$$

Each of these equations behaves as shown in Fig. 29. However, it is instructive to visualize the behavior of the system according to the representation in Fig. 30.

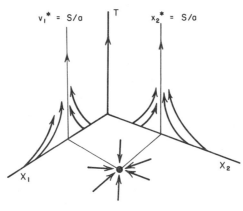

Fig. 30

In the figure we see the X_1–T plane and the X_2–T plane, in which the solutions to the two independent equations are displayed as in Fig. 29. However, we can now use these pairs of solution curves in the two planes to obtain projections (the system trajectories) on the X_1–X_2 plane. This plane, the points of which represent precisely the states of our two-cell system, is the state plane, or state space, for the system, and the properties of the system trajectories in this plane indicate in a particularly straightforward pictorial way the properties and behavior of the system.

If we now look at the steady-state behavior of the system of Eq. (38), we find that there is a single steady state, the coordinates of which are given by

$$x_1{}^* = x_2{}^* = S/a.$$

It is further clear that this steady state is stable, and is approached from any initial state. Further, since $x_1{}^*$, $x_2{}^*$ represent the steady-state concentrations of morphogen in the first and second cell, respectively, it is clear that this system will move according to our thermodynamic intuitions, from any initial inhomogeneous state (that is, one in which the morphogen concentration in the two cells is different) to a homogeneous state (in which the morphogen concentration in the two cells is the same).

Now let us suppose that we complicate our system somewhat by allowing a coupling between the two cells, that is, allow an exchange of morphogen between them. We shall suppose that this coupling is proportional to the concentration difference between the two cells, according to the following equations:

$$\begin{align} dx_1/dt &= -ax_1 + S + D(x_1 - x_2), \\ dx_2/dt &= -ax_2 + S + D(x_2 - x_1) \end{align} \tag{39}$$

or

$$\begin{align} dx_1/dt &= (D - a)x_1 - Dx_2 + S, \\ dx_2/dt &= -Dx_1 + (D - a)x_2 + S. \end{align} \tag{40}$$

To solve these equations, if we write

$$\begin{align} u_1 &= x_1 + x_2 - 2S/a, \\ u_2 &= x_1 - x_2, \end{align} \tag{41}$$

then the equations take the form

$$du_1/dt = -au_1, \tag{42a}$$

$$du_2/dt = (2D - a)u_2, \tag{42b}$$

which again look like a pair of uncoupled (independent) linear equations like Eqs. (38). The single steady state available to the system in the u_1, u_2 coordinates is at the origin $(0, 0)$. However, the stability properties of this steady state now depend entirely on the sign of the coefficient $(2D - a)$ in

Eq. (42b). If this sign is negative (that is, if $D < a/2$), then the steady state
at the origin is approached by all solutions regardless of initial conditions,
as shown in Fig. 31. However, if $(2D - a) > 0$ (that is, if the coupling D
between the two cells is greater than $a/2$), then the solutions of Eq. (42b)
are *growing* exponentials, yielding the hyperbolic trajectories shown in Fig.
32. The relation of the U_1–U_2 axes to the original X_1–X_2 axes, which specify
the morphogen concentrations, are shown in Fig. 33. The origin of coordi-

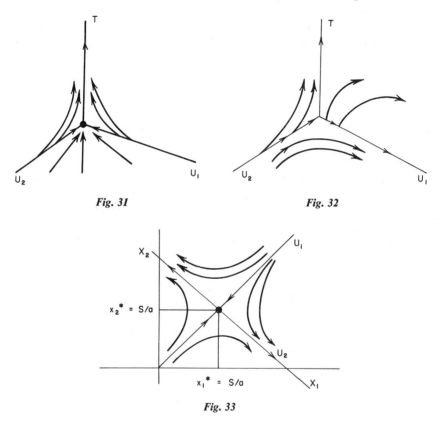

Fig. 31   Fig. 32

Fig. 33

nates in the U_1–U_2 axes is just the (homogeneous) steady state for the system
in the X_1–X_2 coordinates. However, when $(2D - a) > 0$, this steady state
is now *unstable*. Any perturbation away from this steady state, and off the
line $u_2 = 0$ (that is, any perturbation for which $x_1 \neq x_2$ such that the
initial amount of morphogen in the two cells is different) will determine a
system trajectory which moves along a hyperbolic path farther and farther
from that steady state. However, since the morphogen concentrations are

intended to represent concentrations of a real chemical substance, we are always restricted to the region of the X_1–X_2 plane for which $x_1 \geqslant 0$, $x_2 \geqslant 0$. This means that the trajectory must stop when it intersects either the X_1 axis ($x_2 = 0$) or the X_2 axis ($x_1 = 0$). But any point on the X_1 axis corresponds to a situation in which there is no morphogen in the second cell, by definition; likewise any point on the X_2 axis corresponds to a situation in which there is no morphogen in the first cell. Thus, the trajectories shown in Fig. 33 may be described verbally as follows. Equations (39) still admit a homogeneous steady state, just as did Eqs. (38), which were predicated on the absence of coupling between the two cells. However, if the coupling is sufficiently strong, this steady state can become *unstable*. When this happens, any deviation from exactly equal morphogen concentration in the two cells becomes progressively magnified, until the cell which initially had less morphogen eventually becomes entirely free of morphogen, while the cell which initially had more morphogen ultimately contains all the morphogen.

Thus, as may be seen from Eq. (42b), the difference of morphogen concentration acts autocatalytically, via a positive feedback loop, to automatically generate an inhomogeneity or polarity from an initially homogeneous steady state. The conceptual similarity with the Rashevsky model of the preceding section should now be apparent. However, the Turing model is in many ways considerably simpler; the equations involved are ordinary differential equations (instead of partial differential equations), and they are linear (though supplemented with a boundary condition or constraint on the positivity of the morphogen concentrations which in fact makes their behavior quite nonlinear indeed). Furthermore, it provides a simple example of a system which violates the thermodynamic intuition that a system will necessarily tend to move from a state of inhomogeneity to one of homogeneity.

Although the simple one-morphogen system we have constructed above has the advantage that the basic behavior of the system is easily visualizable, it is unsatisfactory in a variety of ways. Perhaps the most unbiological aspect of the one-morphogen system is that the coupling between the two cells which we have introduced is of an active character, rather than being simple, passive diffusion. However, it is possible to remedy this defect, if we allow the state of each of our cells to be specified, not by the concentration of a single morphogen x, but by the concentrations of two distinct morphogens x and y. Thus, the state space of each individual cell now becomes two dimensional, and the state space of a two-cell system becomes four dimensional, instead of the corresponding one- and two-dimensional state spaces we employed before. The higher dimensionality clearly involves the loss of visualizability, which is why it is best to proceed initially with the less plausible but more transparent system.

For two morphogens x, y the corresponding Turing equations for a single cell, analogous to Eq. (37), take the form

$$dx/dt = ax + by + S_1,$$
$$dy/dt = cx + dy + S_2. \tag{43}$$

The reactive part of these equations is no longer a simple decay (the only reaction available in the one-morphogen case), but may now be any first-order reaction involving the two morphogens. For a two-cell system, interacting now through passive diffusion, the equations analogous to Eqs. (39) will be of the form

$$dx_1/dt = ax_1 + by_1 + S_1 + D_1(x_2 - x_1),$$
$$dy_1/dt = cx_1 + dy_1 + S_2 + D_2(y_2 - y_1),$$
$$dx_2/dt = ax_2 + by_2 + S_1 + D_1(x_1 - x_2),$$
$$dy_2/dt = cx_2 + dy_2 + S_2 + D_2(y_1 - y_2). \tag{44}$$

Here D_1, D_2 represent the diffusion constants for the morphogens x, y, respectively.

It is clear that if we take a pair of uncoupled cells [that is, put $D_1 = D_2 = 0$ in Eqs. (44)], the steady state of this system will not be affected by the introduction of diffusive coupling, although of course the stability properties of the steady state may be changed. These stability properties will now depend on the relation between the various parameters a, b, c, d, S_1, S_2, D_1, D_2. Such an analysis is complicated; we shall content ourselves, following Turing, with exhibiting a specific example of a system of the form (44) for which the homogeneous steady state is unstable, analogous to the situation shown in Fig. (33). Indeed, let us put

$$a = 5, \quad b = -6, \quad c = 6, \quad d = -7,$$
$$S_1 = S_2 = 1, \quad D_1 = 0.5, \quad D_2 = 4.5. \tag{45}$$

This system clearly has a homogeneous steady state at the point $x_1^* = y_1^* = x_2^* = y_2^* = 1$. But this steady state is unstable, with the parameter values we have chosen. For if we take the initial conditions as

$$x_1^0 = 1.06, \quad x_2^0 = 0.94, \quad y_1^0 = 1.02, \quad y_2^0 = 0.98 \tag{46}$$

(that is, an initial configuration with more of both morphogen in the first cell than in the second) then it is readily verified that both morphogens will tend to accumulate in the first cell (that is, *against* the initial concentration gradient) at initial rates of 0.12 and 0.04, respectively. Thus, this more realistic model, built from only first-order reactions and passive diffusion, has the same property as the simpler one-morphogen system, namely, to magnify any random departure from a homogeneous but unstable steady state, and thereby spontaneously generate the appearance of inhomogeneity or polarity in the system.

These Turing models can be generalized in a variety of ways. For one thing, we can use a larger number of morphogens, with more general reaction or diffusion kinetics connecting them. Another kind of generalization involves the use of more complicated configurations of cells. Indeed, once we have specified that cells interact with each other through diffusions across common boundaries, we can write down the dynamical equations for any topological array of cells; the topology of the array will be mirrored in the form of the diffusional terms (since these terms tell us essentially which cells are diffusionally interacting; hence which cells are neighbors). For instance, let us consider with Turing the case of a ring of N cells, as shown in Fig. 34. Here each

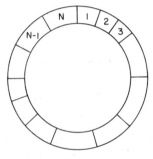

Fig. 34

cell has exactly two neighbors, and it is clear that, in the two-morphogen case, the system as a whole is represented by the $2N$ equations

$$dx_i/dt = ax_i + by_i + D_1(x_{i+1} + x_{i-1} - 2x_i) + S_1,$$
$$dy_i/dt = cx_i + dy_i + D_2(y_{i+1} + y_{i-1} - 2y_i) + S_2,$$

$$i = 1, \ldots, N, \quad (47)$$

provided we identify cell N with cell 0, and denote the concentrations of morphogen x and y in the ith cell by x_i, y_i, respectively.

Following Turing, let us investigate the behavior in such a ring of cells more closely. As we noted above for the simplest Turing system, if a single isolated cell has a (stable) steady state $x = x^*$, $y = y^*$, then any array whatever will have a steady state with the concentration of x morphogen in each cell equal to x^* and the concentration of y morphogen in each cell equal to y^*. We can get rid of the source terms (the constants) in Eqs. (47) above by moving the origin of coordinates in the $2N$-dimensional state space of the system to this steady state; in the new variables

$$\bar{x}_i = x_i - h, \qquad \bar{y}_i = y_i - k,$$

Eqs. (47) will have the form

$$d\bar{x}_i/dt = a\bar{x}_i + b\bar{y}_i + D_1(\bar{x}_{i+1} + \bar{x}_{i-1} - 2\bar{x}_i),$$
$$d\bar{y}_i/dt = c\bar{x}_i + d\bar{y}_i + D_2(\bar{y}_{i+1} + \bar{y}_{i-1} - 2\bar{y}_i). \qquad i = 1, \ldots, N, \qquad (48)$$

Turing solves these equations by making the linear coordinate transformation,

$$x_r = \sum_{s=0}^{N-1} \exp(2\pi irs/N)\, \xi_s,$$
$$\qquad\qquad\qquad\qquad\qquad r = 1, \ldots, N, \qquad (49)$$
$$y_r = \sum_{s=0}^{N-1} \exp(2\pi irs/N)\, \eta_s.$$

The inverse transformation is given by

$$\xi_r = N^{-1} \sum_{s=1}^{N} \exp(-2\pi irs/N)x_s,$$
$$\qquad\qquad\qquad\qquad\qquad r = 1, \ldots, N, \qquad (50)$$
$$\eta_r = N^{-1} \sum_{s=1}^{N} \exp(-2\pi irs/N)y_s.$$

It is now readily verified that, in the new state variables ξ_r, η_r, the dynamical Eqs. (48) assume the form

$$d\xi_s/dt = (a - 4D_1 \sin 2\pi s/N)\xi_s + by_s,$$
$$\qquad\qquad\qquad\qquad\qquad s = 1, \ldots, N, \qquad (51)$$
$$d\eta_s/dt = c\xi_s + (d - 4D_2 \sin 2\pi s/N)\eta_s;$$

that is, in these variables, the $2N$ equations break down into N noninteracting pairs of the form (51). Each such pair can be solved by ordinary linear methods; indeed, if $\lambda_1^{(s)}$, $\lambda_2^{(s)}$ are the (distinct) eigenvalues of the matrix

$$\begin{pmatrix} a - 4D_1 \sin 2\pi s/N & b \\ c & d - 4D_2 \sin 2\pi s/N \end{pmatrix}, \qquad (52)$$

that is, roots of the characteristic equation

$$\begin{vmatrix} [a - 4D_1 \sin 2\pi s/N] - \lambda & b \\ c & [d - 4D_2 \sin 2\pi s/N] - \lambda \end{vmatrix} = 0, \qquad (53)$$

then we can write the solutions of Eqs. (51) in the form

$$\xi_s = A_s \exp(\lambda_1^{(s)}t) + B_s \exp(\lambda_2^{(s)}t),$$
$$\qquad\qquad\qquad\qquad\qquad (54)$$
$$\eta_s = C_s \exp(\lambda_1^{(s)}t) + D_s \exp(\lambda_2^{(s)}t),$$

and, substituting Eqs. (54) back into Eqs. (51) and (49), we find, in the original state variables,

$$x_r = h + \sum_{s=1}^{N} [A_s \exp(\lambda_1^{(s)}t) + B_s \exp(\lambda_2^{(s)}t)] \exp(2\pi irs/N),$$
$$\qquad\qquad\qquad\qquad\qquad r = 1, \ldots, N, \qquad (55)$$
$$y_r = k + \sum_{s=1}^{N} [C_s \exp(\lambda_1^{(s)}t) + D_s \exp(\lambda_2^{(s)}t)] \exp(2\pi irs/N).$$

The stability properties of the system will clearly depend on the properties of the $2N$ numbers $\lambda_1^{(s)}$, $\lambda_2^{(s)}$. The sums (55) will be dominated by those terms for which these numbers have the largest real part. If we assume that these numbers are all real, and that the largest of them is the s_0th, then we can write (55) asymptotically as

$$x_r = 2A_{s_0} \exp[2\pi i s_0 r/N + \lambda^{(s_0)} t] + h,$$
$$y_r = 2C_{s_0} \exp[2\pi i s_0 r/N + \lambda^{(s_0)} t] + k. \tag{56}$$

From this expression it follows that, as a function of r, the concentration of each morphogen has s_0 maxima, which may be regarded as the peaks of a standing wave around the ring; there are likewise s_0 minima, which may be regarded as the troughs of the standing wave. The situation in this case is as diagramed in Fig. 35. If there is an instability, then the amplitude of this standing wave grows with time. It was suggested by Turing that arguments of this sort might account for the kinds of tentacled patterns which are found in coelenterates, for example, around the head of a hydra.

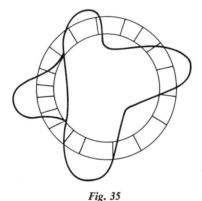

Fig. 35

There are a variety of other cases [for example, those in which the eigenvalues of the matrix (52) are complex], but it is not necessary for us to examine them here; the interested reader is referred to Turing's original paper [Turing, 1952].

It is clear that systems like that of Turing can spontaneously generate many interesting kinds of patterns. Moreover, since they do represent examples of systems which move spontaneously from a homogeneous to a nonhomogeneous state, they are of interest to scientists interested in foundations of thermodynamics, particularly in the "far-from-equilibrium" condition. For this reason, the Turing systems have recently been extensively investigated from this thermodynamic viewpoint by Prigogine and his co-workers [see, for example, Prigogine, 1969] as examples of a class of behaviors which, for obvious reasons, he calls "symmetry-breaking instabilities." The interested reader is referred to these treatments for further details.

D. The Keller–Segel Models

In this section we shall consider yet a third variation on the theme of instabilities as a model for the generation of inhomogeneities in biological systems. The model to be considered [Keller and Segel, 1970] involves the phenomenon of *aggregation* which occurs at one stage in the life cycle of the cellular slime molds. These organisms are most interesting in themselves, and have more recently been studied with still greater interest as simple biological models of more complex morphological and developmental processes that occur in multicellular organisms. The portion of the slime mold life cycle which interests us here concerns the transition from a population of free-living single amoebas uniformly distributed on the surface of a medium to a highly aggregated form in which the amoebas are localized within a relatively small number of compact units, or slugs, which behave as entire new organisms. Eventually these slugs differentiate to form a fruiting body and spores; the spores give rise to new amoebas and the cycle repeats. We shall be concerned here only with that phase in which the amoebas cease to be uniformly distributed over the surface of the medium on which they live and begin their aggregation.

It was postulated some time ago that the aggregation process was chemotactic in nature, depending on the concentration of a hypothesized substance, *acrasin*, secreted by the individual amoebas. The cells also produced an extracellular enzyme, acrasinase, which inactivated acrasin. They were considered positively chemotactic, moving toward regions of greater acrasin concentration. For details on these hypotheses and the biological nature of the slime molds, see Bonner [1967].

To model this system mathematically, we shall suppose that we can treat the density of a population of amoebas, distributed over a plane region A, as a continuous function of position and time. This in itself, of course, is a substantial idealization, but it is no different from the sorts of arguments used in treating chemical kinetics or population dynamics by analytical methods. Thus, we shall assume that the amoeba density, measured, for example, in mass per unit area, is represented by a continuous function $a(x, y, t)$. We can obtain a relationship for this function from the same kinds of conservation arguments applicable to general diffusion problems, regarding the amoebas essentially as a diffusing chemical species. In particular, we can write

$$(\partial/\partial t) \int_A \int a(x, y, t)dx \, dy = \int_A \int Q_a \, dx \, dy - \int_S \mathbf{J}_a \cdot \mathbf{\eta} \, ds, \qquad (57)$$

where Q_a is the rate of production of amoebas per unit volume per unit time, $\mathbf{J}_a$ is the flux of amoeba mass, and $\mathbf{\eta}$ is the unit normal vector to the boundary S of the region A. If the hypotheses of the divergence theorem are satisfied, we can rewrite Eq. (57) as

$$(\partial/\partial t) \int_A \int a(x, y, t) \, dx \, dy = \int_A \int Q_a \, dx \, dy - \int_A \int \mathbf{V} \cdot \mathbf{J}_a \, dx \, dy \qquad (58)$$

or, if we can differentiate under the integral sign,

$$\int_A \int [\partial a/\partial t - Q_a + \mathbf{V} \cdot \mathbf{J}_a] \, dx \, dy = 0 \qquad (59)$$

from which it follows, since A is arbitrary, that $a(x, y, t)$ must satisfy the differential equation

$$\partial a/\partial t = Q_a - \mathbf{V} \cdot \mathbf{J}_a. \qquad (60)$$

We shall suppose that the amoebae are not dividing; that is, that $Q_a \equiv 0$.

Now the flux $\mathbf{J}_a$ of amoeba mass depends on two factors: the active chemotaxis along a positive gradient of acrasin, and a random diffusive movement due to the motility of the individual cells. Qualitatively, the first of these will cause the amoebas to tend to accumulate towards regions of high acrasin concentration; the second will cause the amoebas to accumulate towards regions of low amoeba concentration. Assuming that both these movements obey the laws of standard diffusion, we can write

$$\mathbf{J}_a = D_1 \mathbf{V}\rho - D_2 \mathbf{V}a, \qquad (61)$$

where $\rho = \rho(x, y, t)$ is the acrasin density in the region A, and D_1, D_2 are in general functions of ρ and a. In particular, D_1 measures the influence of the acrasin gradient on amoeboid motion, while D_2 measures the influence of the random diffusive movement to the overall amoeboid motion. Putting all these results into Eq. (60), we find

$$\partial a/\partial t = -\mathbf{V} \cdot (D_1 \mathbf{V}\rho) + \mathbf{V} \cdot (D_2 \mathbf{V}a). \qquad (62)$$

This expression involves, of course, the acrasin density ρ. We can obtain a similar expression for ρ by using the same technique we have just employed for a. We have

$$\partial \rho/\partial t = Q_\rho - \mathbf{V} \cdot \mathbf{J}_\rho. \qquad (63)$$

Just as before,

$$\mathbf{J}_\rho = -D_\rho \nabla \rho, \tag{64}$$

where we may assume that D_ρ is a constant. The production term Q_ρ, involving sources and sinks of acrasin in the region A, may be split into an intrinsic rate of production $f(\rho)$ of acrasin per amoeba per unit time, and the destruction of acrasin through interaction with acrasinase. Denoting acrasinase by E, and assuming standard Michaelis–Menten kinetics of the form

$$\rho + E \underset{k_{-1}}{\overset{k_1}{\rightleftharpoons}} C \xrightarrow{k_2} E + \text{products} \tag{65}$$

we can write

$$Q_\rho = k_1 \rho E + k_{-1} C + a f(\rho). \tag{66}$$

Putting all these results together, we have

$$\partial \rho / \partial t = -k_1 \rho E + k_{-1} C + a f(\rho) + D_\rho \nabla^2 \rho. \tag{67}$$

We can by similar reasoning also obtain the equations

$$\partial C / \partial t = k_1 \rho E - (k_{-1} + k_2) C + D_C \nabla^2 C \tag{68}$$

and

$$\partial E / \partial t = -k_1 \rho E + (k_{-1} + k_2) C + a g(\rho, E) + D_E \nabla^2 E, \tag{69}$$

where $g(\rho, E)$ is the rate of acrasinase production per amoeba per unit time. The four equations (62), (67), (68), and (69) completely determine the system.

We can simplify the relation (67) by using the classical ideas of enzyme kinetics as applied to the acrasin–acrinase interaction (65). The "steady state" assumption is that

$$k_1 \rho E - (k_{-1} + k_2) C = 0 \tag{70}$$

and the conservation condition (which we may suppose to be at least approximately satisfied) states that

$$E + C = E_0 = \text{const} \tag{71}$$

for which it follows that

$$E = E_0 / (1 + K\rho), \tag{72}$$

where K is the Michaelis constant $k_1 / (k_{-1} + k_2)$. Substituting this in (67), and writing

$$k(\rho) = E_0 k_2 K / (1 + K\rho),$$

we have

$$\partial \rho / \partial t = -K(\rho)\rho + a f(\rho) + D_\rho \nabla^2 \rho. \tag{73}$$

This last equation, together with Eq. (62), constitute a simpler pair of equations characterizing the system.

We now suppose that the system admits a steady state, that is, solutions

$$a(x, y, t) = a_0, \qquad p(x, y, t) = p_0, \tag{74}$$

where a_0, p_0 are constants; such a solution would correspond to constant density of amoebas and acrasin in A. To investigate the stability of such a solution, we shall use the standard trick of linearization. We expand the right-hand sides of Eqs. (62) and (73) around the steady state (a_0, p_0), and discard all terms of higher degree than the first. Thus, we find

$$\partial \bar{p}/\partial t = [a_0 f'(p_0) - \bar{k}]\bar{p} + f(p_0)\bar{a} + D p \nabla^2 \bar{p}, \tag{75}$$

$$\partial \bar{a}/\partial t = -D_1(a_0, p_0)\nabla^2 \bar{p} + D_2(a_0, p_0)\nabla^2 \bar{a}, \tag{76}$$

where

$$\begin{aligned}
\bar{a}(x, y, t) &= a(x, y, t) - a_0, \\
\bar{p}(x, y, t) &= p(x. y, t) - p_0, \\
\bar{k} &= k(p_0) + p_0 k'(p_0)
\end{aligned} \tag{77}$$

and

$$|\bar{a}| \ll a_0, \qquad |\bar{p}| \ll p_0.$$

If we assume solutions to Eqs. (62) and (73) of the form

$$\bar{a} = \bar{a}_0 \cos(\alpha x + \beta y)e^{\sigma t}, \qquad \bar{p} = \bar{p}_0 \cos(\alpha x + \beta y)e^{\sigma t}, \tag{78}$$

where $\bar{a}_0$, $\bar{p}_0$, x, β, σ are constants, and substitute these presumed solutions into the linearized Eqs. (75), (76), we find the conditions

$$\begin{aligned}
\{[f'(p_0)a_0 - \bar{k} - (\alpha^2 + \beta^2)D_p] - \sigma\}\bar{p}_0 + f(p_0)\bar{a}_0 &= 0, \\
D_1(p_2, a_0)(\alpha^2 + \beta^2)\bar{p}_0 - [D_2(p_0, a_0)(\alpha^2 + \beta^2) + \sigma]\bar{a}_0 &= 0,
\end{aligned} \tag{79}$$

which is a pair of linear equations for $\bar{p}_0$, $\bar{a}_0$. These equations will have a nontrivial solution if and only if the determinant of the coefficients vanishes. If we now write

$$F = f'(p_0)a_0 - \bar{k} - q^2 D_p, \qquad q^2 = \alpha^2 + \beta^2, \tag{80}$$

this condition becomes

$$\begin{vmatrix} F - \sigma & f(p_0) \\ D_1 q^2 & -(D_2 q^2 + \sigma) \end{vmatrix} \tag{81}$$

or that σ must be a root of the equation

$$\sigma^2 - \sigma(F - q^2 D^2) - (q^2 f(p_0)D_1 + q^2 D_2 F) = 0. \tag{82}$$

But from the assumption (78), σ measures the *stability* of the steady state (a_0, p_0); if $\sigma < 0$, all solutions damp back toward the steady state, while if $\sigma > 0$, the solutions will generally grow away from the steady-state solution. It is now easy to verify that the condition that the roots of Eq. (82) be negative is

$$D_1 f(p_0) + D_2 f(p_0)a_0 < D_2(\bar{k} + D_p q^2). \tag{83}$$

If this inequality is not satisfied, then the steady-state solution is *unstable*, and exactly as we have seen before, a random perturbation of the system away from this steady state will generally be increasingly amplified by the system. Since that steady state represents the homogeneous distribution fo amoebas and acrasin, a departure from this steady state represents an aggregation.

E. Morphogenesis and Epigenesis

The morphogenetic models we have presented so far, particularly those dealing with the generation of polarities and inhomogeneities, are of a rather general character. As noted earlier, they represent demonstrations of the *capability* of a class of biologically plausible systems to exhibit a certain kind of biologically significant behavior. However, such demonstrations of capability, while important, are not entirely satisfactory to many biologists, nor do they even theoretically represent a complete solution to the problem of understanding how form and pattern are generated in specific biological systems. For in order to have such a complete understanding in a specific system, we would need to identify those substances which play the role of morphogens (although this is not an entirely trivial kind of problem even conceptually; see Rosen, 1968) and show that the kinetics of these substances are in accord with the kind of dynamics postulated by Turing or Rashevsky (or at least behaves qualitatively similarly to them). Even if we do not have such detailed information available, it would be important to relate these models to the overall and rather detailed picture presently available regarding the genetic control of cellular events. For clearly, differentiation is a genetically controlled process, while the mechanisms we have considered for the generation of gradients and inhomogeneities seem independent of any genetic mechanism.

In Chapter 2, Volume II, we consider in detail a number of mechanisms for the control of gene expression in specific biochemical terms, through pairs of coupled excitation and inhibition mechanisms (particularly induction and repression of genes, and activation and inhibition of catalytic protein). These controls on the expression of genes were what we called *epigenetic*, and we pointed out at that time that such epigenetic mechanisms find one of their major applications in an understanding of the genetic control of differentiation. The question, then, is how we are to relate the specific epigenetic mechanisms described in Chapter 2, Volume II to the rather more general differentiation models considered herein.

Perhaps the simplest connection between the two kinds of models can be made by looking upon a morphogen (in the sense of the Turing models) as an inducer or repressor of a particular gene. The generation of an inhomogeneity (or gradient) in the concentration of such a morphogen in a population of initially chemically and genetically identical cells will, if the inhomo-geneity is sufficiently pronounced, lead to a derepression of a previously repressed gene, or the repression of a previously induced gene, in one local-ized region of the aggregate. This in turn leads either to the production of a new catalytic protein, or the loss of some previously present catalytic protein. In the first case, it is easy to suppose futher that the reaction catalyzed by the new protein produces a product which is itself a repressor or derepressor of other genes in the cell, and so on. In the second case, the loss of a catalytic protein may result in the loss of a metabolite (the product of the reaction catalyzed by the protein) which was itself a repressor for another gene. Thus in all circumstances, the initial inhomogeneity in a morphogenetic substance (which is itself not necessarily of a genetic character) generated through a Turing-like mechanism, can be amplified and cascaded through an aggregate of initially identical cells through a sequence of repressions and derepressions. This kind of coupling between epigenetic mechanisms, involving as they do a specific picture of the relationships between gene expression and cellular activities, and arguments based on the generation of instabilities in initially homogeneous populations, gives us the grounds for a much more complete picture of morphogenesis through differentiation than either mechanism could do alone. This is because epigenetic mechanisms are basically of an *intracellular* character; they tell us little about how events within cells deter-mine the kinds of gross phenomena which we observe in populations of cells (such as the generation of polarities); on the other hand the instability models, while geared to a discussion of populations, seem too crude in terms of what we know of individual cells. By combining the two kinds of mecha-nisms we can extract the best features of both for an overall understanding of differentiation mechanisms and their control.

Of course, we have barely been able to scratch the surface of complex phenomena like differentiation in these pages. There is an enormous litera-ture, dating back to the turn of the century, on the control of differentiation through gradients of concentration of morphogenetic substances; empirical concepts like the evocation of a particular pattern of differentiation in response to an organizing stimulus from another cell type or tissue; or the competence of a cell to respond to such a stimulus, could be formulated and, to a certain extent, understood in terms of the generation and maintenance of material gradients. The kinds of models we discussed above, based on instabilities, represent only the most recent developments in this long history of study of differentiation; a detailed historical treatment would be beyond the scope of this text, but the reader is strongly recommended to consult some of the older literature [for example, Child, 1941] for a fuller picture.

It must also be remembered that any picture of differentiation based on gradients being established and maintained, and on the existence of threshold concentrations of substances required to elicit a particular developmental effect, are open to certain kinds of general criticisms, which have been perhaps best articulated by Wolpert [1969]. Nevertheless, it is certainly true that we have at hand the formal basis for understanding at least certain important aspects of morphogenesis, and that much remains to be done in exploring simply the properties of these.

General References

Anfinsen, C. B., Jr. [1970]. "Aspects of Protein Biosynthesis." Academic Press, New York.

Berrill, N. J. [1961]. "Growth, Development and Pattern." Freeman, San Francisco, California.

Bonner, J. T. [1967]. "The Cellular Slime Molds," 2nd. ed. Princeton Univ. Press, Princeton, New Jersey.

Caspar, D. L. D. [1963]. Assembly and stability of the tobacco mosaic virus particle, *Advan. Protein Chem.* **18**, 37–121.

Caspar, D. L. D., and Klug, A. [1960]. The structure of small viruses, *Advan. Virus Res.* **7**, 225–325.

Caspar, D. L. D., and Klug, A. [1962], Physical principles in the construction of regular viruses, Cold Springs Harbor Symposia XXVII.

Caspar, D. L. D., and Klug, A. [1963]. Structure and assembly of regular virus particles, *Viruses, Nucleic Acid and Cancer, Symp. Fundam. Cancer Res., 17th*, 27–39.

Child, C. M. [1941]. "Patterns and Problems of Development." Chicago Univ. Press, Chicago, Illinois.

Crane, H. R. [1950]. "Principles and problems of biological growth, *Sci. Mon.* **70**, 376–389.

Crick, F. H. C., and Watson, J. D. [1956]. The structure of small viruses, *Nature* **177**, 473–475.

Goel, N., and Leith, A. G. [1970]. Self-sorting of anisotropic cells, *J. Theor. Biol* **28**, 469–482.

Goel, N., Campbell, R. D., Gordon, R., Rosen, R., Martinez, H. M., and Ycas, M. (1970). Self-sorting of isotropic cells, *J. Theor. Biol.* **28**, 423–468.

Grover, J. W. [1961]. The enzymatic dissociation and reproducible reaggregation *in vitro* of 11-day embryonic chick lung, *Develop. Biol.* 3, 555–568.

Grover, J. W. [1962]. The influence of age and environmental factors on the behaviour of re-aggregated embryonic lung cells in culture, *Exp. Cell Res.* **26**, 344–359.

Grover, J. W. [1963]. Reaggregation and organotypic redevelopment of dissociated embryonic chick lung cells in short-term culture, *Nat. Cancer Inst. Monogr.* **11**, 35–50.

Keller, E. F., and Segel, L. A. [1970]. Initiation of slime mold aggregation viewed as an instability, *J. Theor. Biol.* **26**, 397–415.

Monod, J., Wyman, J., and Changeaux, J. P. [1965]. On the nature of allosteric transitions, *J. Mol. Biol.* **12**, 88–118.

Moscona, A. A. [1965]. Recombination of dissociated cells and the development of cell aggregates, *in* "Cells and Tissues in Culture" (E. N. Willmer, ed.), Vol. I. Academic Press, New York.

Perutz, M. F. [1965]. Structure and function of haemoglobin I. A tentative atomic model of horse oxyhaemoglobin, *J. Mol. Biol.* **13**, 646–668.

Perutz, M. F., Kendrew, J. C. and Watson, H. C. [1965]. Structure and function of haemoglobin. II. Some relations between polypeptide chain configuration and amino acid sequence, *J. Mol. Biol.* **13**, 669–678.

Prigogine, I. [1969]. Structure, dissipation and life, *in* "Theoretical Physics and Biology" (M. Marois, ed.), pp. 23–52. North-Holland Publ., Amsterdam.

Rashevsky, N. [1961]. "Mathematical Biophysics," 3rd ed. Dover, New York.

Rosen, R. [1968]. Turing's morphogens, two-factor systems and active transport, *Bull. Math. Biophys.* **26**, 493–499.

Schroedinger, E. [1944]. "What is Life?" Cambridge Univ. Press, Cambridge, Massachusetts.

Steinberg, M. [1963]. Tissue reconstruction by dissociated cells, *Science* **141**, 401–408.

Steinberg, M. [1964]. The problem of adhesive selectivity in cellular interaction, *in* "Cellular Membranes in Development" (M. Locke, ed.), p. 344. Academic Press, New York.

Thompson, D'Arcy, W. [1961]. "On Growth and Form." Harvard Univ. Press, Cambridge, Massachusetts.

Toth, F. [1964]. "Regular Figures." Macmillan, New York.

Townes, P. L., and Holtfreter, J. [1944]. Directed movements and selective adhesion of embryonic amphibian cells, *J. Exp. Zool.* **128**, 53–120.

Trinkhaus, J. P. [1969]. "Cells into Organs." Prentice-Hall, Englewood Cliffs, New Jersey.

Turing, A. M. [1952]. The chemical basis of morphogenesis, *Phil. Trans. Roy. Soc. B* **237**, 5–72.

Wartiovaara, J. [1966]. Cell contacts in relation to cytodifferentiation in metanephrogenic mesenchyme *in vitro*, *Ann. Med. Exp. Biol. Fenn.* **44**, 469–503.

Weyl, H. [1952]. "Symmetry." Princeton Univ. Press, Princeton, New Jersey.

Wilson, H. V. [1907]. On some phenomena of coalescence and regeneration in sponges, *J. Exp. Zool.* **5**, 245–258.

Wolpert, L. [1969]. Positional information and the spatial pattern of cellular differentiation, *J. Theor. Biol.* **25**, 1–47.

Wood, R. L. [1965]. An electron microscope study of developing bile canaliculi in the rat, *Anat. Rec.* **151**, 507–530.

Chapter 2

MECHANICS OF EPIGENETIC CONTROL

Robert Rosen

Center for Theoretical Biology
State University of New York at Buffalo
Amherst, New York

I. Introduction

We do not need to emphasize the dominant role which the cell theory has played in biology. The methodological reasons for the preeminence of the cell concept, however, are generally not stressed, but they are both subtle and important. The first basic fact, now regarded as too trivial to mention, is that all organisms are in fact composed entirely of cells; that is, they can effectively be regarded as being constructed in what is nowadays called a *modular* fashion, or as complicated arrays composed of a small number of basic kinds of units. In this way we can regard the complex physiological behavior of any multicellular organism as being caused by, and reflected in, the corresponding behavior of the cells of which it is composed. Second, despite vast quantitative differences, all cells seem to share the same overall qualitative characteristics of organization; this indeed is why we recognize such diverse entities as a bacterial cell, a protozoan, an alga, or the many kinds of cells found in complex multicellular organism as in fact being cells. Third, the properties of cells seem relatively simpler to understand than the properties of complex organisms. From these three simple observations stems the overwhelming importance of the cell concept: It suggests a strategy whereby we may decompose or analyze complicated biological activities into relatively simple parts (cells) that may be studied and understood in isolation; from an understanding of these parts we may then synthesize an understanding of the complex behaviors which were our original interest. Thus it is that cell biology has come to play a predominant role in all of the biological sciences.

Furthermore, organization at the cellular level seems to offer us a closer contact with the quantitative and highly developed sciences of physics and chemistry than can be hoped for at higher organizational levels in biological systems. Thus, the basic tools of cell biology, besides those of conventional observation, have come to include more and more of the tools of the physicist and chemist, and the basic data of cell biology to be more and more related to the organization of physical matter within cells. Indeed, the upshot of this entire train of development, culminating in the molecular biology of the past two decades, is this: *The properties of any cell are most conveniently studied in terms of the specific molecular species of which the cell is composed.* This allows us to make close contact on the one hand with the molecular theory of the physicist, and on the other hand with well-developed kinetic and dynamic analysis of the mathematician. It is the dynamical properties of cells, studied in terms of temporal variations of their molecular constituents (both their species and their concentration) with which we shall be concerned in the present chapter.

The most striking characteristics of cells, both in themselves and in the roles they play in the development and physiology of multicellular organisms, have to do with the regulatory and control properties they manifest. The basic problems of cell biology, and by inference, all those problems of physiology that may be studied in terms of cells, may then be regarded as revolving around the question of how it is that a cell is able to control the species and concentrations of the molecules of which it is composed and how it can modulate the interactions between them. We shall want to understand the principles whereby these control mechanisms act to modify the state of the cell, both as a function of its own initial state, and as a function of the influences which impinge on the cell from its environment.

Modern molecular biology and its ancestral disciplines of biochemistry and genetics have provided a convenient framework in terms of which the various control mechanisms that play a role in cell biology may be analyzed.

Perhaps the earliest relevant observation was that most biological reactions occurring in cells are mediated through the activity of exceedingly specific catalysts, or enzymes, which by 1940 had been identified with a specific family of macromolecules (proteins) present in all cells. The specificity of these catalysts provided the first important regulatory principle, since clearly the kinds of enzymes present in a cell, and their respective concentrations, determined which metabolic reactions would occur in that cell and at what rate. In line with the basic philosophy of cell biology we have enunciated above, then, the questions of cell biology can be reduced to questions regarding the kinds and concentrations of catalytic proteins present in a cell.

We push the question back one step: What is it which determines the kinds of catalytic proteins that will be present in a cell and in what proportions? The answer here was sought from genetics, and by 1960 the answer was: The role of the genome in a cell is precisely to provide the information required to specify the structure of catalytic protein; the genome of a cell is thus the full library of catalytic potentialities which the cell can exhibit. This library is preserved unchanged (barring mutations) by means of an autocatalytic replication of the genes themselves, while the genes can "catalyze," in a sense, the formation of specific enzymes. The unraveling of the molecular details of these processes, which are now too well known to require elaboration here, represents one of the great scientific advances of the century.

In this framework we can thus understand a large percentage of the kinds of regulatory properties that cells exhibit, at least qualitatively. But this cannot be the entire story of cellular regulation. For one thing, the problems of differentiation and development stand in apparent contradiction to this story. For all the cells of a multicellular organism constitute a *clone*, descended in a genetically conservative way from a single parent cell (zygote), and therefore contain the same genome; yet in the adult organism there exist literally hundreds of kinds of cells of the utmost diversity. At a simpler level, bacterial cells grown on different nutrients, or protozoa grown under different conditions, may exhibit equally wide differences in morphology and chemical composition. Furthermore, the phenomena of differentiation and polymorphism in cells are not arbitrarily variable, but are clearly integrated and coordinated into discrete and sharply separated types; a differentiated cell is either a nerve cell or a muscle cell or a liver cell, and never exhibits some of the characteristics of each. From the facts of differentiation, then, it is clear that there are regulatory mechanisms superimposed on the mechanisms of genetic control of enzyme structure, and the enzymatic control of metabolic reactions. These are mechanisms which on the one hand can be regarded as modulating genetic *expression* (that is, the determination of which

genes shall actually be actively engaged in synthetic activities at a particular
time), and also the rates at which catalytic protein shall actually operate.
These we shall call *epigenetic mechanisms*, and these will be the specific subject matter of the present chapter.

We noted earlier that a fruitful formulation of all problems of cellular
regulation and control was in terms of the kinds and concentrations of the
reactants present in a cell at a particular instant (where by reactant we
comprise both the small molecules, or metabolites, and the macromolecules,
including DNA, RNA, and protein, which are involved in catalysis of
cellular processes). If we identify each such species with a coordinate axis in a
Euclidean space of sufficient dimension, and assign to each a number,
representing the concentration of that species in the cell at an instant of time,
then we can identify the *state* of the cell at that instant as a single point in
the Euclidean space. We shall call that space the *state space* of the cell, when
the cell is regarded as a chemical system of interacting molecules. Typically the cell will change state in the course of time, as a consequence of the
chemical reactions taking place within the cell, and of the materials entering
and leaving the cell from its environment; as this happens, the representative
point in the state space will trace out a curve, parameterized by time, which
we shall call a *trajectory* of the system. These curves, or trajectories, represent the solutions of the rate equations obtained according to ordinary
chemical kinetics from a knowledge of the specific reactions occurring within
the system. If we have N reactants $x_1, \ldots, x_N$, these rate equations take on
the familiar dynamical form

$$dx_i/dt = f_i(x_1, \ldots, x_N), \qquad i = 1, \ldots, N. \qquad (1)$$

where $x_1, \ldots, x_N$ represent the concentrations of substances $x_1, \ldots, x_N$.†
The questions of regulation with which we shall be concerned all fall within
this framework, and involve following the change of state of a cell in time
by moving along the corresponding trajectory. We shall basically be concerned with the following kinds of problems: (a) How can a cell stably manifest different stable states in the same environment? (b) How can a cell with
the same library of catalytic specificities (that is, the same genome) manifest
different stable states? Sometimes we shall be primarily interested in these
stable end states, especially if the transition between them is rapid (as in the
adaptation of bacteria to different substrates); sometimes we shall be interested in the details of the trajectory connecting two such end states (so-called
"transient behavior"), especially if the transition is slow, as in many differentiating systems. But as will be clear from the subsequent development, the

---

† Concentrations of all substances throughout this chapter will be denoted by the italic
counterpart of the substance; that is, the concentration of x (or A, B, C) is x (or A, B, C).

language we develop is the same in all cases, and the distinction between them involves merely a change in emphasis.

II. Regulatory Behavior in Open Chemical Systems

In line with our general principles of parsimony (see Chapter 3, Volume I) we are now going to investigate the kind of regulatory phenomena which are exhibited by the most general kinds of chemical systems, that is, those for which the fewest special structural assumptions have been made. Clearly, if a particular kind of regulatory activity is already implicit in such systems, there is no need to postulate special mechanisms to achieve it. And as we shall see, a substantial number of suggestive properties are already manifested by the most general systems. In this section, we undertake an examination of these properties and their interpretations.

According to the principles laid down in the preceding section, we wish to examine the manner in which a system governed by equations of the form of Eq. (1) changes state in time. Let us introduce some familiar terminology, which will be very helpful to us in organizing our ideas. Let us suppose that the states of our system are determined by the concentrations of reactants $x_1, \ldots, x_N$ at an instant of time, and suppose that at our initial time t_0 the concentrations of these reactants are $x_1^0, \ldots, x_N^0$. We shall say that our system is *closed*, if for every state $(x_1(t), \ldots, x_N(t))$ of the system on the trajectory passing through the initial state $x_1^0, \ldots, x_N^0$ we have

$$x_1(t) + \cdots + x_N(t) = x_1^0 + \cdots + x_N^0. \tag{2}$$

Intuitively, this means of course that none of the reactants are entering or leaving the system. The total reactants are thus *conserved* [Eq. (2) is often called a conservation condition], and the system reactions serve only to interconvert the reactants. A system that does not satisfy Eq. (2) will be called an *open system*. Open systems are thus characterized by the fact that there are fluxes of matter through the system, with reactants entering from the system environment on the one hand and leaving the system into the environment on the other. A set of dynamical rate equations like Eq. (1) thus typically refer to an open system; to obtain a closed system we must add an additional constraint condition like Eq. (2). Geometrically, the effect of the conservation equation is to require that the trajectories of the closed system be constrained to lie on an $(N-1)$-dimensional hyperplane in the state space, whose Eq. (2) is uniquely determined by the total amount of initial reactant $x_1^0 + \cdots + x_N^0$. Thus, closed systems composed of the same reactants but with different initial amounts of total reactants will have their trajectories constrained to lie on *distinct and nonintersecting hyperplanes*.

It is always of interest to consider those states of the system for which the rates of change of all of the reactant concentrations simultaneously vanish. If $(x_1{}^*, \ldots, x_N{}^*)$ is such a state, then from Eq. (1) it follows that

$$f_i(x_1{}^*, \ldots, x_N{}^*) = 0 \tag{3}$$

for each $i = 1, \ldots, N$. Intuitively, since the rate of change of each reactant concentration vanishes in such a state, it follows that once the system arrives in the state $(x_1{}^*, \ldots, x_N{}^*)$, it can never leave it as a result of autonomous system activity. Such a state for a closed system is called a state of *equilibrium*; for open systems it is typically called a *steady state*. Equilibria and steady states thus represent *entire trajectories* of closed and open systems, respectively.

We shall also be much concerned with the stability characteristics of equilibria and steady states. Stability refers to the behavior of the system trajectories in the vicinity of these states. We shall say that such a state is *stable* if all of the trajectories in the vicinity approach this state arbitrarily closely as time increases; or, in other words, if a small displacement of the system away from this state causes the system to tend back to the state as time increases. Examples of such stable states are shown in Fig. 1. Conversely, such a state is called *unstable* if at least one trajectory in every neighborhood of the state moves away from that state. The instability is called absolute if every trajectory near the state moves away from the state; examples of such instabilities are shown in Fig. 2. An important example of an instability which is not absolute, but conditional, is the saddle-point behavior shown in Fig. 3. In any case, a system which has been moved away from an unstable steady state or equilibrium typically will not return to it, but will be carried farther and farther away from it by the intrinsic dynamics of the system.

Finally, we have the case of *neutral stability*, in which the nearby trajectories neither approach nor diverge from the state, but remain forever roughly equidistant. This means that the nearby trajectories are closed curves, as shown in Fig. 4. As we shall see later, closed trajectories are very important, since they correspond to *periodic solutions* of the rate equations.

Let us give a very simple example of the distinction between the behavior of closed and open systems in terms of the system trajectories. The simplest closed system of chemical reactions is the monomolecular interconversion of two species:

$$A \underset{k_{-1}}{\overset{k_1}{\rightleftharpoons}} B.$$

The rate equations are

$$dA/dt = k_{-1}B - k_1 A, \qquad dB/dt = k_1 A - k_{-1}B, \tag{4}$$

from which the closure condition follows immediately:

$$d(A + B)/dt = 0$$

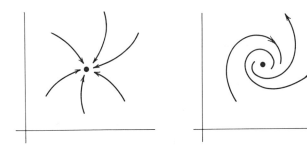

Fig. 1

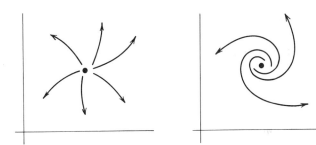

Fig. 2

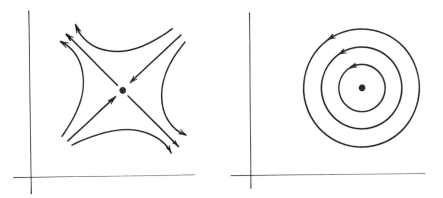

Fig. 3   *Fig. 4*

or

$$A + B = \text{const.} \qquad (5)$$

The state space here is two dimensional and the trajectories are constrained to lie on the lines determined by Eq. (5); the particular line on which any trajectory will lie is determined by the constant, which is the sum of the initial concentrations of A and B. The situation is indicated in Fig. 5. The equilibria of Eq. (4) are those points for which dA/dt and dB/dt simultaneously vanish. Clearly every point on the line

$$k_{-1}B - k_1 A = 0 \qquad (6)$$

will have this property. Thus, as Fig. 5 indicates, each line $A + B = \text{const}$ will intersect the line $k_{-1}A - k_1 B = 0$ in exactly one point, the *equilibrium point* for the system determined by the initial amount of reactants $A_0 + B_0$.

It is easy to verify that these equilibrium points will all be stable if and only if $-(k_1 + k_{-1}) < 0$, which is always the case in chemical systems; hence, the trajectories will be oriented as shown.

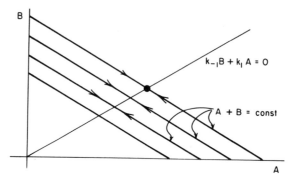

Fig. 5

EXERCISE: Using the diagram in Fig. 5, verify *le Chatelier's principle* for the system (4): If one of the state variables of a closed system in equilibrium is perturbed, the system will move to a new equilibrium in such a way as to oppose the direction of perturbation. Show that the principle fails if we allow several of the reactant concentrations to be perturbed at once.

Let us note one further point regarding the kinetics of the closed system. If the initial reactant concentrations A_0, B_0 are such that $A_0 + B_0 = R$, then the equilibrium concentration A^*, B^* corresponding to these initial conditions will satisfy the simultaneous equations

$$A^* + B^* = R, \qquad k_1 A^* - k_{-1}B^* = 0.$$

If we write $K = k_1/k_{-1}$, we find explicitly for the equilibrium concentrations

$$A^* = KR/(K+1), \qquad B^* = R/(K+1).$$

These relations illustrate the point that for closed systems the equilibrium concentrations do not depend on the individual rate constants, but only on their ratios; the constant K we have defined is the *equilibrium constant* of the reaction. Moreover we note how these equilibrium concentrations depend explicitly on the initial choice of the constant R.

On the other hand, let us consider a typical open system involving the same reactants, but now allow the reactants A and B to enter and leave the system at definite (constant) rates. Thus, our most general reaction system is

$$\overset{k_0}{\searrow} A \underset{k_{-1}}{\overset{k_1}{\rightleftharpoons}} B \overset{k_2'}{\nearrow}$$
$$\underset{k_0'}{\nearrow} \qquad\qquad \underset{k_2}{\searrow}$$

and the corresponding rate equations are

$$
\begin{aligned}
dA/dt &= (k_0 - k_0') + k_{-1}B - kA_1, \\
dB/dt &= (k_2 - k_2') + k_1A - k_{-1}B.
\end{aligned}
\tag{7}
$$

Now in general there is no conservation condition, and the system will have a (unique) steady state at the point

$$
\begin{aligned}
A^* &= [(k_0 - k_0') + (k_2 - k_2')]/(k_{-1} + k_1K), \\
B^* &= [(k_0 - k_0')(K - k_{-1} - k_1K) + K(k_2 - k_2')]/(k_{-1} + k_1K)
\end{aligned}
$$

which will have the character of that shown in Fig. 2a; that is, it will be stable. Le Chatelier's principle is no longer true for such a system, but in fact far stronger regulatory behavior is observed: If the system is perturbed away from a stable steady state, it will return to that state. Note carefully that the steady state in question no longer depends on initial conditions, as was the case for the corresponding closed system, but is automatically reached from any initial state. Furthermore, the steady state will be seen to depend explicitly on the individual rate constants, and not just on equilibrium constants as was the case in the closed system. Let us sum up the distinctions we have found in Table I.

TABLE I

Closed system	Open system
1. Equilibrium depends on initial conditions.	1. Steady state is independent of initial conditions.
2. Equilibrium is independent of individual rate constants.	2. Steady state depends on individual rate constants.
3. Perturbation generally drives the system to a new equilibrium.	3. Perturbation will result in return to the same steady state.

Although we have illustrated these results only in the simplest open and closed systems, the reader should have no trouble in convincing himself that they are generally true for arbitrary open and closed systems.

Thus, if we are interested in regulatory behavior, we see that open systems (at least around stable steady states) exhibit far stronger intrinsic regulatory properties than do closed systems. Indeed, their stubborn return to a stable steady state amounts to a built-in homeostat. There is in fact an exact equivalence between homeostatic behavior, such as that manifested, for example, by the thermostatically controlled temperature in a room governed negative feedback, and the behavior of an open system in the vicinity of a stable steady state. In the thermostat case, the open system in question consists of the room to be controlled, the source of heat, and the thermostatic coupling between them; the setpoint on the thermostat determines where the steady state will be, and the negative feedback principle is geared precisely to ensure the stability of that steady state. Thus, it can be said that the concepts of regulation (or control) and of stability are exactly equivalent ways of formulating the same kind of system behavior, although the precise formulation of this equivalence would take us far beyond the scope of this chapter.

An immediate corollary of this kind of stable behavior manifested by open systems in the neighborhood of a stable steady state is as an explanation for the biological phenomenon of *equifinality*. This is a general name given to biological phenomena in which a perturbation (usually a mutilation or excision of a portion of the system) is followed by a recovery or repair of the system in which its initial conformation or chemical properties are reestablished. Such phenomena are found very often in development and in regeneration; they were often considered magical or an evidence of vitalistic, nonphysical principles governing these phenomena. However, as we can see, properties of equifinality are automatically manifested by an open system in the vicinity of a stable state, and require in fact no specific and separate mechanism for their accomplishment.

We have thus seen that open systems allow a built-in homeostasis around stable steady states, which could in principle account for much of the constancy of cellular properties in fluctuating environments. However, most of the questions we raised in the Introduction refer to ordered *changes* of state of the cell, both automatically and as a result of environmental interactions. Thus, we must now turn to the question of whether and how a general open system, unconstrained by further structure, can manifest such ordered changes of state, and how these changes come about.

The first crucial point which must be understood is that *open systems in general admit a multiplicity of steady states*. This fact often comes as a surprise to those used to dealing with simple physical systems (like gases) or with closed systems or simple linear open systems like Eqs. (7); these all have a single steady state which is globally stable and is approached from any possible initial state. With general nonlinear open systems, however, multiple steady states generally do exist. It is difficult to give a nontrivial or non-artificial simple example with just two reactants; perhaps the best-known biological example is the system

$$\longrightarrow A \longrightarrow B \longrightarrow$$

in which each of the reactants catalyzes its own formation as indicated; the rate equations of this system are the well-known Volterra equations (with the rate constants set to unity)

$$dA/dt = A - AB, \qquad dB/dt = AB - B, \tag{8}$$

which admits two steady states; one at the origin and one at the point (1, 1). As is the rule in such systems, these two steady states have quite different stability properties; the origin is a saddle point (as in Fig. 3), while the point (1, 1) is neutrally stable (as in Fig. 5); the trajectories of this system are shown in Fig. 6. Perhaps the simplest nontrivial example of multiple steady

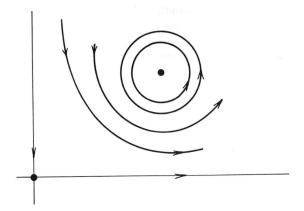

Fig. 6

states, and one which we shall see much of in one form or another, is the following variant of the Volterra system: Consider the pair of Volterra systems

$$\longrightarrow A \longrightarrow B \longrightarrow,$$
$$\longrightarrow C \longrightarrow D \longrightarrow,$$

coupled in the following way: The product D of the second system combines with the reactant A of the first system to form an inactive complex, and the product B of the first system likewise combines with the reactant C of the second system to form another inactive complex. These are indicated by the dotted arrows indicated in the diagram. The dynamical equations are then

$$dA/dt = A - AB - AD, \qquad dB/dt = AB - B - BC,$$
$$dC/dt = C - CD - CB, \qquad dD/dt = CD - D - AD, \qquad (9)$$

and the steady states are the solutions of the simultaneous algebraic equations

$$A - AB - AD = 0, \qquad AB - B - BC = 0,$$
$$C - CD - CB = 0, \qquad CD - D - AD = 0. \qquad (10)$$

As can be seen, there are a number of steady states available to the system. However, two are of particular interest, since these are the only two stable steady states. One of these is characterized by the product values $B = 0$, $D = 1$, while the other is given by the product values $B = 1$, $D = 0$. Every trajectory of the system approaches one of these steady states or the other (except for some unimportant saddle-point behavior); the entire state space can thus be partitioned up into two subsets, representing the states lying on trajectories which move to one or the other of the stable steady states. Although it is impossible to diagram the trajectories in a four-dimensional

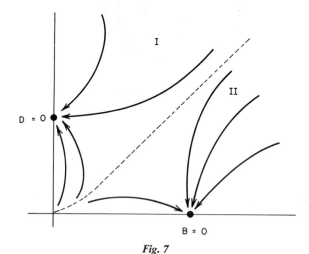

Fig. 7

state space, the situation is roughly as indicated in Fig. 7. This is a figure we shall also see a good bit more of as we progress. The important thing to note is that *the system can be shifted from one stable mode of behavior to the other by an appropriate alteration of initial conditions*; for this reason systems of this type are often called *biochemical switches*.

These systems automatically exhibit another important property implicit in the stability behavior; namely, they manifest the phenomenon of threshold. If we are in a state falling in the region labeled II in Fig. 7, then this means that we are in a region which determines trajectories corresponding to the steady-state condition $B = 0$; if we manipulate this state (say by adding or removing reactant species), we can do so up to a certain point without modifying the steady state to which the system will tend. However, past a certain point, the modification will be sufficient to force the state of the system out of the region II and into the region I; the system will then move along a trajectory tending to the stable steady state for which $D = 0$; such a perturbation is "above threshold" and causes a "switch" in system behavior from one to the other steady state. Such switches play a dominant role, not only in the study of epigenetic mechanisms, but also in the theory of the central nervous system.

It should be quite clear from these simple examples that an open system with multiple steady states can exhibit quite complex patterns of switching behavior; in general, the more reactions there are in a system, the more complex are the switching patterns to which the system can give rise. If the thresholds implicit in these switching patterns are relatively small, then clearly a small change in reactant concentrations can cause a large change in system behavior. This fact plays an important role in the theory of epigenetic mechanisms, for it means that, without changing the reactants (and most particularly, without changing the specific catalysts present in the system) we can completely shift the mode of activity to which the system will tend. Thus, in particular, systems possessing the same genome can, by these means, exhibit a large variety of different behaviors without any loss or modification of the "genetic information." Once again we stress that these conclusions hold good for arbitrary open systems with multiple steady states, and hence hold *a fortiori* for the kinds of multiply catalyzed reaction systems which we find in cells of every kind.

III. Parametric Changes in Open Chemical Systems

In Section II, we have considered some of the properties of general open chemical systems that may have a bearing on epigenesis, and which are in any case reminiscent of the kinds of homeostatic and regulatory properties found in real cells. The emphasis was on directed changes of state caused by specific alterations in the reactant concentrations. In the present section, we shall consider the manner in which such directed changes of state may be caused in general open chemical systems by the equally important process of parametric variation. In these situations, the changes of state elicited in the system are brought about, not by direct changes in reactant concentrations, but in the dependence of system parameters upon internal and external stimuli.

First, it shall be important to clarify exactly what we shall mean by a system parameter. Dynamical equations of all kinds, and chemical rate equations in particular, typically contain a variety of quantities which are supposed to remain constant throughout the discussion, but the values of which must be determined by conditions or constraints exterior to the equations themselves. In the cases we have been discussing, such quantities represent the individual rate constants of the reactions involved, or the concentrations of chemical species held fixed at a definite value, either through appropriate coupling with an infinite external reservoir, or through the imposition of some kind of conservation equation (as is typical, for example, in enzyme kinetics). Such quantities as these, which may be regarded as reflecting the details of structure of an individual system, but the exact numerical values of which cannot be inferred from the dynamical equations themselves, are what we shall mean by *system parameters*.

As should be clear from the two examples we have already given, the distinction between a system parameter and a fully fledged state variable is not a sharp one; the concentration of a particular reactant may be regarded as a parameter if it is held fixed at some definite value, and as a state variable if it is not. The distinction between the two will blur still further as we proceed with the analysis of the present section. Typically a parameter is regarded as some quantity pertaining to system structure which is variable only with difficulty, and which is independent of the specific state the system happens to be in; such mechanical quantities as the mass of a particle, the stiffness of a spring, or the viscosity of a ponderable medium are further examples of these properties. That the parametric values are independent of the system state does not mean, however, that they are absolutely constant. Typically such parameters can be regarded as functions of other aspects of the system environment; for instance, chemical rate constants, which are typical system parameters, are strongly dependent on temperature. Indeed, it is precisely this fact which we shall exploit to see how ordered changes of state can be manifested in open chemical systems as a response to other kinds of influences than the simple modification of reactant concentrations.

In Section II, we tied potential epigenetic changes of state to the stability properties manifested by an open chemical system in the neighborhoods of its steady states. Thereby we were able to study its regulatory properties as arising from the system's intrinsic dynamical response, embodied in the rate equations, to manipulations of the reactant concentrations. In this section, we shall also employ this strategy to determine how the stability properties of an open chemical system depend upon the specific parametric values assigned to the system.

Let us consider a very simple situation at the outset. Consider a single reactant x, the concentration of which is changing according to the rate law

$$dx/dt = \alpha x - S. \tag{11}$$

There are two parameters in the system; the quantity α is a rate constant, and the quantity S represents an influx or outflow (depending on the sign of S) of the reactant x. In what follows we shall restrict our attention to the behavior of the system as a function of the parameter α, and leave it to the reader to construct the parallel discussion of the manner in which system behavior depends on S. We shall only assume that S and α are always of opposite sign.

First, it is obvious that the system of Eq. (11), whatever the values assigned to the parameters, will always have a single steady state, that has the value

$$x^* = S/\alpha.$$

The rate equation (11) is also immediately integrable; we find

$$x(t) = (x_i - x^*)e^{\alpha t} + x^*, \tag{12}$$

where x_i represents the initial reactant concentration. From this explicit solutions, we can determine the stability properties of the unique steady state. It is quite clear from Eq. (12) that these stability properties depend entirely on the sign of the parameter α. If $\alpha > 0$, and if $x_i > x^*$, then $x(t)$ will grow in the positive direction without bound; that is, it will move ever farther away from the steady state; while if $x_i < x^*$, $x(t)$ will grow without bound in the negative direction, and the same conclusion holds. Thus, if $\alpha > 0$, it follows that the steady state x^* is unstable. Conversely, if $\alpha \leqslant 0$, it follows similarly that the steady state x^* is stable. The actual absolute magnitude of the parameter α determines only the rate at which the trajectories approach or diverge from the steady state.

To graphically illustrate the dependence of the stability properties of this simple system upon the parameter α, we shall introduce a two-dimensional representation, which will graphically exhibit the dependence of system properties on the two relevant quantities (the reactant concentration x, and the rate constant α). The idea is basically the following: We shall assign to each value of the rate constant α a copy of the entire state space of the system (in this case, a line); we thus obtain something like a two-dimensional state space, but fibered into a field of lines parallel to the X axis, one line for each value of α. The situation is as indicated in Fig. 8. From this kind of representation we can graphically see how both the specific steady state, and the behavior of the trajectories in the vicinity of the steady state (that is, the stability properties) depend on *both* the dynamical rate law and on the character of the parameter.

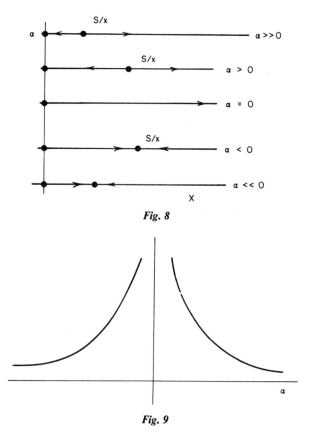

Fig. 8

Fig. 9

A particularly valuable way of representing this data is to plot the coordinate of the steady state, $x^* = S/\alpha$, against α. When we do this, we obtain the hyperbolic relationship shown in Fig. 9 (of course, a different but similar curve will be obtained for each choice of the parameter S).

Thus, it is suggested that, if we wish to move our system from an initial reactant concentration x to a desired reactant concentration $\bar{x}$, and to keep the system there, we merely need choose that value $\bar{\alpha}$ of the parameter α such that (a) $\bar{\alpha} < 0$ and (b) $\bar{x} = S/\bar{\alpha}$. The second condition will ensure us that the desired state behaves as the steady state of the system, while the first condition ensures that this steady state is stable (and hence is approached from all other states). This can be accomplished without any initial modification of reactant concentration, simply by making an appropriate change in some environmental influence on which the parameter depends.

Of course, the simple situation we have just considered does not have sufficient richness to lead to specific interesting consequences; in particular, we have only a single reactant and the rate equations allow only a single steady state. If we have multiple steady states, however, we encounter the possibility of eliciting switching behaviors parallel to those we discussed in Section II, but proceeding through parametric variations and not through alterations in reactant concentrations. These phenomena fall under the general heading of *hysteresis*, although the conventional hysteretic phenomena familiar from physics represent only a very special subclass of what is possible. To fix ideas, we shall consider a very simple example, constructed to parallel the discussion we have already given.

Consider then a system consisting of a single reactant x, but which possesses three steady states with stability properties as shown in Fig. 10. It is difficult to give a chemically meaningful rate equation for such a simple system, though mathematically such a rate equation would be of the form

$$dx/dt = (x - a)(x - b)(x - c)f(x).$$

The three numbers a, b, c here will all be taken as specific functions of a single parameter α, and the function $f(x)$ is chosen so as to ensure that the three steady states at $x = a$, $x = b$, $x = c$ have the required stability properties, but is otherwise arbitrary (except, of course, for the requirement that $f(x)$ can vanish at most at the three points a, b, c). Now we have already seen, in our simple example, that the locations of the steady states depend on α.

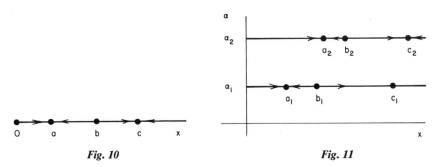

Fig. 10   Fig. 11

Let us construct a diagram analogous to Fig. 8, and suppose that for two distinct parametric values α_1, α_2, the corresponding state spaces are as shown in Fig. 11. We start the system out sitting at the steady state a_1 corresponding to the parametric value α_1. We then increase the parametric value from α_1 to α_2. At this parametric value, the state $x = a_1$ is no longer at the steady state; it will accordingly move autonomously to the nearest available steady state, which as will be seen from the figure is the state $x = a_2$. This in itself results in a directed change of state of our system. But if we now return our parametric value back to its original value α_1, we observe that something quite new happens. The new steady state at $x = a_2$, to which the system now tends, is to the right of the old unstable steady state b_1. Thus, if we return the parametric value to α_1, the system will not return to its original state at $x = a_1$, but will rather move to the entirely new state $x = c_1$, which is now the nearest available stable steady state.

This kind of switch, induced by parametric variation, has many interesting properties which were not present in the switching behavior we discussed in Section II. Prime among these is the fact that the switch was induced by a reversible alteration in parametric value, which led to an irreversible change of state. The actual switch itself may, or may not, be reversible through further parametric variation. It may happen, for instance, that parametric values below α_1 correspond to state space behavior like that shown in Fig. 12 in which case, decreasing the parametric value from α_1 to α_0 will reverse the switch; this is the typical hysteresis situation. But if the situation is as shown

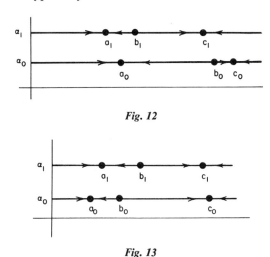

Fig. 12

Fig. 13

in Fig. 13, then no amount of parametric variation will reverse the effect of the first switch. For this reason, such hysteretic phenomena are frequently regarded as endowing the systems which exhibit them with a *memory* of environmental alterations which occurred in the past; a mere knowledge of initial conditions is not sufficient to determine their state at a future time unless the temporal variations of parameters like α is also known. These types of hysteresis phenomena are clearly reminiscent of the kinds of events that occur in cells during differentiation; permanent switches in behavior which are induced by particular environmental circumstances, which cannot be reversed simply by restoring the system environment to its original condition.

It will also be seen that the kind of hysteretic switch we have described possesses threshold properties similar to those we have seen previously.

And of course, although a hysteretic switch need not in general be reversible through further environmental manipulation, it can always be reversed through an appropriate manipulation of reactant concentrations (since the multiple steady states which make the hysteresis possible are also sufficient to enable us to induce switching through alterations in reactant concentrations as well).

Having displayed this many possibilities, which are generally available in a general open system possessing a number of steady states, and containing a number of variable parameters, we are now in a position to make a variety of observations that tie the above results even more closely to the specific kinds of chemical systems which we know to occur in real cells. We have seen above, for example, that the steady states of such a system depend on the individual rate constants (system parameters), and indeed we exploited precisely that fact in our discussion of hysteresis. It therefore follows that any influence which causes a modification in these rate constants thereby exerts an effect, possibly a very large effect, on the steady state behavior of the entire system. However, as we carefully noted earlier, in real cells the reactions which occur are typically mediated by specific catalytic proteins (enzymes); the effect of any catalyst is precisely to cause such modification of the rate constant of the reaction it catalyzes. Indeed, since the rate of a catalyzed reaction is approximately proportional to the concentration of the catalyst, it follows readily that *the steady-state properties of an open system of catalyzed reactions are exquisitely sensitive, in general, to the precise values of the catalyst concentrations,* far more so than the corresponding system of uncatalyzed reactions would be to ordinary reactant concentrations.

We can now begin to discern two distinct levels of epigenetic control in cellular systems. On the one hand, there is the kind of control which specific values of enzyme concentrations exert on the concentrations of metabolites occurring within cells. And, on the other hand, there are the mechanisms whereby enzyme concentrations are themselves controlled. Since the total library of enzymatic specificities resides by definition in the cell genome, and since the genes themselves may be regarded as catalysts involved in the synthesis of specific enzymatic proteins, everything we have said thus far leads to the conclusion that this last level of epigenetic control operates through modifications in the rate constants for the reactions leading from the genes to the finished proteins. Obviously, the most parsimonious way for these rate constants to be modified is to allow them to be specific functions of the concentrations of metabolites present in the system, or in the system environment (just as we control rate constants in an ordinary open chemical system by allowing them to be specific functions of catalyst concentrations). The work of the remainder of the present chapter, then, will be to develop a framework in which questions of this kind can be formulated and their implications studied.

To conclude this section, we should stress again that these last results, regarding the dependence of steady-state values on rate constants, and hence on catalyst concentrations, belong exclusively to open systems. In a closed system, as we have seen, the equilibrium values of the reactant concentrations depend not on the individual rate constants, but on their ratios. By definition, a catalyst (such as an enzyme) modifies both the forward and backward rates of a reversible reaction in the same proportion, and hence leaves equilibrium constants unaffected. As a result, the equilibria of closed catalyzed systems are *independent of catalyst concentrations*; only the rate of approach to equilibrium is affected by the presence of a catalyst in a closed system. Consequently, closed systems simply cannot manifest any of the kinds of behavior we have been describing; behavior on which our subsequent discussions of epigenetic mechanisms will be based.

IV. Reaction Rate Control in Catalyzed Open Systems

We are now in a position to begin our investigation into reaction rate control in open chemical systems, particularly those containing catalysts. The basic strategy is the following: we know quite generally that the individual reaction rates, as system parameters in open systems, determine the steady state properties of the system, and so can result in directed changes of state. We know further that these parameters are expressible as definite functions of the concentrations of specific reactants (catalysts) already present within the system. We are going to carry this program one step further, and explore how the catalyst concentrations themselves can be modified as functions of reactant concentrations (that is, as functions of the state of the system itself). This type of situation, in which information about the present state of a system is employed to determine directed changes of subsequent states, is well known in engineering applications, where it is quite generally described as *feedback*.

Quite generally, let us consider an arbitrary open chemical system, involving the reactants x, y, z, . . . , and satisfying the rate equations

$$dx/dt = f(x, y, z, \ldots),$$
$$dy/dt = g(x, y, z, \ldots),$$
$$dz/dt = h(x, y, z, \ldots),$$
$$\vdots$$

(13)

We shall suppose henceforth that f, g, h, . . . are continuous and have continuous first partial derivatives in the state space (where, as usual, we have denoted a reactant and its concentration by the same symbol). We may then interpret $v_x = dx/dt$ as the net rate at which the reactant x is being produced as a result of the reactions occurring in the system, and similarly for $v_y, v_z, \ldots$.

Suppose we are particularly interested in the reactant x, and are interested in the manner in which alterations in the concentrations of other reactants affect the rate at which x is produced. How would we go about determining the effects of such alterations? The most natural way is the following: Suppose at some instant t_0 the system is in the state

$$(x(t_0), y(t_0), \ldots), \tag{14}$$

and suppose at that instant the system is perturbed by a modification of some other reactant, say z. We then can look at the sign of the partial derivative

$$\partial v_x / \partial z \big|_{t=t_0}$$

evaluated at the state described by Eq. (2). There are three possibilities:

(a) $\partial v_x / \partial z \big|_{t=t_0} > 0$, (b) $\partial v_x / \partial z \big|_{t=t_0} < 0$, (c) $\partial v_x / \partial z \big|_{t=t_0} = 0$.

If (a) holds, then an increase in the instantaneous value of z will cause the rate at which x is being produced to be increased, or, alternatively, a decrease in the instantaneous value of z will cause the rate at which x is being produced to be decreased. If (b) holds, the reverse is true; an increase in the instantaneous value of z will cause the rate of production of x to be decreased, or alternatively, a decrease in the instantaneous value of z will cause the rate of production of x to be increased. If (c) holds, the rate of production of x is independent of alterations in the instantaneous value of z; all relative, of course, to the initial state, Eq. (2). In Case (a), we shall say that z is an *activator* for x; in case (b), we shall say that z is an *inhibitor* for x. It is not excluded that $z = x$; in this case we say that x is a *self-actuator* or *self-inhibitor* in Cases (a) and (b), respectively.

Since by hypothesis all the first partials of production rates are continuous, it follows that if z is an activator for x relative to some initial state, it is an activator for x in some entire neighborhood of that state; likewise, if z is an inhibitor for x relative to some initial state.

Thus, in general, we can decompose the state space Ω of the system into a set of subregions, such that, in any subregion, the sign of all the first partial derivatives $\partial v_y / \partial z$ is the same at all points in the subregion.

We shall say that a subregion of this kind, for which the first partial derivatives of the rates $v_x, v_y, \ldots$ with respect to a reactant z are of constant sign throughout, is a *region of constant character* with respect to the reactant z.

Let us at this point introduce a valuable notation, due to Higgins [1967], as are most of the methods in this section. Let us indicate the fluxes through the system of x and y, respectively, by the large arrows shown in Fig. 14, and indicate each of the possible interactions of the reactants x, y with these fluxes by a small arrow. The interactions of x with its own flux v_x, and of y

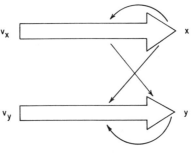

Fig. 14

with its own flux v_y, will be called *self-couplings*; the interaction of x with the flux v_y, and of y with the flux v_x, will be called *cross couplings*.

At each point in the state space we can associate a sign to each of the possible couplings, positive if the coupling is activational, negative if it is inhibitory (if the coupling is zero, we simply omit the corresponding arrow). There will then be such a flux diagram associated with every point in the state space, and hence with each region of constant character. We shall call this diagram the Higgins diagram of the region. These diagrams will be very helpful in our subsequent analysis.

Let us now apply these simple ideas to some familiar reactions, to be sure that the terminology we have introduced is consistent with other usages of the terms "activation" and "inhibition." Since all of these results, and the subsequent applications, will essentially be variations on a single theme (though growing successively more elaborate), we shall emphasize the basic pattern common to all of them.

A. Uncatalyzed Reactions

1. Monomolecular Irreversible Reactions

The typical reaction of this type is governed by the stoichiometric equation

$$A \xrightarrow{k} B \tag{15}$$

and, according to the law of mass action, is governed by the single rate equation

$$dA/dt = -dB/dt = -kA$$

(we shall typically denote a substance and its concentration by the same symbol when there is no danger of confusion). If we measure the rate v of the reaction by the appearance of the product B; that is, if we put

$$v = dB/dt = kA, \tag{16}$$

then it is trivial to verify that $\partial v/\partial A > 0$, and hence that A is always an activator for the reaction. Since B does not appear in the rate equation, B is neither an activator nor an inhibitor.

2. Multimolecular Irreversible Reactions

Here we shall consider the typical reaction,

$$n_1A_1 + n_2A_2 + \cdots + n_rA_r \xrightarrow{k} m_1B_1 + m_2B_2 + \cdots + m_sB_s, \qquad (17)$$

where the n_i, m_i are nonnegative integers (not all zero). This of course contains the preceding as a special case. Once again, the law of mass action allows us to write down the rate equation governing the system:

$$dA_1/dt = dA_2/dt = \cdots = dA_n/dt = -dB_1/dt = \cdots = -dB_s/dt$$
$$= kA_1^{n_1}A_2^{n_2}\cdots A_r^{n_r}. \qquad (18)$$

Once again, if we measure the reaction rate v as the rate of appearance of the products B_i, it is trivial to verify that $\partial v/\partial A_i > 0$ for each index i, and hence each reactant A_i is in this sense an activator for the reaction.

3. Monomolecular Reversible Reactions

For reversible reactions of this type we now have the stoichiometric equation

$$A \underset{k_{-1}}{\overset{k_1}{\rightleftharpoons}} B \qquad (19)$$

now involving two rate constants, one for the forward reaction and one for the backward reaction. The corresponding rate equation now takes the form

$$dA/dt = -dB/dt = k_{-1}B - k_1A. \qquad (20)$$

If we once again measure the rate v of the reaction by the appearance of the product B, then we have

$$\partial v/\partial A = k_1 > 0, \qquad \partial v/\partial B = -k_{-1} < 0.$$

Thus, looking at the forward reaction, it follows that A is an activator for the reaction and B is now an inhibitor for that reaction. Looking at the backward reaction, however, for which A is a product and B is a reactant, it follows that A is an inhibitor of the back reaction while B is an activator.

4. Multimolecular Reversible Reactions

The typical multimolecular reversible reaction has the stoichiometric equation

$$n_1A_1 + \cdots + n_rA_r \underset{k_{-1}}{\overset{k_1}{\rightleftharpoons}} m_1B_1 + \cdots + m_sB_s, \qquad (21)$$

and the corresponding set of rate equations,

$$dA_i/dt = -dB_j/dt = -k_1 A_1^{n_1} A_r^{n_r} + k_{-1} B_1^{m_1} \cdots B_s^{m_s}.$$
$$i = 1, \ldots, n, \qquad j = 1, \ldots, m. \tag{22}$$

Just as before, the A_i all activate the forward reaction and inhibit the back reaction, while the B_i all inhibit the forward reaction and activate the back reaction.

B. Catalyzed Reactions

1. Catalyzed Monomolecular Irreversible Reactions

We now proceed to generalize the preceding examples in an important way. We shall assume that the reactant A is transformed irreversibly to the product B under the influence of a catalyst E. Typically this situation is denoted stoichiometrically by

$$A \xrightarrow{E} B. \tag{23}$$

In order to write down a rate equation for such a catalyzed reaction, it is necessary to reexpress Eq. (11) in terms of ordinary mass-action type reactions. This situation was discussed exhaustively in Chapter 2, Volume I, for enzyme-catalyzed reactions, and we shall now recall the high spots of that analysis. We typically break the reaction (23) down into a pair of reactions involving an intermediate reactant (which in enzyme-catalyzed reactions is the enzyme–substrate complex) denoted by C, as follows:

$$A + E \underset{k_{-1}}{\overset{k_1}{\rightleftharpoons}} C, \qquad C \xrightarrow{k_2} B + E. \tag{24}$$

From these, the rate equations for each reactant can be immediately written down:

$$dA/dt = -k_1 AE + k_{-1}C,$$
$$dB/dt = k_2 C,$$
$$dC/dt = k_1 AE - k_{-1}C - k_2 C. \tag{25}$$

It follows that

$$dC/dt = -dA/dt - dB/dt. \tag{26}$$

If we now make the *steady-state assumption* that the intermediate reactant C reaches an essentially constant concentration very quickly, then we can treat the concentration of the intermediate C as a *parameter* of the system, and set $dC/dt = 0$. From Eq. (26), this means that, just as in the uncatalyzed case, we can write the reaction rate v as

$$v = dB/dt = -dA/dt.$$

Making use of the conservation equation

$$E + C = E_t = \text{const},$$

where E_t is the total concentration of catalyst E and the steady-state hypothesis $dC/dt = 0$, it then follows from the second and third equations of (25) that

$$v = k_2 A E_t/(K + A), \tag{27}$$

where we have written $K = (k_2 + k_{-1})/k_1$, which is, of course, the typical Michaelis–Menten expression for the rate of an irreversible enzyme-catalyzed reaction.

We shall now compare this expression (27) with the corresponding expression (16) for an uncatalyzed reaction of the same type. As before, it is evident that $\partial v/\partial A > 0$, so that A is an activator for the reaction; and as before, since B does not appear in the final rate equation, B is neither an activator nor an inhibitor. But now we have an additional variable in the problem; namely the total catalyst concentration E_t. If we now express the manner in which the rate v depends on this variable, we again find always $\partial v/\partial E_t > 0$; that is, the catalyst itself activates the reaction, as we should expect if our terminology is to be in accord with conventional usage.

There is now one more most important obaservation to be made. Let us suppose that the reaction (15) can proceed spontaneously, and that to the reaction mixture we add some quantity E_t of catalyst; what happens to the rate of the reaction? This is easy to determine; we have the system of stoichiometric equations

$$A \xrightarrow{\ k_1\ } B$$

$$A + E \underset{k_{-1}}{\overset{k_1}{\rightleftharpoons}} C \tag{28}$$

$$C \xrightarrow{\ k_2\ } B + E$$

to which correspond the rate equations

$$\begin{aligned} dA/dt &= -k_0 A - k_1 AE + k_{-1} C, \\ dC/dt &= k_1 AE - k_{-1} C, \\ dB/dt &= k_0 A + k_2 C. \end{aligned} \tag{29}$$

On the basis of what we have already done, it is easy to see that the overall reaction rate equation for the system (28) is

$$v = \frac{dB}{dt} = \left(k_0 + \frac{k_2 K E_t}{1 + KA}\right) A. \tag{30}$$

Let us now compare Eq. (30) with the rate equation (16) for the uncatalyzed reaction. We find the two are of the same form if we make the formal identification

$$k = k_0 + k_2 K E_t/(1 + KA).$$

That is, we in effect express the rate "constant" of the uncatalyzed system as in fact a function of the new reactant E (or E_t). It is clear that E_t is *an activator of the uncatalyzed reaction* (as well as, trivially, of the catalyzed reaction), and specifically that it acts *through an increase of the reaction rate*; this increase in rate rests upon a specific functional dependence between the rate and the concentration of the reactant E. It is this crucial observation, which we shall find again and again, that provides the basis for a deep formal understanding of the manner in which epigenetic controls are exerted in biological systems. In a nutshell, we shall find that *epigenetic controls consist in general of modifications of reaction rates, implemented through the functional dependence of rate-determining parameters on reactant concentrations*, and consisting of activations, inhibitions, or a combination of the two.

2. Catalyzed Irreversible Multimolecular Reactions

Reactions of this type are of the stoichiometric form,

$$n_1A_1 + n_2A_2 + \cdots + n_rA_r \xrightarrow{E} m_1B_1 + \cdots + m_sB_s, \tag{31}$$

which, as with the irreversible catalyzed monomolecular reactions, must be reduced to a series of mass-action type relations before the rate equations can be written down. However, the matter is no longer as simple as it was in the monomolecular case, because there are a vast number of different stoichiometric schemes which are compatible with Eqs. (28), each differing from the others in specific mechanisms and each leading to rather different rate equations. We can see this if we consider the very simplest bimolecular reaction system

$$A_1 + A_2 \xrightarrow{E} B_1 + B_2. \tag{32}$$

The following different stoichiometric equations can be assumed just for this simple reaction:

$$A_1 + E \underset{k_{-1}}{\overset{k_1}{\rightleftharpoons}} C$$
$$C + A_2 \xrightarrow{k_2} B_1 + B_2 + E; \tag{33}$$

$$A_1 + E \underset{k_{-1}}{\overset{k_1}{\rightleftharpoons}} C_1$$
$$C_1 + A_2 \underset{k_{-2}}{\overset{k_2}{\rightleftharpoons}} C_2$$
$$C_2 \xrightarrow{k_3} B_1 + B_2 + E; \tag{34}$$

$$A_1 + E \rightleftharpoons C_1,$$
$$A_2 + E \rightleftharpoons C_2,$$
$$C_1 + C_2 \longrightarrow B_1 + B_2 + E; \tag{35}$$

$$A_1 + A_2 \rightleftharpoons C_1,$$
$$C_1 + E \rightleftharpoons C_2,$$
$$C_2 \longrightarrow B_1 + B_2 + E; \tag{36}$$

there are, of course, many other possibilities as well. For each such possibility, the corresponding rate equations may be written down using the
methods employed in the preceding paragraphs, but naturally the forms of
these equations will be rather different.

To illustrate the basic ideas, let us treat the cases denoted by (33) and (34)
in detail, utilizing the same format we applied to the monomolecular case.
For the system (33), we have the reactant rate equations

$$dA_1/dt = k_{-1}C - k_1A_1E, \qquad\qquad dA_2/dt = -k_2CA_2,$$
$$dC/dt = kA_1E_1 - (k_{-1} + k_2A_2)C, \qquad dB_1/dt = dB_2/dt = k_2CA_2. \tag{37}$$

Since

$$dC/dt = -dA_1/dt - dB_1/dt = -dA_1/dt + dA_2/dt,$$

the steady-state assumption $dC/dt = 0$ implies that the rate of reaction v
is given by

$$v = dB_1/dt = dB_2/dt = -dA_1/dt = -dA_2/dt.$$

Solving the relation $dC/dt = 0$, using the conservation condition $E + C
= E_t = $ const to eliminate E from this relation, and substituting in the equa-
tion for dB_1/dt gives us the rate equation

$$v = k_2E_tA_1A_2/(K_1 + A_1 + K_2A_2) \tag{38}$$

(where we have written $K_1 = k_{-1}/k_1$, $K_1 = k_2/k_1$) which we will view in
more detail in a moment.

If the stoichiometry of system (32) obeys the scheme (34) instead of (33),
the rate equations are more complicated, since five rate parameters are now
involved. We have

$$dA_1/dt = k_{-1}C_1 - k_1A_1E,$$
$$dA_2/dt = k_{-2}C_2 - k_2A_2C_1,$$
$$dC_1/dt = k_1A_1E - k_{-1}C_1 - k_2C_1A_2 + k_{-2}C_2, \tag{39}$$
$$dC_2/dt = k_2C_1A_2 - k_{-2}C_2 - k_3C_2,$$
$$dB_1/dt = dB_2/dt = k_3C_2.$$

As before, the steady-state conditions $dC_1/dt = 0$, $dC_2/dt = 0$ imply that
the reaction rate v is given by

$$v = dB_1/dt = dB_2/dt = -dA_1/dt = -dA_2/dt.$$

We apply these steady-state conditions, together with the conservation
relation $E + C_1 + C_2 = E_t = $ const. as follows:

1. From the equation for $dC_2/dt = 0$, apply the steady-state condition to
obtain

$$C_2 = C_1A_2/K_2',$$

where $K_2' = (k_{-2} + k_3)/k_2$.

2. Substitute this into the steady-state relation for C_1, to obtain

$$C_1 = \frac{EA_1}{k_{-1}/k_1 + k_3 A_2/k_1 K_2}.$$

3. Substitute these relations for C_1 and C_2 into the conservation equation, solving this for E. Using this and the equation for dB_1/dt, we finally find

$$v = \frac{k_3 F_t A_1 A_2}{K_1 K_2' + K_2' A_1 + K_3' A_2 + A_1 A_2}, \tag{40}$$

where we have written

$$K_1 = k_{-1}/k_1, \qquad K_3' = k_3/k_1.$$

Now let us compare the two rate equations (38) and (40) with the corresponding rate equation for an uncatalyzed bimolecular irreversible reaction, which we obtain from Eq. (18) by putting $r = s = 2$, $n_1 = n_2 = m_1 = m_2 = 1$. This gives simply, for the rate of the uncatalyzed reaction,

$$v = k A_1 A_2. \tag{41}$$

If we now append the uncatalyzed reaction to the reaction schemes (33) and (34), we find that the respective reaction rate equations are given by

$$v = \left(k_0 + \frac{k_2 E_t}{K_1 + A_1 + K_2 A_2}\right) A_1 A_2 \tag{42}$$

and

$$v = \left(k_0 + \frac{k_3 E_t}{K_1 K_2' + K_2' A_1 + K_3' A_2 + A_1 A_2}\right) A_1 A_2, \tag{43}$$

respectively. From this we find once again, on evaluating for these two rate equations, that the catalyst E activates the uncatalyzed reaction, and does so through the introduction of a specific functional relation involving the rate-determining parameter k_0 of the uncatalyzed reaction and the catalyst concentration E (or E_t). The form of this functional relation may differ for different stoichiometries, as is seen by comparing Eqs. (41), (42), and (43), but the overall activation effect is the same. The reader should, as an exercise, carry out the corresponding calculation of the rate equations for the stoichiometric schemes (35) and (36), and verify our assertions for these cases as well.

3. Catalyzed Reversible Monomolecular Reactions

We can pass from the irreversible catalyzed monomolecular reaction (24) by making the dissociation of the intermediate C into B catalyst reversible. We then have the stoichiometric relations

$$A + E \underset{k_{-1}}{\overset{k_1}{\rightleftharpoons}} C$$
$$C \underset{k_{-1}}{\overset{k_2}{\rightleftharpoons}} B + E \tag{44}$$

to which correspond the rate equations

$$dA/dt = k_{-1}C - k_{-1}AE,$$
$$dC/dt = k_1AE + k_{-2}BE - (k_{-1} + k_2)C, \qquad (45)$$
$$dB/dt = k_2C - k_{-2}BE.$$

On the basis of the steady-state assumption and the conservation condition for the catalyst, it is readily verified that the equation for the (forward) reaction velocity v is of the form

$$v = \frac{K_0 AE_t}{K_1 K_2 + AK_2 + BK_1} - \frac{K_0' BE_t}{K_1 K_2 + AK_2 + BK_1}, \qquad (46)$$

where K_0, K_0', K_1, K_2 are specific constants.

Just as before, it is immediate to verify that A activates the forward reaction, while B inhibits that reaction; conversely, B activates the reverse reaction, while A inhibits it. With respect to E_t, the situation is more complicated. We have

$$\frac{\partial v}{\partial E_t} = \frac{K_0 A - K_0' B}{(K_1 K_2 + AK_2 + BK_1)^2}, \qquad (47)$$

which may be positive or negative for the forward reaction, depending on the relative concentrations of A and B. Roughly, Eq. (35) says that if there is an excess of A over B, then the catalyst is an activator of the forward reaction, while if there is an excess of B over A, then the same catalyst is an inhibitor of that reaction. (Converse statements are true for the reverse reaction.) This is another most important observation, for it indicates that *the same reactant can activate a reaction in one set of circumstances and inhibit the same reaction in another set of circumstances*, depending on initial conditions. This becomes still clearer if we adjoin the uncatalyzed reaction (19) to the catalyzed reaction (44); in this case we obtain for the overall rate equation

$$v = \left(k_1 + \frac{K_0 E_t}{K_1 K_2 + AK_2 + BK_1} \right) A - \left(k_{-1} + \frac{K_0' E_t}{K_1 K_2 + AK_2 + BK_1} \right) B, \qquad (48)$$

explicitly exhibiting the functional dependence of the rate-determining parameters for the uncatalyzed reaction on the catalyst E. Again, whether E will activate or inhibit the forward reaction will clearly be seen to depend upon the initial conditions; this circumstance again finds its origin in the specific form of the functional dependence of these rate-determining parameters upon the reactant E.

4. Catalyzed Reversible Multimolecular Reactions

On the basis of what has already been said, we can now essentially repeat the discussion of catalyzed irreversible multimolecular reactions given in

Section IV.B.2 above, except that in every case we allow the breakdown of the various complexes involved to be reversible. As in Section IV.B.2, the specific kinetics depends upon the specific reaction mechanisms which are assumed. The reader may find it a useful exercise to repeat the calculations based on reaction mechanisms (33)–(36) above, with all reactions assumed reversible, to obtain the catalyzed reaction rates for the catalyzed systems, and compare these with those of the uncatalyzed reactions, in the same manner we have employed previously.

C. Activation and Inhibition of Catalyzed Reactions

1. Catalyzed Irreversible Monomolecular Reactions with Activator

The next few reaction mechanisms we shall consider introduce a further complexity into the kinetic analysis. Namely, we shall assume the existence of reactants in the system which modulate the overall rate of the catalyzed reaction, just as the catalyst itself modulates the rate of the uncatalyzed reaction. In case the catalyst itself is an enzyme, such reactants are typically called *activators* and *inhibitors* of the enzyme. In our study of the kinetics of such systems, we shall have two concerns: (a) to understand the quantitative characteristics of catalytic systems involving such activators and inhibitors, and (b) to show that our general terminology, introduced at the beginning of this section, is consistent with the intuitive behavior of these systems.

Let us suppose then that we begin with the basic reaction (15), upon which we have superimposed a specific catalyst E, satisfying the stoichiometric equations (24). We shall now introduce a further reactant F (read "facilitator") which will act in the fashion of an activator in the conventional sense of enzyme kinetics. As in previous situations, we can make a variety of assumptions about the stoichiometry of the involvement of F in the system. Some typical ones are

$$
\begin{aligned}
E + F &\underset{k_{-3}}{\overset{k_e}{\rightleftharpoons}} E_a \\
E_a + A &\underset{k_{-4}}{\overset{k_4}{\rightleftharpoons}} C_a \\
C_a &\overset{k_5}{\longrightarrow} B + E_a;
\end{aligned}
\tag{49}
$$

$$
\begin{aligned}
C + F &\underset{k_{-4}}{\overset{k_3'}{\rightleftharpoons}} C_a \\
C_a &\overset{k_4'}{\longrightarrow} B + E_a;
\end{aligned}
\tag{50}
$$

$$
\begin{aligned}
A + F &\rightleftharpoons A_a \\
A_a + E &\rightleftharpoons C_a \\
C_a &\longrightarrow B + E_a.
\end{aligned}
\tag{51}
$$

The first of these schemes (49) represents, in standard enzyme-kinetic terms, what is typically called simply "activation." It involves the direct reaction of the catalyst with the facilitator to form a new "activated complex," here denoted by E_a, which then behaves exactly as did E in the simple scheme (24). The scheme (50) is different, in that it involves the "activation" of the substrate A. It cannot, of course, be excluded that two or more of these mechanisms are simultaneously involved (together with still other plausible schemes by which the facilitator F can be involved in the reaction); this greatly complicates the computations (because of the large number of rate constants arising from the mechanisms involved), but the underlying idea of the computations remains in all cases the same.

Let us carry out the computation in detail for the simple scheme (49). The dynamical rate equation is

$$v = dB/dt = k_5 C_a. \tag{52}$$

In this situation, the steady-state assumptions mean that the two complexes C_a, E_a reach their "equilibrium" value much more rapidly than the other reactants; that is, we have

$$dC_a/dt = k_4 E A_a - (k_4 + k_5)C_a = 0, \tag{53}$$

$$dE_a/dt = k_3 EF - k_3 E_a + k_{-4} C_a - k_{-4} E_a A + k_5 C_a = 0. \tag{54}$$

After combination of Eqs. (53) and (54), the steady-state equation (54) becomes just

$$0 = k_3 EF - k_{-3} E_a. \tag{55}$$

From Eq. (53),

$$C_a = k_4 E_a A/(k_{-4} + k_5) = AE_a/K, \tag{56}$$

where $K = (k_{-4} + k_5)/k_4$. From Eq. (55),

$$E_a = EF/K', \tag{57}$$

where $K' = k_3/k_{-3}$. From the conservation condition

$$E_t = E + E_a + C_a; \tag{58}$$

together with Eqs. (56) and (57), it follows that

$$E = E_t/(KK' + K'F + AF), \tag{59}$$

and finally, substituting (59) in (57) and the result in (52), we get the final overall rate equation

$$v = dB/dt = k_5 AE_t F/(KK' + K'F + AF). \tag{60}$$

This rate equation can be regarded as a special case of the more general reaction system obtained by combining the system (24) with the system (49), when we put $k_1 = k_{-1} = 0$; that is, the catalyst does not function in the

absence of the facilitator F. However, if we now allow these rate constants
to be nonzero, we can obtain in like fashion the effect of the facilitator
F upon the catalyzed reaction (24). For the combined system, the rate
equation is

$$v = dB/dt = k_5 C_a + k_2 C \qquad (61)$$

and the conservative equation is

$$E_t = E + E_a + C_a + C. \qquad (62)$$

Solving as before, and making the appropriate rearrangements, the rate
equation (61) becomes

$$v = \frac{k_5 E_t A(F + k_2 KK'/k_5 \bar{K})}{KK' + K'F + AF + KK'A/\bar{K}}, \qquad (63)$$

where $\bar{K} = (k_2 + k_{-1})/k_1$. When $k_1 = k_{-1} = 0$, we have $\bar{K}^{-1} = 0$, so Eq.
(63) reduces to Eq. (60), as it should. Likewise, if $k_3 = k_{-3} = 0$, Eq. (63)
reduces to Eq. (27).

As we have done before, let us compare the rate equation (63) involving
the facilitator F with the corresponding rate equation (27) for the same
reaction without the facilitator. The comparison is most easily made if we
express Eq. (27) directly in terms of the specific rate constants instead of the
Michaelis constant. In these terms Eq. (27) becomes

$$v = \frac{k_2 A E_t}{(k_{-1} + k_2)^2 + k_1(k_{-1} + k_2)A}, \qquad (64)$$

which is of the form (63) if we identify

$$k_2 \sim k_5(F + k_2 KK'/k_5 \bar{K}),$$
$$(k_{-1} + k_2)^2 \sim KK' + KF,$$
$$k_1(k_{-1} + k_2) \sim F + KK'/K.$$

That is, if we again appropriately choose the individual rate "constants"
of the simple reaction to be appropriate functions of the new reactant. Thus
we see reappearing, at each successive level of complexity, the same pattern
of rate control through the system parameters which we noted before.

It is now of interest to evaluate the partial derivatives

$$\partial v/\partial F, \qquad \partial v/\partial A$$

for Eq. (63). Let us first consider $\partial v/\partial F$. If we perform the indicated computation, it will be seen that the sign of this partial derivative depends only on the
expression

$$k_5 - k_2. \qquad (65)$$

If $k_5 > k_2$, or in other words, if the overall rate of the reaction scheme (49)
is greater than that of (24), then the sign of the expression (65) will be posi-

tive, and by our general definition, F will be an activator of the reaction. But the positivity of (65) is, intuitively, precisely the circumstance in which we would call F an activator of the reaction. On the other hand, if (65) is negative, then F *inhibits the reaction* (24). This too is in accord with our intuitive expectations; it means effectively that the reactant F is binding the catalyst E into a less reactive material than would be the case if F were absent from the system. We shall consider inhibition of catalysts in more detail below; for the moment it is sufficient to observe that the ordinary intuitive ideas of activator and inhibitor are both subsumed under the general formalism we have developed, as special cases of a single concept.

A similar situation obtains with respect to $\partial v/\partial A$, though the computation is somewhat more complicated. Upon performing the indicated differentiation and simplifying, we find

$$\partial v/\partial A = Kk_sF + k_sF^2 - (K^2\bar{K}k_2/K'^2)A - (k_2K/\bar{K})AF.$$

This is a rather complicated relationship, depending on the relative values of the various rate constants, and also on the relative amounts of the reactants A and F present in the system. This is a simple indication of how complex the question of activation or inhibition can really become. However, it is verified fairly readily that, if the reaction scheme (49) has a greater overall rate than the reaction scheme (24), then A *is actually an inhibitor of the total reaction* (intuitively, it competes with the facilitator F for the catalyst E, binding it into a less reactive complex).

We shall draw one final result before we turn to other matters. If in the reaction scheme (49) we replace the first equation by the equation

$$E + pF \underset{k_{-3}}{\overset{k_3}{\rightleftharpoons}} E_a \qquad (66)$$

(that is, we assume the stoichiometry is such that p molecules of the facilitator F are required to combine with a molecule of catalyst E to form a single molecule of "activated" catalyst), then all of the derivations go through as before, except that each factor F in (60) is replaced by F^p. Specifically, the rate equation for the reaction scheme in which the first reaction of (49) is replaced by (66) can be written

$$v = \frac{k_sAE_tF^p}{KK' + (K' + A)F^p}. \qquad (67)$$

To indicate how complex these kinetics can become, we may expand (66) into a series of reactions

$$E + F \rightleftharpoons E_a$$
$$E_{a_1} + F \rightleftharpoons E_{a_2}$$
$$E_{a_{p-1}} + F \rightleftharpoons E_{a_p}$$

and suppose that each of the complexes E_{a_i} has some catalytic activity through reactions of the form

$$E_{a_i} + B \rightleftharpoons C_{a_i} \longrightarrow B + E_{a_i}.$$

However, these types of kinetics can occur, and are in fact typical of allosteric systems. We shall discuss such systems shortly.

We leave it to the reader to carry out analogous analyses for the alternate mechanisms (50), (51), and draw the corresponding conclusions.

2. *Catalyzed Irreversible Monomolecular Reactions with Inhibitor*

We suppose again that we begin with the basic system (24), to which has been added a further reactant I. As before, various assumptions can be made of the manner in which I can be involved in the reaction scheme; for example, we can have the scheme

$$A + E \underset{k_{-1}}{\overset{k_1}{\rightleftharpoons}} C$$

$$C \overset{k_2}{\longrightarrow} B + E \tag{68}$$

$$E + I \underset{k_{-3}}{\overset{k_3}{\rightleftharpoons}} E_i$$

(inhibition of catalyst), or

$$A + E \underset{k_{-1}}{\overset{k_1}{\rightleftharpoons}} C$$

$$C \overset{k_2}{\longrightarrow} B + E \tag{69}$$

$$C + I \underset{k_{-3}}{\overset{k_3}{\rightleftharpoons}} C_i$$

(inhibition of the complex C), or

$$A + E \underset{k_{-1}}{\overset{k_1}{\rightleftharpoons}} C$$

$$C \overset{k_2}{\longrightarrow} B + E \tag{70}$$

$$A + I \underset{k_{-3}}{\overset{k_3}{\rightleftharpoons}} A_i$$

(inhibition of substrate). Here the new substances E_i, C_i, A_i are considered inactive, in that they do not participate in the catalyzed reaction. Since these schemes are rather easier to analyze than the corresponding activator systems, we shall to analyze than the corresponding activator systems, we shall derive the rate equations for all of them, and compare them.

We derive the rate equations in all cases in the familiar pattern, using the steady-state hypothesis for the complexes C and A_i, C_i, or E_i as is appropriate, and the conservation condition for total catalyst E_t. For the scheme (68), for instance, the conservation condition reads

$$E_t = E + C + E_i.$$

The rate equation corresponding to scheme (68) is then

$$v = \frac{dB}{dt} = \frac{k_1 k_2 A E_t}{1 + K_1 A + K_2 I},$$ (71)

where we have written

$$K_1 = k_1/(k_2 + k_{-1}), \qquad K_2 = k_3/k_{-3}.$$

Likewise, the rate equation for the scheme (69) is

$$v = \frac{dB}{dt} = \frac{k_1 k_2 A E_t}{k_{-1}[1 + (k_1/k_{-1})A + (k_3/k_{-3})(k_1/k_{-1})AI]},$$ (72)

while for the scheme (70) the rate equation is

$$v = \frac{dB}{dt} = \frac{k_1 k_2 (A_t - A_i) E_t}{1 + K_1 (A - A_i) + K_2 I}.$$ (73)

In all cases, it is immediately verified that

$$\partial v/\partial I < 0$$

so that the reactant I is indeed inhibitory in the general sense, in accord with the conventional terminology. The same is of course true if the participation of I in the reaction (24) can occur through all three schemes (68), (69), (70) simultaneously; the details are left to the reader.

It is to be noted that exactly the same equations of rate obtain if the stoichiometry of the reactant I obeys the following multimolecular scheme:

$$R + pI \rightleftharpoons R_i,$$

where R is either E, C, or A, except that we replace I in the corresponding rate equation by I^p. It is also to be noted that in each case, on comparing the rate equation for the system containing I to the rate equation (27) for the system in the absence of I that the effect of I is again manifested by making the rate "constants" of the uninhibited reaction specific functions of the inhibitory reactant I.

3. Catalyzed Irreversible Monomolecular Reactions with Activator and Inhibitor

We now turn to the most interesting situation, which is indeed, the goal of our previous discussions; namely, the effect of the presence of both facilitators (or activators) F and inhibitors I on the basic reaction (24). As should be apparent, there are a very large number of specific reaction mechanisms which could be envisaged for the simultaneous effect of activator and inhibitor on the reaction (24), each with its own kinetics as embodied in its own rate equation. We shall here consider in detail only three possible mechanisms, both as illustrations of the general properties of such schemes, and because these are particularly important for the applications to be developed. These schemes are the following:

$$
\begin{aligned}
E + F &\rightleftharpoons E_a \\
E_a + A &\rightleftharpoons C_a \\
C_a &\longrightarrow B + E_a \\
E + I &\rightleftharpoons E_i;
\end{aligned}
\tag{74}
$$

$$
\begin{aligned}
E + F &\rightleftharpoons E_a \\
E_a + A &\rightleftharpoons C_a \\
C_a &\longrightarrow B + E_a \\
C_a + F &\rightleftharpoons C_i;
\end{aligned}
\tag{75}
$$

and, most interesting of all,

$$
\begin{aligned}
E + F &\rightleftharpoons E_a \\
E_a + A &\rightleftharpoons C_a \\
C_a &\longrightarrow B + E_a \\
F + I &\rightleftharpoons F_i.
\end{aligned}
\tag{76}
$$

This last case (76) involves the direct interaction of the facilitator F with the inhibitor I, to form an inactive complex F_i. This situation has become of dominant interest because of its controlling role in genetic catalytic systems according to the Jacob–Monod picture of the operon (see subsequent sections). We shall accordingly consider only this case in detail; rate equations for more complicated versions of this basic idea, and for systems like (74) and (75), may be set up and studied similarly, in what by now should be the familiar pattern.

The relevant rate and conservation equations are

$$
\begin{aligned}
dC_a/dt &= k_2 E_a A - (k_{-2} + k_3)C_a = 0, \\
dE_a/dt &= k_1 EF + (k_3 + k_{-2})C_a - (k_2 A - k_{-1})E_a = 0, \\
E_t &= E + E_a + C_a, \qquad F_t = F + E_a + F_i + C_a, \\
I_t &= I + F_i, \qquad\qquad v = dB/dt = k_3 C_a.
\end{aligned}
\tag{77}
$$

Writing $K = k_2/(k_2 + k_3)$, we find first that

$$
C_a = KAE_t/(1 + K) - KAE/(1 + K) - KAE_a/(1 + K),
$$

and

$$
E = (k_2 A - k_1)E_a - (k_3 + k_2)C_a/(k_{-1}F).
$$

Thus

$$
C_a = \frac{KAE_t}{1 + K} - \left[\frac{KA(k_2 A - k_1)}{k_{-1}F} - \frac{KA}{1 + K}\right]E_a + \frac{KA(k_3 + k_2)}{k_1 F}C_a.
\tag{78}
$$

But

$$
E_a = (F_t - F - F_i) - C_a
$$

and substituting this in (78), and solving for C_a, we find

$$C_a = \frac{[KAE_t/(1+K) - KA(k_2A - k_1)/k_{-1}F - KA/(1+K)](F_t - F - F_i)}{1 - KA(k_2A - k_1)/k_{-1}F + KA/(1+K) - KA(k_3 + k_2)/k_{-1}F},$$
(79)

which can then be substituted into the rate equation for $v = dB/dt$.

The rate equation arising from Eq. (79) can be seen to be of a rather complex form, even in this relatively simple situation. Nevertheless, there is one important point to be observed. The partial derivatives

$$\partial v/\partial F, \qquad \partial v/\partial I,$$

which express the dependence of the reaction rate on the facilitator and inhibitor concentrations, may be interpreted as formally exhibiting a system of this kind as a biochemical "switch" of the type we have discussed previously. Whether or not the reaction will go depends entirely upon the initial relative concentrations of F and I. The exact expressions for the relative rates in terms of the concentrations of F and I may be obtained from Eq. (79).

V. Examples of Epigenetic Controls

In the present section, we are going to briefly examine some actual (and for the most part, well known), examples of the manifestations of the kinds of control mechanisms we have discussed in some actual cellular systems. Naturally, we cannot go too deeply into the experimental details in this short space, and we refer the reader to the bibliography at the end of the chapter for more complete expositions. The point of this section is to show how widespread the state-transition mechanisms we have discussed above actually are in cellular systems, and to point up some of their metabolic consequences, both for epigenesis, and for ordinary cell physiology and homeostasis. Our discussion will be subdivided into two main parts: those controls exerted at the level of enzymatic catalysis which manifest themselves through effects on reaction rates of enzyme-mediated reactions, and those controls exerted at the genetic level, which manifest themselves through the control of catalyst concentrations.

A. Rate Control in Enzyme Systems

1. *"Feedback Inhibition"*

Simple cells, such as bacteria, are able to function effectively in the simplest environments, typically containing nothing more than a carbon source (generally a carbohydrate), a nitrogen source (ammonia), and a few inorganic ions. From these simple materials they are able to manufacture all of the metabolites necessary for growth and multiplication. In particular, they are able to synthesize all of the amino acids required for incorporation into cellular proteins.

It was early observed that if amino acids were supplied exogenously to such cells, then the exogenous amino acids were preferentially incorporated into protein, and the internal synthetic pathways leading to these amino acids were shut down. Clearly this is an intelligent (or "adaptive") response of the cells to the alteration in environment; it is certainly more efficient to use metabolites that are freely available than to waste resources and energy synthesizing them from simpler constituents. It therefore became a matter of considerable interest to see how this adaptive response comes about.

One's first thought would be the following: We have seen above that any stable steady state (such as that of a bacterium in a stationary phase in a culture medium) has built-in regulatory properties, derived from the rate equations which govern the system. In the case of chemical systems, these rate equations ultimately rest on the principle of mass action. One of the well-known consequences of this principle is that, if the end product of a reversible reaction (or reaction sequence) is added to the system, the reaction is slowed down, or even reversed, depending on the amount of end product added. Thus, any chemical system might be expected to exhibit this type of adaptive behavior.

There were two lines of evidence, however, which indicated that this explanation was not adequate.

1. In *Escherichia coli*, the amino acid isoleucine is synthesized from the amino acid threonine via the following sequence of reactions:

threonine $\longrightarrow$ α-ketobutyrate $\longrightarrow$ acetohydroxybutyrate $\longrightarrow$ isoleucine.

It so happens that the first step in this pathway, catalyzed by the enzyme threonine deaminase, is *irreversible*.

2. It is possible to obtain mutants which are unable to synthesize one or another of the enzymes catalyzing particular steps in the above sequence. In the absence of exogenous isoleucine, the appropriate intermediate (the product of the enzyme which precedes the missing one in the mutant) accumulates, but when isoleucine is exogenously supplied, the synthesis of the intermediate is inhibited.

The only interpretation for such experimental facts was that the exogenous metabolite acted so as to *inhibit the first important enzyme in the biosynthetic pathway*. That is, the effect of isoleucine in the above pathway would be to decrease the rate at which the first enzyme, threonine deaminase, catalyzed its reaction. The effect of this inhibition would be, in effect, to shut down the entire pathway leading to isoleucine. This effect, which is termed *end-product inhibition*, has subsequently been observed in many enzyme systems, and in many different kinds of cells belonging to organisms from many phyla.

It is clear that this method of control of reaction rates may be considerably more general than simple end-product inhibition; we can imagine that the end products of one particular synthetic pathway will inhibit enzymes involved in quite different pathways. Indeed, precisely such a mechanism underlies the "switch" described above, whose rate equations we gave as Eq. (9). Thus, we can imagine the enzymic catalysts in a cell as arrayed into a network (or networks) with coupling between the different enzymes in the network mediated through the inhibitory effects of one enzyme product upon the rate constant of another. The net effect of such activity will manifest itself as a well-coordinated, or *integrated*, response of the cell as a whole to exogenous environmental situations.

A good example of such an integrated response, involving many enzymes interacting in this fashion, is found in the control of energy metabolism. Most cells possess two main pathways by which high-energy phosphate bonds (ATP) may be formed: the tricarboxylic acid cycle, which functions under aerobic conditions, and the Emden–Meyerhof pathway, which typically functions under anaerobic conditions. It was discovered by Pasteur in 1860 that free oxygen inhibits glycolysis (that is, the entire chain of events leading to ATP through the Emden–Meyerhof pathway). Conversely, there is a converse effect, called the Crabtree effect, or the glucose effect in which aerobic respiration is inhibited by the addition of exogenous carbohydrates. Thus, these two enzymic systems, aerobic and anaerobic, are capable of functioning as a switch, through the inhibition of enzymes in the other. These systems, however, are extremely complicated, and the details would be far beyond the scope of our present discussion; for a discussion of such details we refer the reader to the references, particularly those by Chance, Garfinkel, and their co-workers [Chance, 1960; Chance and Hess, 1959a, b, c, d, 1961; Chance *et al.*, 1960, 1962; Garfinkel and Hess, 1964, 1965].

2. *"Activation"*

If we can imagine the products of one enzyme or enzyme pathway exerting an inhibitory effect on the rates of enzymes, it is of course equally easy to imagine the converse effect; the activation (or augmentation of reaction rate) of particular enzymes by the products of other enzymes. Indeed, as we saw in the preceding section, such an effect can arise in principle from many causes, including the binding of an inhibitor into an inactive complex by a specific enzyme product. Clearly too, the net effect of such activations between enzyme pathways will also manifest itself in the integrated behavior of cells in response to particular environments.

An early example of this type of activation was found in a sharp increase in DNase activity in *E. coli* when the cell is invaded by bacteriophage DNA. This increase in activity (that is, reaction rate) was shown not to be due to *de novo* enzyme synthesis, but rather due to the release of an inhibition of preexisting enzyme. Thus, it represents a true "substrate activation," in the sense in which we have employed that term.

Many other examples of such substrate activations are known, in which a preexisting enzymatic protein is activated by the substrate of the enzyme, or by another metabolite. However, since such systems closely resemble enzyme inductions (the *de novo* synthesis of the enzyme brought about by the substrate; see below) the number of clear-cut examples is not as large as for the various feedback inhibition mechanisms we have discussed. Indeed, many mechanisms that were originally considered as examples of inductions have turned out on closer examination to involve activation mechanisms (perhaps the best-known of these is the activation of tryptophan pyrrolase by tryptophan in liver fractions).

Furthermore, there are many specific cofactors (for example, metabolic products of other pathways) that some enzymes require to become fully active. And, of course, there are equally well-known cases in which the enzyme is synthesized as an inactive zymogen, requiring the mediation of another enzyme before it can become fully active. Such situations are discussed in any textbook of biochemistry or enzymology. The main point here is that in the area of enzyme activations we find an inverse mechanism to that of enzyme inhibitions, and as we saw above, the antagonism of these two opposite processes can lead on the one hand to an exquisitely finely tuned homeostat, and, on the other hand, can account for at least some of the controlled changes of state which we have regarded as epigenetic.

3. Allostery

We have seen abundantly that activations and inhibitions of enzymatic catalysts can arise from interactions of metabolites, not directly involving the enzyme itself. The most interesting cases, however, are those in which the enzyme interacts directly with the activator or inhibitor. Intuitively, such interactions can be regarded as causing conformational changes in the catalytic protein itself, thereby modifying the characteristics of the active catalytic site and consequently altering (increasing or decreasing) the rate of the catalyzed reaction. As usual, inhibition effects have been the most thoroughly studied.

The conventional ideas of enzyme inhibition, until relatively recently, clustered around the concept of competitive inhibition. That is, the inhibitor was regarded as "competing" for the active site with the enzyme substrate, binding the enzyme into an inactive form. This is the familiar situation described in simplest terms by Eqs. (68) and (71) above. Many cases of competitive inhibition have been thoroughly studied and are classical; such as the competition of carbon monoxide for the active sites on hemoglobin; once again we refer to the Notes and General References for further literature [specifically Webb, 1963]. On the other hand, it was early apparent that certain types of inhibition were noncompetitive; as we have seen above, this can be determined directly from observing the kinetics involved; the well-known Lineweaver–Burke plot of the Michaelis–Menten equations display most clearly the difference between competitive and noncompetitive inhibitor kinetics. Aside from these kinetic differences, it was also observed that competitive inhibitors are generally closely similar to the natural substrates in their chemical structure, whereas noncompetitive inhibitors need not be, and indeed in general are not.

Naturally, when end-product inhibition was discovered, the question was raised as to whether the inhibition was competitive or not. Chemical comparisons indicated that the inhibiting end product was substantially different from the substrate in its chemical conformation, making it difficult to understand how both the substrate and inhibitor could compete for the same site. In 1961, the first definite piece of evidence was obtained by Gebhart and Pardee; they found that they could modify the structure of an enzyme exhibiting feedback inhibition (aspartic transcarbamylase) by mild heating in such a way that the enzyme retained its catalytic activity but could no longer be inhibited by the end product (cytidine 5′-phosphate). This was evidence that at least two distinct sites were involved, one of which bound the substrate, the other the inhibitor (significantly, the catalytic activity of the treated enzyme was higher than normal). Such observations were repeated on many other enzymes subject to feedback inhibition, and indicated that the mechanisms of feedback inhibition were in themselves carefully controlled and highly evolved.

Such behavior was termed by Jacob and Monod to be *allosteric* (to be distinguished from the *isosteric* kind of inhibition arising from closely similar compounds competing for the same active site). These authors put forward the suggestive view, implicit in what we have already said above, that

> An allosteric enzyme is a chemical transducer, allowing interactions to occur between compounds which would not otherwise react or interact in any way. In fact, allosteric proteins appear to constitute the type of 'universal' interconnecting elements required for the construction of physiological circuits [Monod and Jacob 1961, p. 399].

There is thus at hand a rich abundance of evidence of various kinds that demonstrates the existence of the formal controls we have introduced in the preceding sections. So far, however, we have dealt only with activation and inhibition effects entirely at the level of the enzymatic catalysts themselves, and the manner in which their rates of activity could be modified by the state of the system. We must now turn to the second, and perhaps more important, level of epigenetic control, namely that which concerns the manner in which the rates of *genetic* catalysts is controlled. This is, of course, the problem of *gene expression*, formulated in a garb calculated to emphasize the close relationship between all types of epigenetic mechanisms, from the simplest to the most complex.

B. Rate Control of Genetic Catalysis

Once again, we shall cast the discussion of rate control of genetic systems in a parallel fashion to the one of rate control in enzymatic catalysts. As before, there will be a pair of antagonistic processes at work: that corresponding to an increase in rate, which we shall call *induction* in this context, and that corresponding to a decrease in rate, which we shall call *repression*.

1. Repression

It was discovered in the early and middle 1950s that if bacterial cells were grown in media containing particular metabolites, such as amino acids, then the *synthesis* of enzymes in the biosynthetic pathway leading to these metabolites was greatly reduced. After it had been determined that such effects did not arise because of activators or inhibitors of preexisting protein, it was recognized that this effect takes place directly at the genetic level. As in the case of enzyme feedback inhibition, there was a clear teleological virtue in this behavior, the utilization of exogenous metabolite and the consequent conservation of energy and material that could best be employed elsewhere.

Furthermore, this effect is of great generality. It subsequently has been shown that many enzymes are repressed, in this sense, by the products of the reactions which they catalyze. Thus, alkaline phosphatase is often repressed by inorganic phosphate, and isocitratase, an important enzyme in the glucolytic pathway, is repressed by succinate. Further, evidence is accumulating that such effects are generally found throughout the biological world.

It turns out, as should perhaps be intuitive from all that we have said, that repression phenomena from both the epigenetic and homeostatic viewpoints are most interestingly considered when coupled with inverse, or inductive, effects. Therefore, we shall now briefly turn to a discussion of induction before proceeding to a discussion of the manner in which rates of genetic activity are epigenetically controlled.

2. Induction

It was recognized long ago (well into the 19th century, in fact) that bacteria grown under different kinds of growth conditions (for example, utilizing different carbohydrates as carbon sources) exhibit differences in the kinds of enzymes they produce; further, these differences are *adaptive* (in the sense that the enzymes present under any particular set of environmental circumstances tend to be precisely those involved in the metabolism of the carbon source which is present). For this reason, enzymes of this type came to be called *adaptive enzymes*, to be distinguished from the so-called *constitutive enzymes* that are produced by the cells under all environmental conditions. Over the years, a great many cases of this phenomenon were discovered, and it was discovered that, in many cases at least, the adaptive enzyme formation was accompanied by the *de novo* synthesis of active enzymatic protein (although a number of cases turned out to be selection effects in rapidly growing cultures, substrate activations of previously inactive zymogens, and the like). Such *de novo* synthesis of enzymatically active protein, typically brought about as an adaptive response to the enzyme substrate or a related compound, has come to be called *enzyme induction*. It represents an increase in the rate at which a specific enzymatic protein is synthesized, as a function of the concentration of a metabolite (which is therefore called an *inducer* of the enzyme). Such enzyme inductions have been of the greatest interest for many years, because of their clear epigenetic implications, and their implications for related problems, such as tumor induction, antibody formation, and the like.

Enzyme induction is not a simple affair, as is shown by a glance at perhaps the most thoroughly studied case, the induction of β-galactosidase in *E. coli* in the presence of a β-galactoside. In general, the response of the cell to a substrate like a β-galactoside is an integrated one, involving the coordinated induction of many enzymes. For instance, in addition to β-galactosidase, a group of enzymes (permeases) are induced which are involved in the active transport of β-galactosides into the cell. Indeed, in many cases, the protein synthesis induced by specific substrates involves the formation of an inducible permease, the specific enzyme catalyzing a reaction involving the substrate being in fact constitutive but inaccessible to the substrate because of permeability problems.

Because of the great interest in adaptive enzyme formation as a mechanism for modifying cellular properties without any change in the genome, many ingenious hypotheses have been proposed over the years to account for the phenomenon. Only during the past 15 years or so, however, have data been collectible that bear directly on this problem. Briefly, it typically appears that induction and repression are inseparably linked together in the control of enzyme synthesis, that is, in the rate at which a particular gene is expressed. An inducible enzyme like β-galactosidase can be regarded as normally repressed when cells are grown in the absence of β-galactosides, the repressor in question being a molecule or metabolite always present in the cell. The effect of the β-galactoside involves its combination with the repressor to form an inactive complex, thereby removing the repression. This is the situation whose kinetics we have studied above [see Eq. (76)].

The detailed study of the interaction between inducers and repressors, mostly in bacterial systems, has been reduced to a coherent form by the well-known *operon hypothesis*, introduced by Jacob and Monod. Since this concept has come to occupy a dominant role in cellular biology in the years since it was introduced, it is only necessary here to touch on the main points; for further details and the experimental evidence on which the scheme is based, we refer the reader to the literature cited in the bibliography.

3. The Operon Concept

In 1961 Jacob and Monod presented a detailed mechanism for the mechanisms involved in the induction and repression of enzymatic protein, which at the same time provided a new way for looking at the organization of the genetic material itself. This mechanism was based on the following four specific hypotheses, each suggested by a large (and increasing) number of data drawn from bacterial systems.

1. The information which determines the primary structure of a specific protein (polypeptide) resides in an entity termed the *structural gene* for the polypeptide. This is conventionally regarded as a particular DNA sequence embedded in the chromosome.

2. Associated with a structural gene is a region responsible for the initiation of synthesis of the primary gene product of the structural gene (messenger RNA). This region is called the *operator region* for the structural gene. It is not excluded that the same operator region may be associated with more than one structural gene. The complex consisting of an operator region and all of the structural genes associated with it is termed an *operon*, viewed by Jacob and Monod as a natural unit of genetic transcription.

3. The activity of any particular operator region is controlled by a distinct (and often remote) genetic entity called a *regulator gene*. A (direct or indirect) product of the activity of such a regulator gene is a metabolite (generally not protein) called *repressor*, denoted by the symbol R. The repressor R combines specifically and reversible with its corresponding operator, preventing the transcription of the associated structural genes.

4. The repressor R can interact with metabolites F (called effectors) in a variety of ways. Typically we have a reversible combination into a complex:

$$R + F \rightleftharpoons RF.$$

It may be, for example, that R is the active repressor and that the complex RF is inactive. In this case F functions as an inducer. Or it may be that the complex RF functions as the active repressor, in which case the structural genes continue to be transcribed until F is exogenously supplied; thus, F functions in this case as a simple repressor. A number of other interesting situations are possible, as should be clear from the discussion of the preceding sections.

Clearly the various metabolites F which may function as inducers and repressors in such a scheme may be (and in general are) produced through the activity of other operons. We thus once again have the possibility of decomposing a system organized in this way into one or more *networks*, built of operon units, and wired together through the interaction with the metabolic products of one such unit with others. Upon these networks are of course to be superimposed the kinds of interactions which modify the rates at which specific enzymatic catalysts operate, which we have already described.

We thus see that all of the directed state transitions in cells which we have discussed may be understood in a common framework. They represent an elaboration of the kind of behavior that occurs in any kind of open system, but which becomes more and more subject to careful control, and to integrated activity, when the systems involve specific catalysts, and even more so when the production of these catalysts occurs within the system through a genetic mechanism. A most important methodological point also emerges in such systems: The idea that a complicated biochemical system can be regarded as a network built of essentially similar subunits interacting through induction and repression mechanisms provides us with important conceptual advantages. First, it recaptures at a lower level of organization (which, as noted at the outset, is one of the main attractions of the cell theory itself) the idea of modular construction. Second, it allows us to view the behavior of the overall system in terms of stimulus and response of definite units, with specific and definable functional properties, instead of being faced at all times with the complexity of a large set of analytically opaque rate equations. Third, it displays the behavior of such a large biochemical system in terms of an integrated pattern of *information flow*, in terms of the familiar block-diagram concepts of engineering and control theory. Finally, and perhaps most important, it exhibits the exceedingly close relationships that exist between epigenetic control phenomena in biochemical and genetic systems and other kinds of network theories, originally developed to explain biological phenomena at a completely different level. In particular, the theory of such metabolic and genetic networks is seen as essentially identical to the well-developed theory of *neural nets*, developed to understand the behavior of the central nervous system and its behavioral correlates. It suffices only to point out that neural networks are also built out of essentially similar units, wired together in a definite pattern, and which interact with each other through excitation (induction) and inhibition (repression). This is an exceedingly important observation, which will play a role in biology analogous to the role which the discovery of equivalent minimization principles has played in unifying different areas of physics. We shall see many indications of the fruitfulness of this unification in subsequent developments (see Arbib, Chapter 3, Volume II, Chapter 3, Volume III of this treatise).

VI. Some Specific Epigenetic Circuits

In the present section, we shall describe a few simple epigenetic circuits and describe some of their properties. We shall do this by combining the specific results and criteria of Section IV with the mechanisms described in Section V. Unfortunately, a fully detailed exposition of the mathematical techniques involved would be far beyond the scope of the present chapter; for a rigorous discussion of these matters the reader is urged to consult the original papers listed in the bibliography, or the author's text, *Dynamical System Theory in Biology.*

Consider the very simplest enzyme-catalyzed (reversible) reaction (44), to which correspond the rate equations (45). We know how to set up the rate equations for such a system in case the reaction involves a facilitator F [according to stoichiometric schemes analogous to Eqs. (49), (50), and (51)] or an inhibitor I [using stoichiometric schemes analogous to Eqs. (68), (69), and (70)], and when both an activator and an inhibitor are involved [analogous to Eqs. (74), (75), and (76)]. As the reader can readily observe by trying his hand at an exact development of the rate equations involved, the complexity of these systems increases very rapidly. The situation becomes even more complex if, following the intuitive discussion of Section V above, we allow the role of the facilitator F, or the inhibitor I, to be played by one of the reactants A, B already in the system. The reader is urged to try his hand at writing down some of the rate equations for these cases, and see just how complex they do become, even in the case (which is by no means the commonest situation in practice) where all the reactions are monomolecular.

We shall take a different approach to obviate the detailed kinetic complexities and manifold special cases that arise in this type of analysis, motivated by the general discussion of activation and inhibition in arbitrary chemical systems given in Section IV. Namely, we shall do the following: Notice that in the simple stoichiometric scheme (44) there are a number of parameters (four rate constants, together possibly with conservation conditions for the catalysts and/or the reactant A) at our disposal. For simplicity, we shall concern ourselves solely with the rate constants. We have abundantly seen that large directed changes of cell state can be governed by the specific values assigned to these parameters, and that in general, the effect of facilitators and inhibitors of arbitrary reactions is in fact manifested through their effect on these rate constants. What we shall do, then, to assess the effect of an arbitrary reactant M on a catalytic system like (44) is to make the rate constants in (44) depend upon M in specific ways. Of course, the specific stoichiometry of a definite kind of interactions between M and a catalytic system will induce such a relationship automatically; but since these relationships are complex, and since they are quantitatively different for different stoichiometries even when the overall qualitative effect is the same, we shall replace these complicated relationships by simple ones, inducing the same overall qualitative behavior, but which are immediately amenable to study. This type of approach, though not mathematically rigorous, is certainly plausible, and can be made rigorous by means of a mathematical discussion of the "robustness" of this type of dynamical argument (that is, by its independence of the specific form of the rate equations, providing a few simple and general qualitative conditions are satisfied).

Accordingly, let us assume that the rate constant k_1 in the system (44) is a specific function of the metabolite M. This means that the rate at which the active complex C is formed, and hence the rate at which the entire reaction $A \rightleftharpoons B$ proceeds, depends on the concentration of M. For simplicity we assume that the other rate constants are independent of M, but the discussion readily generalizes to the case where all of the other rate constants depend on M in particular ways. We can assess the effect of M on the system, according to our previous discussion, by evaluating the manner in which the rate of the overall reaction depends upon M; that is, by evaluating the partial derivative

$$\partial v / \partial M.$$

This rate v is given by Eq. (46). If we evaluate the partial derivatives in question, writing the constants in terms of the original rate constants, and placing $k_1 = k_1(M)$, it is readily seen that

$$\partial v / \partial M = [\partial k_1 / \partial M] \, F(A, B, E_t) \tag{80}$$

where $F(A, B, E_t)$ is a positive function of its arguments. If now $k_1(M)$ is a monotonic increasing function of M, it follows that

$$\partial k_1 / \partial M > 0 \qquad \text{and hence} \qquad \partial v / \partial M > 0$$

so that M, according to our general definition, is always an activator (or facilitator) of the system. Conversely, if $k_1(M)$ is a monotonic decreasing function of M, it follows that $\partial k_1 / \partial M < 0$ and hence that M is an inhibitor of the system. Thus, if we wish to study merely the qualitative effects of activators or inhibitors on catalytic systems, it suffices to make $k_1(M)$ as simple a function as possible, subject only to the monotonicity required to elicit the behavior under study.

This strategy remains valid in the case for which M is already a reactant of the system; that is, when M = A or M = B. In this case, even evaluating the partial derivatives

$$\partial v/\partial A, \quad \partial v/\partial B$$

becomes more complex.

EXERCISE: Is simple monotonicity of the partial derivatives $\partial k_1(A)/\partial A$, $\partial k_1(B)/\partial B$ still sufficient to guarantee that A or B will always either activate or inhibit?

The general principle remains: In order to guarantee, under these circumstances, that A is an activator of the reaction, it is sufficient to choose $k_1(A)$ such that

$$\partial v/\partial A < 0.$$

Likewise, to make A an inhibitor, it is sufficient to choose $k_1(A)$ such that

$$\partial v/\partial A < 0.$$

To make B an activator, we need only choose $k_1 = k_1(B)$ such that

$$\partial v/\partial B > 0,$$

while to make certain that B is an inhibitor, we need only choose $k_1 = k_1(B)$ so that

$$\partial v/\partial B < 0.$$

The cases of greatest interest, as might be guessed from the discussion of Section V, are those in which B is an inhibitor of the system (44), and in which A is an activator of the system. The cases in which B inhibits the reaction which produces it may be interpreted in a variety of ways. If E is an enzymatic catalyst, then the inhibition of E by the product B of its own activity is a special case of the *feedback (or end-product) inhibition*, which we discussed extensively in Section V; if *E* represents the catalytic activity leading from the primary genetic material to the metabolite B which is the product of an enzymatic protein encoded by the primary genetic material, then the situation is one of *end-product repression*. In either case, we shall refer to this process as being one of *back inhibition*, and denote the existence of such a back inhibition in the system (44) by the diagram

$$A \xrightarrow{\;-E\;} B \tag{81}$$

Conversely, in the case in which A is an activator, we shall speak of *forward activation*, and denote the situation by

$$A \xrightarrow{\;+E\;} B \tag{82}$$

EXERCISE: Discuss the biological interpretation and significance of the two remaining possibilities: *forward inhibition*

$$A \xrightarrow{-E} B \qquad (83)$$

and *back activation*

$$A \xrightarrow{+E} B. \qquad (84)$$

The cases we have just discussed, involving the simplest reaction scheme (44), can obviously be generalized considerably. For example, we can consider a chain of catalyzed reactions of the form

$$A_1 \xrightarrow{E_1} A_2 \xrightarrow{E_2} A_3 \xrightarrow{E_3} \cdots \xrightarrow{E_{n-1}} A_n. \qquad (85)$$

If the product A_m at any stage is an inhibitor of any enzyme E_k in the system, where $k < m - 1$ (that is, if one or more of the rate constants involved in the kth reaction are suitable functions of A_m) then we speak of *back inhibition* in the chain, and denote the situation, analogous with the simplest case,

$$A_1 \longrightarrow \cdots \longrightarrow A_k \xrightarrow{E_k^-} \cdots \longrightarrow A_m \longrightarrow \cdots \longrightarrow A_n. \qquad (86)$$

If A_m inhibits one of the subsequent enzymes, we have *forward* inhibition, denoted by

$$A_1 \longrightarrow \cdots A_m \longrightarrow \cdots \xrightarrow{-E_{m+k-1}} A_{m+k} \longrightarrow \cdots A_m. \qquad (87)$$

Likewise, if A_m activates an enzyme E_{m+k}, $k > 0$, we have *forward activation*,

$$A_1 \longrightarrow \cdots \longrightarrow A_m \longrightarrow \cdots \xrightarrow{+E_{m+k}} A_{m+k+1} \longrightarrow \cdots \longrightarrow A_n, \qquad (88)$$

while if A_m activates a preceding enzyme, we have *back activation*,

$$A_1 \longrightarrow \cdots \longrightarrow A_k \xrightarrow{E_k^+} \cdots \longrightarrow A_m \longrightarrow \cdots \longrightarrow A_n. \qquad (89)$$

Clearly this kind of analysis can be extended still further, to arbitrarily branched networks of catalyzed reactions, or even to those containing closed cycles (such as the Krebs cycle), though in these cases the concepts of "forward" and "backward" lose much of their cogency.

We will now turn attention to one very interesting cellular phenomenon, of the greatest interest for epigenetic control and many other biological processes; namely, the capacity of certain forward- and backward-activation and inhibition systems to exhibit *oscillations*. We recall that oscillations in reactant concentrations (see Fig. 4) are associated with neutrally stable steady states, corresponding to periodic solutions of the rate equations. The oscillations in question are *autonomous* oscillations inherent in the system dynamics, and not *forced oscillations* corresponding to the response of a system to a periodically fluctuating input from the environment.

Why are such autonomous oscillations of interest? For one thing, they represent a departure from the conventional view that any reaction system, left to itself, will approach some constant steady-state situation. Oscillatory behavior does indeed represent a kind of steady state, since in mathematical terms both the constant steady state and the closed trajectories corresponding to periodic behavior represent the possible *limiting behaviors*, as time goes to infinity, of system trajectories near them. But the overwhelming importance of oscillations is that they exhibit at least the possibility of *internal timekeeping within the system*. They thus, for example, could provide in principle an intrinsic biochemical basis for the numerous cyclical or rhythmical processes found in biology. Or, still closer to the epigenetic problems which are our main interest in this chapter, they could in principle keep time for the carefully controlled *temporal organization* of developmental phenomena. In differentiation, for example, specific developmental events, occurring quite remotely from one another in the developing system, are carefully synchronized, even in the absence of direct coupling between them. This kind of temporal organization can be interpreted by supposing that there is an intrinsic clock, which provides temporal signals determining when these processes are to be initiated and terminated, that can, in some sense, be consulted by these processes. Thus, it becomes a matter of the greatest interest to determine the conditions under which oscillations are possible in chemical and biochemical systems.

Let us consider the situation (86), in which we have a back inhibition. Intuitively, this means that, as the product A_m accumulates, the rate of the reaction leading to A_m is diminished. As the concentration of A_m drops, because of its utilization in subsequent reactions, the inhibition exerted by A_m on E_k is released. Thus, the pathway leading to A_m once again becomes functional and the concentration of A_m once more increases. The cycle can then be visualized as repeating indefinitely. Thus, at least intuitively, it is plausible to see in systems possessing a back inhibition at least the potentiality for oscillations. Indeed, oscillations of this kind are well known in engineering applications of feedback control (where they are a severe nuisance and great pains are taken to avoid them); this fact and the wide occurrence of backward inhibitions in biology, documented in the preceding section, make it plausible to seek oscillatory solutions in such systems.

EXERCISE: Examine, in intuitive terms, whether you would expect systems exhibiting forward activation, forward inhibition, and backward activation to manifest oscillatory behavior.

In general, the determination that a particular set of rate equations does actually admit periodic solutions is a most complicated affair mathematically. It turns out, not surprisingly, that the capacity of a particular system to exhibit autonomous oscillations depends heavily on the values assigned to the various parameters present in the system. In general mathematical terms, the best that can usually be done is to *exclude* certain kinds of rate equations by means of (rather weak) necessary conditions for the existence of oscillations which may not be satisfied by the rate equations under study. The intelligent use of these negative criteria, supplemented by analog and digital simulation of rate equations, thus far has yielded precious, if limited, information on the existence of oscillations in open catalytic systems.

It is of some interest to see how these negative criteria manifest themselves, without going into full mathematical details, in terms of the activation and inhibition patterns present within the system. Let us fix ideas by considering the simplest set of rate equations which do exhibit oscillations:

$$dx/dt = y, \qquad dy/dt = -ax. \tag{90}$$

In chemical terms, these equations state that the reactant y is produced from x, and that x catalyzes the removal of y from the system. Otherwise, these equations are those of the (undamped) harmonic oscillator of mechanics. In this case, x is the displacement and y is the velocity or momentum of the oscillating particle; these are the simplest (linear) equations that exhibit oscillations. To see that the trajectories are indeed closed, we can obtain the differential equation satisfied by the trajectories in the state space (the X–Y plane) by dividing one of the rate equations by the other;

$$dy/dx = -ax/y.$$

These can be immediately integrated, to give the equations of the trajectories themselves:

$$x^2 + ay^2 = R^2.$$

These are clearly ellipses in the X–Y plane, closed curves corresponding to the well-known sinusoidal solutions for $x(t)$ and $y(t)$.

Now let us look at this system from the standpoint of the general self- and cross-activation and inhibition characteristics we developed early in Section IV. We have here two reactants x and y, and we wish to determine the effect of each of these reactants on the rate of production of these reactants. Evaluating the partial derivatives required, we find that

$$\partial v_x/\partial x = 0, \qquad \partial v_x/\partial y = 1 > 0,$$
$$\partial v_y/\partial x = -a > 0, \qquad \partial v_y/\partial y = 0,$$

or in other words, in this oscillatory system, *the cross couplings are always of opposite sign,* and retain this relationship throughout the state space. This corresponds to a Higgins diagram of the form

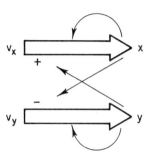

Let us look at another system known to exhibit oscillatory behavior; namely, the Volterra system (8). This is a more complex nonlinear system. Once again, to determine the self- and cross couplings we evaluate the requisite partial derivatives,

$$\partial v_A/\partial A = 1 - B, \qquad \partial v_A/\partial B = -A,$$
$$\partial v_B/\partial A = B, \qquad \partial v_B/\partial B = A - 1.$$

In the first quadrant of the state space (to which we are restricted by the non-negativity of the reactant concentrations) it follows that the cross couplings are again of opposite sign throughout the first quadrant (though the magnitudes of the cross couplings varies with the state of the system, unlike the previous case). The self-couplings here can be of four types, as shown in the diagram of the first quadrant of the A–B plane in Fig. 15a. Thus, each of the four regions of the first quadrant indicated in the figure has its own Higgins diagram; these are, respectively, of the form shown in Fig. 15b.

In all cases the cross couplings are of opposite sign. It turns out from the general mathematical theory (see Rosen [1970] for details) that this property of cross couplings is indeed a necessary condition for oscillations; thus any system for which the cross couplings fail to be always of opposite sign for a pair of reactant concentrations can never show oscillations in these reactant concentrations.

Now let us see what this negative criterion tells us in the various activation and inhibition schemes (86)–(89) we have considered. Consider the case of feedback inhibition, in the general case in which $k = 1$; that is, the first enzyme of the sequence is inhibited by the mth metabolite. This situation has been studied by quite a number of authors, often including, as an extra bit of generality, that the stoichiometry of the inhibition reaction is of the form

$$E_1 + pA_n \rightleftharpoons C_i. \qquad (91)$$

The number p is regarded as an additional system parameter, and turns out

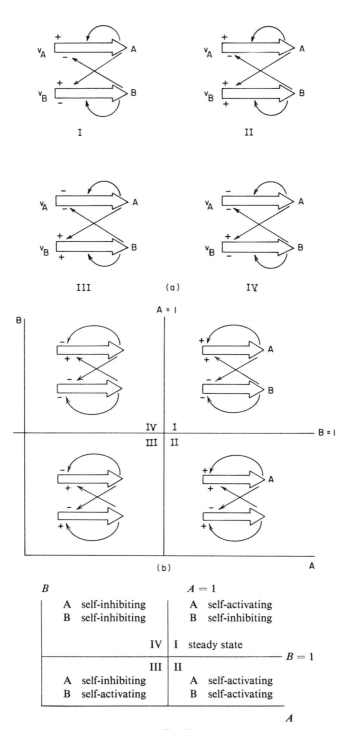

Fig. 15

to play an important role. In any case, the rate equations for such a system, under the steady-state assumption we always utilize, and assuming that A_1 is held constant through contact with an infinite external source, are of the form

$$dA_2/dt = k_1 A_1/(1 + \alpha A_n{}^p) - k_2 A_2,$$
$$dA_i/dt = k_{i-1}A_{i-1} - k_i A_i, \qquad 3 \leqslant i \leqslant n, \qquad (92)$$

where the parameter α represents the ratio of free catalyst E_1 to total initial catalyst concentration E_{1t}, and the "rate constants" k_i are the ratios of the maximal velocity of the ith reaction to the Michaelis constant K_i of the ith enzyme.

From these equations, it easily follows that the cross couplings between A_2 and A_n are of the form

$$\partial v_2/\partial A_n = -p\alpha k_1 A_n^{p-1}/(1 + \alpha A_n{}^p)^2 < 0, \qquad \partial v_n/\partial A_2 > 0, \qquad (93)$$

so that the negative criterion is satisfied in this case, and the system can in principle exhibit oscillations in reactant concentration. The actual capacity of such a system to oscillate may be studied further by a combination of mathematical stability analysis of the trajectories and simulation studies. In Figs. 16 and 17 are shown the result of one such simulation study, where the parameters were chosen as follows:

$$n = 6; \quad p = 4; \quad \alpha = 0.08;$$
$$k_1 A_1 = 5.1; \quad k_i = 1, \quad i = 1, \ldots, 5; \quad k_6 = 0.6;$$
$$A_i(0) = 0, \quad i = 2, \ldots, 6.$$

In Fig. 16 is shown the time course of the two solutions $A_2(t)$ and $A_n(t)$, while in Fig. 17 is shown the (projection) system trajectory in the A_2, A_6 plane. Thus, when the parameters are chosen properly, these systems can indeed exhibit oscillations. For a fuller discussion, including circumstances in which oscillations are not possible (for example, if $n = 2$ and p is too small) we must refer the reader to the original literature.

The case $n = 2$, $p = 1$ mentioned above, for which it was stated that no oscillations are possible, is easy to anlayze directly for its stability properties. The equations are

$$dA_2/dt = k_1 A_1/(1 + \alpha A_3) - k_2 A_2,$$
$$dA_3/dt = k_2 A_2 - k_3 A_3, \qquad (94)$$

which are the equations of a two-dimensional system, like the harmonic oscillator. The analysis proceeds by expanding the right-hand sides of these rate equations in Taylor series around the steady state of the system, and

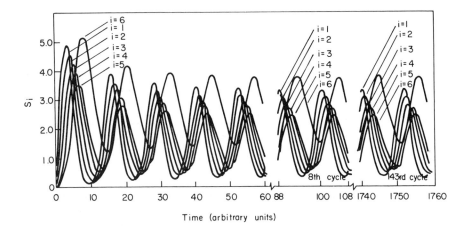

Fig. 16

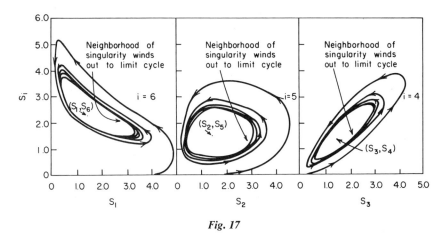

Fig. 17

retaining only the linear terms. In order for the nonlinear system to exhibit oscillations around this steady state, it turns out to be necessary that the resulting linear equations arising from the Taylor expansion be precisely those of the harmonic oscillator. Carrying out this procedure with Eqs. (94) leads to linear systems that are not the harmonic oscillator, to systems the trajectories of which behave as in Fig. 16 or Fig. 17. Hence, these systems cannot oscillate. This fact points up the importance of parametric values in these systems, and the sensitivity of qualitative system behavior to parametric variation.

The equations (94) are of interest for another reason. As noted above, we can equally well interpret such catalytic schemes as pertaining to end-product repression mechanisms as pertaining to enzymic catalytic (end-product inhibition) mechanisms. Variants of these equations, which do indeed admit periodic solutions, were studied at length by Brian Goodwin in an attempt to find autonomous oscillations in repressor systems; that is, in the kinds of genetic control circuits implicit in the Jacob–Monod picture of genetic regulation discussed in the preceding section. Goodwin's equations are of the form

$$dA_2/dt = k_1 A_1/(1 + \alpha A_3) - k_2,$$
$$dA_3/dt = k_2 A_2 - k_3, \qquad\qquad (95)$$

which, it will be seen, differ from Eqs. (94) in that the rates of removal of the reactants A_2 and A_3 from the system are assumed to be *constant* (that is, not dependent on reaction concentration). This assumption, which is rather unreasonable from a biochemical viewpoint, does indeed lead to a system of equations the linearization of which, in the sense described above, does yield the harmonic oscillator, and which can indeed be shown to admit oscillatory solutions. For the details of this analysis, and a mode of further analysis similar to that used by physicists for studying large systems (statistical mechanics) we refer the reader to Goodwin [1963].

One further application of Goodwin's analysis may be mentioned here to motivate the next step in our discussion. As we stated above, his analysis proceeds from an interpretation of a catalyzed sequence of reactions exhibiting backward inhibition in terms of end-product repression of a gene by the product of the reaction catalyzed by the enzyme determined by the gene. This can be represented, in block-diagram form, as in Fig. 18. If we have now a *pair* of such block diagrams, interacting with each other in such a way that the metabolic product of each pathway inhibits the gene of the other pathway, as shown in Fig. 19, then it turns out (as the reader may verify, using the techniques we have already developed) that this system too can exhibit the same kinds of oscillations that each single system exhibited [or else it can behave like a switch; notice the essential similarity between this system and that exhibited on page 89, the rate equations of which are given by Eq. (9)]. Thus, a way is opened to the study of the interaction between different pathways in a metabolic net. To be sure, we have spoken only of inhibitory effects, but exactly the same reasoning can be used to take into account the other kinds of interactions which are possible.

To conclude this chapter, we shall mention a few specific networks, interacting in terms of the elementary processes we have discussed, which have been proposed to account for a few specific biological processes.

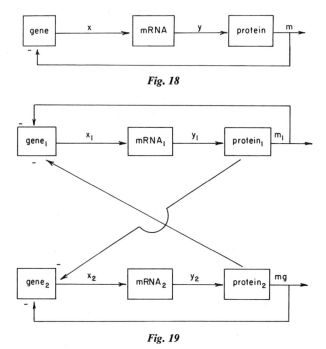

Fig. 18

Fig. 19

A. "Discrimination Network"

The prototype of this network is shown in Fig. 20. It is clearly a special case of the network shown in Fig. 18, but one in which the switching aspect rather than the oscillatory aspect is emphasized. It can manifest at any time one of two possible but mutually exclusive behaviors, depending upon the

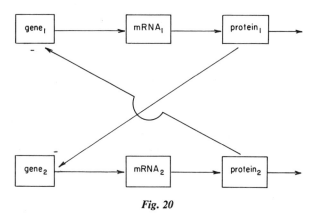

Fig. 20

initial conditions, or upon the environmental stimuli that set those conditions in the net. In another terminology, this was one of the networks proposed by Jacob and Monod, in their original development of the operon concept, to show how phenotypic changes could occur without any change or loss in the primary genetic information. We have called it a *discrimination network*, because of the role such networks have played in the study of the central nervous system in terms of neural nets; networks like the one in question were developed in the 1930s by Landahl and Raskevsky to account for the firing of mutually exclusive behavioral pathways as a response to the larger of two environmental stimuli, and hence to explain behavioral discrimination phenomena.

B. "Learning Network"

The prototype for this network is shown in Fig. 21. This network, in intuitive terms, may initially produce either of two possible responses. However, once the lower pathway in Fig. 21 has become activated, the system will

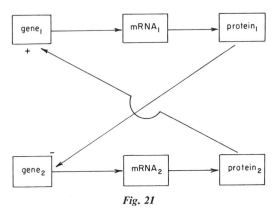

Fig. 21

ultimately "learn" to produce only the lower response regardless of environmental conditions. This kind of network was also first introduced by Landahl and Rashevsky many years ago, and exhibits many of the properties of what we intuitively call "learning" in higher organisms. Indeed, the combination of excitation and inhibition that produces this behavior has many points of similarity to the positive and negative reinforcements through which learning is produced. It, too, was one of the original networks proposed by Jacob and Monod, who were apparently quite unaware that a formal theory for the study of such networks (in fact, several such theories) had been in existence for a long time. Nor has anyone, thus far, really exploited this fact.

C. "Size Regulation Network"

As a final example of an epigenetic network, drawn from the literature, we offer the mechanism for cell size regulation (which amounts to regulation of total protein synthesis) in cells. This is one of the most important homeostatic mechanisms occurring in the regulation of cellular activities, but it has hardly been explained in molecular terms. A mechanism proposed by Ycas, Sugita, and Bensam [1965] consists essentially of a parallel array of operon units, each interacting with all of the others through repressions. When the average activity of the network becomes too large (that is, when the amount of catalytic protein synthesized by the cell exceeds a certain threshold) then all of the operons are shut down. For details, we refer the reader to the original paper [Ycas *et al.*, 1965], and to the discussion in Rosen [1967].

We hope to have, in the above pages, introduced the reader to the fundamental concepts which play a role in epigenetic mechanisms, and with an inkling of one set of mathematical tools whereby these mechanisms may be systematically studied. It should be evident, however, that the surface has been scratched (even though, apart from matters of detail, the exposition was reasonably complete at the time of writing). We hope to have motivated the reader to try his hand in this relatively virgin and most exciting area.

Notes

Section I. The term "epigenesis" as we are using it in this chapter has a rather specialized meaning. The term itself is an exceedingly old one in biology, and typically refers to the hypothesis that each biological individual is developed *de novo* by a dynamical process, implemented by the initial fusing of germ cells. The alternative, that each organism preexists essentially fully formed in the germ, was termed *performation*, and the early biological literature was dominated by essentially sterile arguments between preformationists and epigeneticists. The issue is far from dead, although naturally nowadays the statement of the issues is rather more subtle than in the 17th century. The manner in which we are using the term, namely in the context of those control mechanisms which must be superimposed on purely genetic controls in order to essentially regulate genetic expression, seems to be originally due to Nanney [1958].

For full details on the dynamical description of general rate processes, including the chemical processes treated in this chapter, see *Dynamical System Theory in Biology* [Rosen, 1970]. This reference is appropriate for all discussions of stability throughout the chapter.

Section II. The fundamental distinction between equilibria in closed systems and steady states in open systems was recognized only slowly. Perhaps the earliest recognition of this distinction in biological systems appeared in Hill [1930]. An attempt at a definitive treatment may be found in Burton [1939]. A more comprehensive discussion of the biologically relevant properties of open systems may be found in von Bertalanffy [1953]. There is also a series of papers in this area by Kacser [1957, 1960, 1963].

Section III. There are not many good discussions of hysteresis in the biological literature; for one rather general treatment, see *Mathematical Biophysics* [Rashevsky, 1960].

Section IV. The general treatment of excitation and inhibition in open chemical systems presented in this section is due to Higgins [1967].

There are numerous excellent treatments of the kinetics of (enzyme-) catalyzed reactions. A general one, of great value to the student, is *Behavior of Enzyme Systems* [Reiner, 1959]. Other good treatments can be found in the various classical series devoted to enzymology, such as *The Enzymes* [Boyer, Lardy, and Myrbäck, eds., 1959] (especially Vol. I). Another very good book, with many further references for specific systems, is *Enzyme and Metabolic Inhibitors*, Vol. I [Webb 1963].

Section V. A good review of the early experimental work on induction, repression, activation and inhibition can be found in Vol. 26 of the *Cold Spring Harbor Symposia on Quantitative Biology, Cellular Regulatory Mechanisms* [1961]. In the papers by Jacob and Monod that appear in this volume were also the first appearances of the idea of biochemical or genetic networks.

The kinetic properties of allosteric proteins were first developed by Monod, Wyman, and Changeux [1965].

Section VI. There is an extensive literature dealing with the possibility of oscillations in open chemical systems. An early negative result is found in J. Z. Hearon [1953]. A hypothetical catalytic system which did admit oscillatory solutions was constructed by Spangler and Snell [1961]. The earliest attempt to identify oscillations in a circuit involving inhibitions or repressions is to be found in the book by Goodwin, *Temporal Organization in Cells* [1963]. Since then, numerous studies have been made of the capacity of oscillations in such systems, beginning with the note by Morales and McKay [1967]. A detailed examination of such systems, together with further references, may be found in a series of papers by Walter [1969].

For a fuller discussion of the relationship between neural nets and biochemical or genetic networks, together with further references, see the review by Rosen [1968] and also, *Dynamical System Theory in Biology* [Rosen, 1970].

Another treatment, based on automata theory, but also exploiting the idea of the identity between the networks arising in the study of the nervous system and those arising from cellular controls, is *Cybernetics and Development* [Apter, 1966].

General References

Apter, M. J. [1966]. "Cybernetics and Development." Pergamon, Oxford.
Boyer, P. D., Lardy, H., and Myrbäck, K. [1959]. "The Enzymes." Academic Press, New York.
Burton, A. C. [1939]. The properties of the steady state compared to those of equilibrium as shown in characteristic biological behavior, *J. Cell Comp. Physiol.* 14, 327–349.
Chance, B. [1960]. Analogue and digital representations of enzyme kinetics, *J. Biol. Chem.* 235, No. 8, 2440.
Chance, B., and Hess, B. [1959a]. Metabolic control mechanism. I. Electron transfer in the mammalian cell, *J. Biol. Chem.* 234, 2404.
Chance, B., and Hess, B. [1959b]. Metabolic control mechanisms. II. Crossover phenomena in mitochondria of ascites tumor cells, *J. Biol. Chem.* 234, 2413.
Chance, B., and Hess, B. [1959c]. Metabolic control mechanisms. III. Kinetics of oxygen utilization in ascites tumor cells, *J. Biol. Chem.* 234, 2416.
Chance, B., and Hess, B. [1959d]. Metabolic control mechanisms. IV. The effect of glucose upon the steady state of respiratory enzymes in the ascites cell, *J. Biol. Chem.* 234, 2421.
Chance, B., and Hess, B. [1961]. Metabolic control mechanisms. VI. Chemical events after glucose addition to ascites tumor cells, *J. Biol. Chem.* 236, 239.
Chance, B., Garfinkel, D., Higgins, J. J., and Hess, B. [1960]. Metabolic control mechanisms. V. A solution for the equations respresenting interaction between glycolysis and respiration in ascites tumor cells, *J. Biol. Chem.* 235, 2426.
Chance, B., Higgins, J. J., and Garfinkel, D. [1962]. Analogue and digital computer representations of biochemical processes, *Proc. Fed. Amer. Soc. Exp. Biol.* 21, 75.
Garfinkel, D., and Hess, B. [1964]. Metabolic control mechanisms, *J. Biol. Chem.* 239, No. 4, 971.
Garfinkel, D., and Hess, B. [1965]. Computer simulation in biochemistry and ecology, *in* "Theoretical and Mathematical Biology" (T. H. Waterman and H. J. Morowits, eds.), Chapter 11, pp. 292–310. Ginn (Blaisdell), Boston, Massachusetts.
Goodwin, B. C. [1963]. "Temporal Organization in Cell." Academic Press, New York.
Hearon, J. Z. [1953]. The kinetics of linear systems with special reference to periodic reactions, *Bull. Math. Biophys.* 15, 121–141.
Higgins, J. J. [1967]. The theory of oscillating reactions, *Ind. Eng. Chem.* 59, 18–62.
Hill, A. V. [1930]. Membrane phenomena in living matter: Equilibrium or steady state?, *Trans. Faraday Soc.* 26, 667–694.
Jacob, F., and Monod, J. [1961]. On the regulation of gene activity, *Cold Spring Harbor Symp. Quant. Biol. Cellular Regulatory Mechanisms* 26, 193–211.
Kacser, H. [1957]. Appendix, *in* "The Strategy of the Genes" (C. H. Waddington, ed.), pp. 191–249. Allen & Unwin, London.
Kacser, H. [1960]. Kinetic models of development and heredity, *Symp. Soc. Exp. Biol.* 14, 13–27.
Kacser, H. [1963]. The kinetic structure of organisms, *in* "Biological Organization at the Supercellular Level" (R. J. C. Harris, ed.), pp. 25–41. Academic Press, New York.

Monod, J. and Jacob, F. [1961]. General conclusions: Teleonomic mechanism in cellular metabolism, growth, and differentiation, *Cold Spring Harbor Symp. Quant Biol., Cellular Regulatory Mechanisms* **26**, 389–401.

Monod, J., Wyman, J., and Changeaux, J. P. [1965]. On the nature of allosteric transitions: A plausible model, *J. Mol. Biol.* **12**, 88–118.

Morales, M., and McKay, D. [1967]. Biochemical oscillations in "controlled systems," *Biophys. J.* **7**, 621–625.

Nanney, D. L. [1958] Epigenetic control systems, *Proc. Nat. Acad. Sci.* **44**, 712–717.

Rashevsky, N. [1960]. "Mathematical Biophysics" (3rd ed.), Chapter II, Volume II. Dover, New York.

Reiner, J. M. [1959]. "Behavior of Enzyme Systems." Burgess, Minneapolis, Minnesota.

Rosen, R. [1967]. Two-factor models, neural nets and biochemical automata, *J. Theor. Biol.* **15**, 282–297.

Rosen, R. [1968]. Recent developments in the theory of control and regulation of cellular processes, *Int. Rev. Cytol.* **23**, 25–88.

Rosen, R. [1970]. "Dynamical System Theory in Biology." Wiley, New York.

Spangler, R., and Snell, F. M. [1961]. Sustained oscillations in a catalytic chemical system, *Nature* **191**, 457–458.

von Bertalanffy, L. [1953]. "Biophysik der Fliessgleichgewichts." Vieweg, Berlin.

Walter, C. [1969]. The absolute stability of certain types of controlled biological systems, *J. Theor. Biol.* **23**, 39–52.

Webb, J. L. [1963]. "Enzyme and Metabolic Inhibitors," Volume I. Academic Press, New York.

Ycas, M., Sugita, M., and Bensam, A. [1965]. A model of cell size regulation, *J. Theor. Biol.* **9**, 444–470.

Chapter 3

AUTOMATA THEORY IN THE CONTEXT OF THEORETICAL EMBRYOLOGY

Michael A. Arbib

Department of Computer and Information Science
University of Massachusetts
Amherst, Massachusetts

A biological system may be analyzed at many levels: biochemical, mechanical, and informational, to name a few. Automata theory gives us insight into the information processing of organisms; here we shall briefly discuss models which bear on the way in which genetic information modulates the development of an organism.

To avoid later misunderstanding, let me stress that our emphasis will be on extremely abstract models, designed to foster understanding of whole classes of systems, rather than being more specifically tailored to the solution of specific biological problems. To clarify this point, we may make a rough tripartite classification of models in theoretical biology:

1. Models designed to help explore experimental situations. These range from general data-processing and statistical techniques which help the experi-

mentalist reduce his data to comprehensible form, to specific models based on experimental data which lead to further experiments which lead to refinements of the model, and so on. Two dangers in this approach are that of deforming experiments to yield data that is tractable for the computer irrespective of its real interest, and that of basing a theory on variables which are available simply because the experimentalist has techniques for measuring them.

2. Models which treat abstract systems in great generality, in the hope that deep insight will be gained into the properties of specific biological systems by viewing them as embodiments of general principles, and the hope that the general theory will yield guidelines to help the theorist working at level 1 to select his variables other than at happenstance. The great danger of this approach is that the particular properties that make a system biological may be lost completely if the analysis is too general and qualitative.

3. Pseudomodels which fill the journals, but tell us nothing. The danger of this approach is that one may not realize one is taking it!

We shall emphasize, then, models at level 2. Occasionally we may lapse into level 3, but I hope that the reader will find, even here, that what we say is of interest mathematically, even when he doubts its relevance to biology.

The chapter is essentially self-contained save for a minimal knowledge of such computer terminology as "program" and "subroutine" and an acquaintance with some of the elementary ideas about finite automata presented in the chapter, "Automata Theory in the Context of Theoretical Neurophysiology," Chapter 3, Volume III in the present treatise.

I. Programs in Biological Systems

Before discussing the role of programs in biological systems we remind the reader of current man-made computers, though mainly to insist that when we import such terms as "program" and "subroutine" into our discussion of a biological system, we are trying to convey a feel for the complexity and flexibility of structures built up from simple elements which contain "tests," and ask the reader not to expect the structure of current hardware to resemble the biochemical processors of the cell. The "memory" of man-made computers is divided into a number of "pigeonholes," each with a specified address. Each pigeonhole can contain a string of symbols. Such a "word" may be decoded by the control unit either as data, or as one of a finite repertoire of instructions. Typically, the memory will contain a sequence of instructions for some desired series of computations, as well as some data and intermediate results. The control unit decodes one instruction at a time, and depending on the nature of the instruction, can cause the following actions:

1. cause the input/output unit to read information out of designated storage locations and transmit it to the environment;
2. cause the input/output unit to read information from the environment into designated storage locations;
3. cause data to be transferred from designated storage locations into appropriate registers in the arithmetic unit;
4. cause the arithmetic unit to compute some function of the data in its registers;
5. cause the arithmetic unit to store the result of a computation in a designated storage location;
6. cause the fetching of the next instruction from store (the location may depend on the result of some test of a condition in the computer).

The point to be made here is that all the subtlety of modern computing applications can be obtained by combinations of these simple elements.

We believe that most multicellular organisms on the earth grow from a single cell, called the zygote, and that the overall structure of the organism is specified by a program, coded in the genes, which program is "decoded" by the cytoplasm of this cell and of its descendants. However, many parts of the program are active at once and serve to regulate ongoing activity throughout the cell, as distinct from the "cue in one operation at a time" of the man-made computer of the above paragraph. Many of the details of the organism may be a result of maternal influences and environmental factors, but nonetheless we believe that, *given the nature of the cytoplasmic decoders*, most of the essentials of form and structure are specified by the genes. We can expect these programs to be hierarchically organized, so that it is misleading to associate the different "instructions" of the program with different characters of the organism; many of the instructions may be at a very high level of control and command.

Let us consider ways in which cellular programs can be modified. Perhaps best known are the point mutations whereby individual "letters" of individual "instructions" are modified at random. In general we may expect either that the change is smoothed over by redundancy in the program, or else that the resultant instruction is such nonsense, relative to its place in the program, that the cell will not be viable. Although redundancy thus allows mutations that appear "neutral" in their immediate effect, this does not preclude that their cumulative effect may be significant indeed; the mathematical notion that comes to mind here is that of a random walk with an absorbing barrier, where nothing significant happens until the barrier is reached. However, an occasional rare change may actually yield an immediately "improved" instruction, and this is one of the basic mechanisms of evolution. Another mechanism of program change comes when a cell in splitting maintains the doubled genetic complement in a single daughter cell. This double program may well yield extra possibilities for dynamic interaction. Similarly, two bacterial cells may pool their genetic complement; or a virus, in invading a cell, may, instead of taking over the cytoplasm of that cell to produce more copies of the virus, actually become "trapped" and add new genes to the cell, thus providing new metabolic possibilities. Note that this mechanism allows much faster writing of long programs than the process of simply randomly adding or deleting a single letter to existing programs, since we increase the size of the program dramatically without diminishing its metabolic competence.

These considerations suggest an interesting theory about the evolution of multicellular organisms. We would suggest that single-cell organisms evolved in different ways, each with their own metabolic efficiencies. Cells deficient in metabolite A would naturally thrive in a region rich in that metabolite. Similarly for cells deficient in metabolite B. Now suppose that the first cell produced an excess of metabolite B which it excreted, whereas the second type of cell produced an excess of metabolite A. Then, it would be likely for the cells to migrate into the neighborhood of each other and after a while a partnership or symbiosis might occur between these two types of cells. In such an environment, the sharing of genetic complements would become highly probable, and so after a while we might well end up with a single cell which had the genetic instructions for both the first and the second type of activity. In some cases the result would be a new unicellular organism which was able to produce adequate amounts of both metabolites. In other cases, the result would be a multicellular organism, which could develop from a single cell with the doubled genetic complement, but in which each cell of the mature organism only expressed one of these genetic complements. Here we see the possible beginnings of the evolution of differentiation. In any case, we now appreciate that there are many different ways of building up complex genetic programs, besides the most frequently cited mechanism of point mutations. A question related to this discussion of the evolution of symbiosis is: "What distinguishes one collection of cells from another in the way that a collection of bacteria is distinguished from a human?" Presumably it will be in terms of some measure of functional dependence, but will certainly entail the "any unit is dispensable but everybody contributes" structure that we see again and again in biological systems.

Considering the plasticity of the embryo, we may relate differentiation to progress to different parts of a program (as well as modifications of the cytoplasm) the execution of which involves *tests* of the environment, internal and external, of the cell. Thus, a cell may be executing the same program during early stages of division from the zygote, and if separated from the zygote *during those early stages* may act as a zygote to produce a complete organism (perhaps smaller because the cytoplasm is smaller—an interesting point, stressing that the execution of a biological program proceeds by relative, rather than absolute, instructions to the organism; for example, when a cell divides, the daughter cells are sized relatively to the parent, rather than specified absolutely), whereas some portions of the program may possess no capability of return to the initial portion, no matter what tests are executed.

It is difficult at our current stage of knowledge to understand how a specific change in the genetic program which is translated into a somewhat different distribution of amino acids will yield a particular change at the level of the overall organism. I think that the clue here is to stress once again the hierarchical nature of the genetic program, so that as evolution progresses what is being modified may often be the relation between complex subroutines rather than low-level modification of chemical activity *per se*.

This suggests that viable mutations are more likely to be at high levels of the hierarchy than at low levels; since the program is hierarchical, "low-level" subroutines will be used in so wide a variety of places in the program that changes there are unlikely to be viable. We thus expect most viable changes to be at recent levels, or to build on at new levels. This would seem to be why, although it is false that ontogeny recapitulates philogeny, it is true that it is highly likely that the phylogenetic sequence will be reflected in the ontogenetic sequence: A human embryo is very early distinguishable as such, and so it does not recapitulate phylogeny, but much of the earlier "program execution" is of subroutines similar to those used in related organisms which evolved earlier.

Note that the presence of "tissues" allows genetic specification in terms of compact (nested) subroutines, rather than detailed connectivity specification for every pair of cells. Note, too, that the complexity of a program is a function of the language in which it is expressed.

Let us now turn to a somewhat more critical analysis of what it is that the genetic "program" has to "do." At any time in any one cell only part of the deoxyribonucleic acid (DNA) is actually active. Upon that active portion ribonucleic acid (RNA) is formed which then moves out to the ribosomes where it is used to control the stringing together of amino acids to form proteins. The result is a change in the balance of enzymes and resultant changes in the rates of various chemical reactions. In a complex multicellular organism, as the cells divide, so do both their cytoplasmic structure change, and the active portion of their DNA changes although the overall DNA content seems to remain the same. This process is called differentiation. As cells divide and differentiate, different potentialities are actualized until in the grown organism most cells can only exhibit one function unless gross damage occurs, as we see in the regeneration of limbs in amphibia.

It does not seem to help to think either in terms of being given an overall blueprint from which we must determine parts and how to put them together, or to think of being given a collection of parts with instructions as to which pairs are to be joined. Perhaps more insight is given by thinking of the following rather feeble and completely hypothetical example: We wish to grow a dumbbell-shaped organism from a single cell. Thus, we have the original cell dividing and dividing and dividing in a way which produces a linear chain. We decree then that a certain chemical gradient is set up from the center so that as we move out towards the periphery of the chain the concentration of this chemical increases. It then might be so arranged that when this chemical concentration reaches a certain critical level, it switches the DNA program from that portion which controls the division into a linear chain into that portion which controls the growth of the spherical object on the end. This idea of increase of concentration of a substance over consecutive generations of cell division until, by some mechanism like end-product inhibition, it can switch control, may give us some idea of how a genetic program may be executed in a way which depends on the number of stages involved. Thus, fairly simple instructions can have, as a result of the nature of chemical reactions, fairly far-reaching consequences in terms of numerical proportioning of different cell types. Thus, the genetic program presumably has different subroutines for different tissues, and some fairly delicate chemical reactions are used to govern the switching-in so that the resulting tissues in the mature animal are appropriately proportioned and relatively placed. We shall discuss this in some generality in Section II.

It must also be stressed that the genetic programs are not programs written by a designer who goes from a mature organism to a specification of how that organism may be assembled. Rather they have resulted from a long process of slight modification of programs which already produce a viable organism and which serve to constrain interactions during growth to yield the form of the mature organism. Thus, the shape of the end of a bone is not to be attributed so much to the genetic instructions regarding growth patterns of a single bone in isolation, but is in fact much more in terms of the position of one bone relative to another bone and the musculature, so that the motion of the two bones relative to each other leads to a spherical interface at the joint. Similarly, the shape of the eyeball depends greatly upon the shape of the surrounding bones. Again, many more cells are formed in embryonic growth than will actually survive to maturity; these are the cells from which we can sculpt out correct structures by future interaction of differing portions of the system [Saunders, 1966]. Thus, the organic way of programming a form is quite different from that we normally think of for machines.

Having thus divested ourselves of too literal an interpretation of the genetic "code," it is nonetheless a striking fact that in both real biological systems and the self-reproducing automata we shall present in Section V, we see a distinction between a description of the system and the active portion of the system.† This goes back to the Weissman assumption underlying the current form of the theory of natural selection, namely, that one needs to distinguish the *genotype* (a set of instructions) from the *phenotype* (their functioning embodiment) and that, whereas different environments produce different phenotypes from a given genotype, this change does not itself produce a change in the genotype. However, changes in genotype do cause marked changes in phenotype and it is the genotype changes that propagate, whereas phenotype changes do not. Thus, selection is of genotypes, but it acts on phenotypes. Let us thus briefly consider the question (the importance of which was emphasized to me by Howard Pattee): "Is a *description* of the object a necessary complication to obtain an interesting self-replicating object?" In other words, must an interesting self-replicating object A made out of our elementary units contain within itself an object $e(A)$, a description of A also made out of these units? (Perhaps one should be somewhat uncomfortable with the word "description" smacking as it does of preformationistic concepts. Note that the system is constantly developing; what then is described? In any case, how are we to make a clear-cut separation between the description and the active part of the system?)

Von Neumann [1966] distinguished two methods of self-reproduction. In the passive method the self-reproducing automaton contains within itself a passive description of itself, and reads this description in such a way that the description cannot interfere with the automaton's operations. This is the method used in Section V for self-reproducing automata and, if we identify DNA with an encoding, it appears to be the method used in living organisms. We may contrast this with the active method, in which the self-reproducing automaton examines itself and thereby constructs a description of itself. Von Neumann has suggested that this method would probably lead to a logical paradox. However, I am not at all convinced of this at the time of writing. It would seem to me that DNA does indeed replicate itself by this active method (though not without the help of certain specialized enzymes) and we must ask not, "Is there a logical paradox inherent in the active method?" but rather, "Is there some well-defined cutoff point at which the active method is no longer applicable?"

Pattee [1968] sees the central biological aspect of hereditary evolution as the fact that "the process of natural selection acts on the actual traits or

†The remainder of this section is taken from "Self-Reproducing Automata—Some Implications for Theoretical Biology" by Michael A. Arbib, in *Towards a Theoretical Biology. 2. Sketches* (C. H. Waddington, ed.), pp. 211–214. 1969 Edinburgh University Press.

phenotypes and not on the particular description of this phenotype in the memory storage which is called the gene." He sees this as essential biologically because "it allows the internal description or memory to exist as a kind of virtual state which is isolated for a finite lifetime, usually at least the generation time, from the direct interaction which the phenotype must continuously face" [Pattee, 1968, p. 74].

Goodwin [1963, pp. 9–11] has pointed out that

> In a study of the dynamic properties of a certain class of biological phenomena . . . it is necessary to extract a manageable number of variables from the very large array which occurs . . . . Variables which are major to the phenomenon being investigated become the quantities that define the systems which one intends to study, while [minor] variables become either parameters of the system, thus defining its environment, or . . . noise.
>
> The *relaxation time* of a system is, roughly speaking, the time required for the variables to reach steady state after a "small" disturbance. The significance of this concept is the fact that if two systems have very different relaxation times, then relative to the time required for significant changes to occur in the "slower" system, the variables of the "faster" one can be regarded as always being in a steady state . . . . On the other hand, the variables of the "slow" system will enter into the equations of motion of the "fast" one as parameters, not as variables.

The description, then, is presumably a portion of the system whose relaxation time is so great relative to the dynamics of the system, that it may serve as a permanent record insulated from the vagaries of the existence of the organism. In terms of this notion of insulation from the vagaries to which the organism is subjected, we might note that Michie and Longuet-Higgins [1966] compare the *germ plasm* (DNA specifications) of a cell to the *program* of a user of a large computer system, and they compare the *soma* (cellular machinery of implementation) to the *monitor* of the computer system (the program that controls input/output devices, assigns priorities to different users, translates programs using various compilers, and so on). They suggest that the genotype–phenotype distinction may be compared to the segregation of user's software and system's software. If a user writes a program that could modify the monitor during execution, difficulties could be created for users of other programs using the same monitor. Perhaps similar difficulties could arise if germ plasm and soma were not segregated.

Perhaps there is even something in the horribly naive thought that if an object is genuinely three–dimensional, it cannot be copied in space without completely cutting it apart. Thus, we copy it in the four dimensions of space–time instead, by using a genotype to *grow* a replica.

Just as larger bodies of a given shape have smaller ratios of surface to volume, so can we expect integrated arrays of cells to have fewer of their cells in direct interaction with the environment, so that consequently the output of the system (its parameters of interaction) can only be a small sampling of the state of the system. Thus, the process of determining the state of the system may be too destructive of the normal function of the system, thus encouraging the strategy of creating a new system and using education to transfer aspects of the state of the adult system.

We end this somewhat inconclusive discussion of the nature of descriptions by noting that Thatcher [1963] and Lee [1963] have shown that there are Turing machines (see Section IV) that can print their descriptions on their tapes when appropriately triggered, although their description is not stored explicitly within their program. Perhaps this means that an automaton need not contain within itself an explicit separate description at all times, but need only retain the ability to generate such a description when it is necessary. In other words, we may visualize an organism in which the genome is *potential* rather than *actual*. The big question here, then, is one of reliability. If we do not explicitly segregate the genotype from the phenotype, can we guarantee that damage to the phenotype will not yield dangerous alterations to the potential genotype, as suggested by Michie and Longuet-Higgins? To the extent that we have redundancy in a description, be it explicit or implicit, to that extent are transcription errors of less concern. This is especially worthy of note since we hold that cytoplasm and cell membrane provide specific decoding procedures without which the genetic information would be meaningless.

II. Positional Information

This section is primarily a critical exposition of the theory of positional information presented by Wolpert [1969]. We may distinguish two interdependent processes: *spatial differentiation* whereby the types of various cells in an array of cells is specified and *morphogenesis* whereby an array of cells is molded into shape. Wolpert sees spatial differentiation as a two-stage process whereby a cell first has its position specified, and then undergoes molecular differentiation on the basis of this *positional information*, though it should be noted that positional information is *not* a logical prerequisite for patterning, for one can imagine circumstances in which two cells in an embryo may undergo the same differentiation with neither knowing "which is which." (See Trinkaus [1969] for an excellent review of the forces that can underlie change of shape in morphogenesis.)

Considering the "universality" of the genetic code—at least in the production of enzymes—it is tempting to expect general principles to exist in the "translation" of genotype into phenotype.

Let us focus, then, on *the problem of assigning specific states to an ensemble of identical cells, the initial states of which are relatively similar, such that the resulting ensemble of states forms a well-defined spatial pattern.*

We distinguish "mosaic" development in which removal of a portion of the system results in a lack of those regions which the removed regions would normally form, and "regulative" development in which a normal pattern (though perhaps smaller) would still be formed. Clearly, *regulation requires intercellular communication*. Spatial and temporal organization are intimately connected in development, and we wish to see how the space–time connection of differing cell states can be realized in terms of the physiology of single cells and their interactions.

Wolpert [1968] offered the French flag problem as a paradigm for size-invariant regulative development of axial patterns: The problem is to design a cell such that a linear array of these cells, no matter which or how many of its units are deleted from either end, will always form a "French flag," that is, with the left-hand third blue, the middle third white, and the right-hand end being red, these three colors representing different states of "molecular differentiation" of the cells. This problem represents abstractly such facts as that the ratio of mesenchyme, endoderm, and ectoderm of the sea urchin embryo remains constant over an eightfold range, while as little as 1 % of a hydra can give rise to an almost complete animal.

Classical field-gradient systems would explain such a phenomenon in terms of a metabolic, say, gradient, that would form autonomously with the apical region serving as a high point or dominant region and exerting an influence on other regions. Wolpert observes, however, that these classical theories fail to provide mechanisms for (a) establishment of the apical region as dominant; (b) maintenance of the gradient; (c) determination of the size of the apical region; or (d) apical control of order or size of adjacent parts (this last is sometimes assigned to "inducer activity" by the dominant region).

Webster and Wolpert [see Wolpert, 1968] use a "balancing principle" to solve the French flag problem: The amount of substance made in one region is altered until it "balances" the amount being made or destroyed in another region. Here the position of a cell within the system is indeterminate.

In other solutions, the cell's position with respect to the two ends of the system is first specified, and then this positional information is used to determine how the cell differentiates.

In Section III we shall give a model in which only the position of a few cells is determined, other cells differentiating on "betweenness," rather than more quantitative positional information.

Wolpert [1968] suggests that all solutions to the French flag problem require a mechanism for specifying polarity, a mechanism for the differential response of the cells, and at least one spontaneous self-limiting reaction. However, while the automaton model we shall present in Section III does require the first two mechanisms, we shall see that the interaction of A and B

pulses suggests that the third mechanism might be replaced by demanding a *family* of *inter*limiting reactions.

Again, we should also note that in an embedded Turing machine, such as we shall discuss in Section IV, *no* cell "knows" where it is. The crucial determinant is "what the neighbors are doing." Thus, for instance, we could form a spatial pattern not by *first* making a separate determination of (relative) position, but by making appropriate changes, for example, by "moving in" from a boundary, making at most an *implicit* simultaneous use of position. For example, some of the mechanisms some theorists provide for setting up positional information could serve to establish spatial patterns directly. Dependence of a cell's activity on that of its neighbors may result in an at most indirect use of position, and cells in different positions may be in functionally identical neighborhoods.

Let us now present the three underlying points of Wolpert's [1969] development of the concept of positional information:

1. He explicates the notion of "field" as a collection of cells in a developing system with their position specified with respect to the same set of points.

2. He views the specification of positional information as in general preceding, and being independent of, molecular differentiation, which latter he views as determined by a process called "*interpretation* of positional information." However, it is not clear how we are to tease apart the local cellular changes signaling positional information and the molecular differentiation that underlies spatial pattern.

3. Polarities are defined to specify the directions in which positional information is measured.

To obtain pattern regulation, one asks that positional information is relative with respect to the reference points, though perhaps in mosaic development, absolute position might play a role. In any case, the theory will try to establish a mechanism that specifies positional information in many different fields in many different organisms.

In the case of specifying positional information on a single axis, we may express the mechanism for positional information in terms of: (a) a reference cell or region α_0, specifying the origin of the axis; (b) polarity, a direction for measurement along this axis; and (c) (in regulating systems) a second reference cell or region α_0' at the opposite end of the axis from α_0, to calibrate a scale-adjustment mechanism which alters the units of measurement along the axis so that the total number of units on this axis is independent of axial length.

It is assumed that these subserve a universal mechanism to provide cells with positional information; it then being species specific how this positional information is "decoded" to control the differentiation of the cell.

Wolpert notes that curve I of Fig. 1 could be obtained by making α_0 and α_0' the source and sink, respectively, of some substance; if the concentrations of the substance in these cells were fixed, the linear gradient resulting from diffusion and transport of the substance would specify a cell's position.

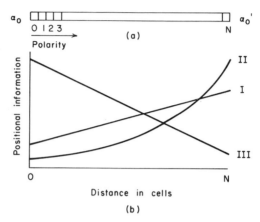

Fig. 1. In (a) a line of cells, N cells long, is shown and the arrow shows the polarity of the system. In (b) three examples of positional information/distance curves are illustrated.
Curve I is a case in which there is a linear increase in positional information with distance from the end, and this could represent the increasing concentration of a substance or the phase angle difference between two-periodic events spreading from the left as suggested by the Goodwin–Cohen [1969] model. Curve II shows another type of relation that could arise for example from active transport of a substance from left to right. Curve III shows a decrease in some property with distance. Note that the value of α_i for the three relationships may be both qualitatively and quantitatively different. [From Wolpert, 1969.]

Goodwin and Cohen [1969], on the other hand, exploit the temporal organization of the cell arising from oscillatory behavior in the cell's physiological control processes and ask how the temporal organization present in the earliest embryo as a result of the cycle of cell division, could form the dynamic basis used to generate spatial organization throughout the developing cell mass. They suggest that two periodic signals of equal frequency, the S event and the P event, are propagated from α_0, but that the S event propagates more quickly than the P event so that the increasing difference between their phase angles can be used as positional information.

Wolpert also notes that the positional information could be *qualitative*, with each of N cells being in different states: the first in state A_1, the second in state $A_2, \ldots$, the last in state A_N. However, we should note that such a scheme seems to belie the positional information idea and seems to go directly to molecular differentiation. Further, it is not well suited for regulation, unless we presume that each cell has access to a measure of the number N of cells.

It is useful, then, to distinguish a *unipolar* axis in which position is determined in a nonregulating fashion, setting out from one reference point in the direction of the polarity, from a *bipolar* axis in which we use the two ends of the axis to regulate position. We shall comment on the mode of this regulation in our discussion of Fig. 3.

Once we have understood such uniaxial models, it is tempting to expect that position in a multidimensional cellular field will be determined by two or three such axes: (α_0, α_0'), (β_0, β_0'), (γ_0, γ_0'). Some examples are shown in Fig. 2.

Wolpert notes that the polarity may have to be defined in some sense before we can establish positional information, and even talks in terms of polarity gradients, raising the specter of an infinite regress of polarity-gradi-

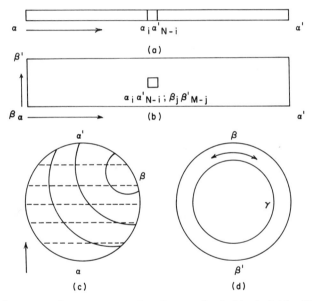

Fig. 2. Diagrams to show some examples of axes and polarities in fields with more than one dimension. In (a) there is a line of N cells and there is one bipolar axis $\alpha\alpha'$. The ith cell, measured in the same direction as the polarity, has positional information $\alpha_i\alpha'_{N-i}$. (b) This is a sheet of cells, N cells long and M cells wide. It is a two-dimensional field having two bipolar axes $\alpha\alpha'$ and $\beta\beta'$. Lines of constant positional information for the $\alpha\alpha'$ axis are at right angles to the long axis; for $\beta\beta'$, parallel to the long axis. Thus, for example, $\beta_0\beta_M'$ is the lower edge of the sheet. (c) This is a spherical sheet of cells having two dimensions: one bipolar axis $\alpha\alpha'$ and one unipolar axis β. The arrow shows the polarity of the $\alpha\alpha'$ axis. Lines of constant positional information with respect to $\alpha\alpha'$ are shown dashed. The solid arcs are the lines of constant positional information with respect to β. Note the radial symmetry of these axes. (d) This shows the cross section of a cylindrical sheet of cells. The $\beta\beta'$ axis is bilaterally symmetrical. One could consider a third dimension along the γ axis, positional information being measured at right angles to the inner sur face [from Wolpert, 1969].

ent-determining gradients, and so forth. The crucial point is that polarity
often may be determined by the outward direction from some dominant
region, but that some initial period of cellular interaction may be required to
determine which of several potentially dominant regions will actually dominate. We might then describe this initial period as the "establishment of a
polarity gradient," whose "crude caricature" may then yield precise positional information by a process of "iterative refinement," just as the output
of a neural network with lateral inhibition undergoes iterative changes even
with constant input as neighbor affects neighbor, and local interactions build
up to global computations.

If we accept positional information as underlying molecular differentiation, we may speak of the *conversion* of positional information into a change,
say, in enzyme activation, which will eventually yield a change in the overall
state of the cell; this end state we may speak of as the *interpretation* of
the positional information, expressing as it does the cell's contribution to
the spatial pattern we have been trying to explain. Of course, developmental
history of the cell and its genome may well affect the interpretation of positional information.

In Fig. 3, we see four different ways in which positional information may
be interpreted to pattern a French flag.

In Fig. 3a we have the unipolar scheme in which cells "count off" their
position starting from α_0 at the left-hand end. There is then some fixed
number a such that the first a cells will be red, the next a white, and those
from α_{2a} to α_{3a} blue. The array can "regulate" for removal of cells from the
left-hand end, provided at least $3a$ cells remain, but can make no adjustment
to excisions at the right-hand end.

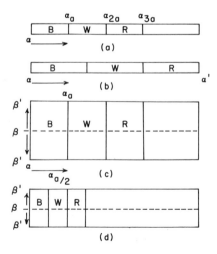

Fig. 3. Diagrams to illustrate the interpretation of positional information
so as to make a French flag. In (a) a
unipolar, one-dimensional system, the
blue region, for example, is between α_0
and α_a. In (b) the axis is bipolar and the
rules for interpretation are such that
the whole axis becomes divided up. In
(c) a sheet of cells has the α axis unipolar
and a bilaterally symmetrical bipolar
$\beta\beta'$ axis. Blue is between α_0 and α_a, and
this should be compared with (d) in
which, due to $\beta\beta'$ axis being shortened,
a is reduced to $a/2$ (B is blue, W is white,
R is red). [From Wolpert, 1969.]

In Fig. 3b we see how a bipolar system is used to obtain size invariance. One mechanism for this suggested by Wolpert is that the ith of the N cells should have positional information α_i starting out from α_0, and positional information α'_{N-i} reckoned from α_0' at the right-hand end. If both these measures are linear of the same slope, Apter [1966] has noted that interpretation could follow the rule

$$\alpha_i/\alpha'_{N-i} < \tfrac{1}{2} \mapsto \text{blue}; \qquad \tfrac{1}{2} \leqslant \alpha_i/\alpha'_{N-i} < 2 \mapsto \text{white}; \qquad 2 \geqslant \alpha_i/\alpha'_{N-i} \mapsto \text{blue}.$$

It is interesting to note the α_i and α'_{N-i}, in this case, can provide each cell with the length of the field. However, it is more interesting that this scheme uses two *scales* of *absolute* position, whereas in Section III we shall present a scheme using two *points* of *relative* position to do the job. I would strongly suggest that

> any scheme obtaining size invariance by "computations" based on
> absolute information obtained by two independent processes rather
> than on relative information obtained as the global result of interactive
> processes is highly suspect.

Our automaton-theoretic scheme does not "overload" the cell with a great deal of absolute information.

On the other hand, it is obvious that if the absolute length is available, it can be used to control the length of the axis under the interpretation that cell division must cease as soon as the length of the axis passes a certain threshold. More standard theories of growth control, more consistent with the tenor of the above paragraph, are based on feedback of some growth-inhibiting substance. Wolpert suggests that such mechanisms, for example, that in which the inhibitor is chalone [Bullough, 1967], can only provide a mechanism for proportionate growth control, though it would be interesting to see a computation taking account of the effect of a changing ratio of cell volume to external surface area.

Anyway, in view of the above criticism, it is reassuring to note that Wolpert's sink–source gradient, presented in our discussion of Fig. 1, does meet our demand for relative information. The phase-shift mechanism of Goodwin and Cohen [1969] also provides relative information, though regulation is by a method that seems at least as *ad hoc* as our automaton model.

The reader may wish to suggest appropriate rules for the biaxial scheme of Fig. 3c, d in which the width of the three stripes is regulated by the height of the "flag."

Wolpert stresses his view of pattern formation as at least a two-step process: First, and independent of molecular differentiation, is the specification of position; second, the cells then differentiate as a result of interpretation of the positional information. I find somewhat disturbing the suggestion

that, as distinct from most other concepts of pattern formation, there is
no interaction between the parts of the pattern *as such*. However, if one
entertains our above suspicion that the two steps are not all that distinct,
then the force of the "as such," and the resulting disturbance, are greatly
diminished.

To the extent, though, that the mechanism for specifying positional information is universal, rather than requiring specific chemical substances for
each field, one might expect interesting implications, as shown in Fig. 4.

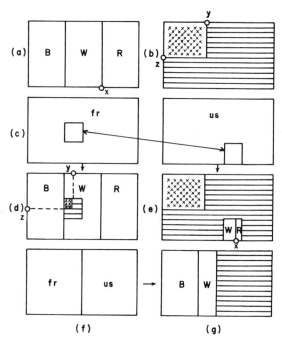

Fig. 4. Some examples to show some possible implications of the universality of positional information. Consider a rectangular field and two different genotypes. Genotype fr
results in the interpretation of the positional information so that a French flag is formed
(a) while genotype us results in the Stars and Stripes (b). If, at an early stage, two pieces
are interchanged as in (c), and if positional information in the two fields is the same, then
the results shown in (d) and (e) will follow; that is, the cells behave according to their
genotype and position and are indifferent to the nature of the surrounding tissue. Similarly, if two halves of different genotypes are joined as in (f) a mosaic as in (g) will form
(B is blue, W is white, R is red). [After Wolpert, 1969.]

Here we see cells from different arrays having their position determined by
the same mechanism within the graft, but with the interpretation of the
positional information depending on the cell's history.

III. Relative Position and Synchronization in Arrays of Automata

A. Synchronization of Arrays of Automata

In 1957, J. Myhill proposed a problem [see Moore, 1964] of synchronization for an array of finite automata which arose in connection with getting all the squares of a Turing machine tape embedded in a tessellation to shift their contents simultaneously to avoid overwriting by a shifting cell of the contents of a cell that was not yet shifting. The problem is:

Design a finite automation with two input lines† and four special states O, E, M, and F such that if we form an array on an *arbitrary* number n of these automata so that each receives as input the state of its two neighbors (we interpret an edge cell as receiving input O from "outside," see Fig. 5) with all in state D, and then set the first cell in state M, there will be a time θ at which all cells *simultaneously* enter state F for the *first* time.

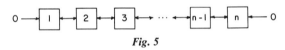

Fig. 5

Note that *the problem would be trivial if the number of states were able to depend on n*; we could then have the ith cell pass on the "message" $i - 1$ and then count down from $i - 1$ to 0.

Moore and Langdon [1967] considered the slight generalization in which any cell can be the initiator and proved the theorem:

Theorem. *If the initiating cell is k units $(0 \leq k \leq n/2)$ from the end of the array, then synchronization must take at least $2n - 2 - k$ time units.*

PROOF: Suppose we can find a cell k_0 in a string of length n_0 such that synchronization occurs at time $m < 2n_0 < 2 - k_0$. This means that cell 1 could not have received a signal from cell n_0 because it takes time $n_0 - k_0 - 1$ for a signal from the initiator to reach n_0 and then time $n_0 - 1$ for it to propagate back to cell 1. Thus, cell 1 fires independently of activity at cell n_0 and so we could add cells to the right of cell n_0 without affecting the firing time of cell 1. But this is absurd, since we could add enough cells for the rightmost cell not to have been affected when cell 1 fires, thus denying synchronization.

In fact we shall see that there do exist cells which attain synchronization in the minimal time. Before looking in moderate detail at such a solution,

---

†A finite automaton with k input lines is a finite automaton whose input set is a (subset of a) k-fold Cartesian product $X_1 \times X_2 \times \cdots \times X_k$, so that an input $(x_1, x_2, \ldots, x_k)$ may be thought of as appearing on k input lines, with x_j on the jth input line. For an explanation of this terminology see Section I, Chapter 3, Volume III of this treatise.

let us first indicate a general notion underlying it, which we shall use in a somewhat biological model. Suppose that we have a line of cells each of the same structure, and that we wish to apportion the cells into different states on the basis of their *relative* positions in the line independent of the length of the line. A general method of doing this is shown in Fig. 6. Here we graph

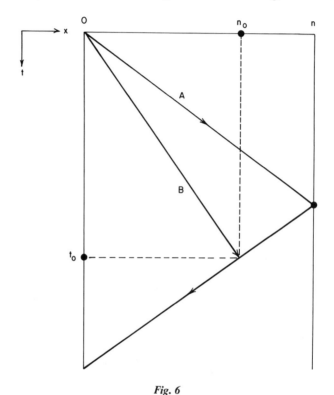

Fig. 6

position along the line on the x axis, and time on the t axis. We show two pulses propagating through the array at constant velocity. The faster A pulse has velocity a and is reflected when it reaches an end. Let us suppose that the B pulse, traveling with velocity b, first intersects the A pulse at position n_0 at time t_0. We then have, clearly,

$$bt_0 = n_0 \quad \text{and} \quad at_0 = n + (n - n_0).$$

Substituting we have

$$2n = at_0 + n_0 = (a/b)n_0 + n_0 = [(a + b)/b]n_0,$$

and we thus have

$$n_0/n = 2b/(a + b),$$

so that we can obtain any desired *relative* intersection point independent of the absolute value of n_0 by appropriate choice of the *ratio b/a*. In fact, if the desired n_0/n is α, and we write b' for b/a, we obtain

$$\alpha = 2b'/(1 + b')$$

so that

$$\alpha + b'\alpha = 2b',$$

whence

$$b' = \alpha/(2 - \alpha).$$

Thus, to divide a line of cells in the ratio α we may use the intersection of two pulses propagating with relative velocity $\alpha/(2 - \alpha)$.

EXAMPLES: If $\alpha = 1/k$, then $\alpha/(2 - \alpha) = 1/(2k - 1)$, so that $\alpha = \frac{1}{2}$ requires $b' = \frac{1}{3}$ and $\alpha = \frac{1}{3}$ requires $b' = \frac{1}{5}$.

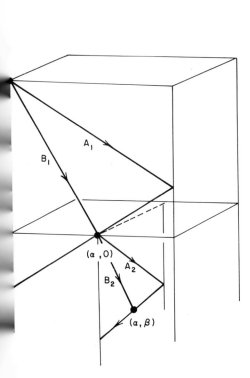

Fig. 7

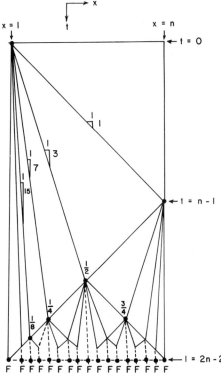

Fig. 8. Pulse propagation for Waksman's [1966] solution of the synchronization problem.

If $\alpha = 1 - 1/k$, then $\alpha/(2 - \alpha) = (k - 1)/(k + 1)$ so that $\alpha = \frac{1}{2}$ requires $b' = \frac{1}{3}$ and $\alpha = \frac{2}{3}$ requires $b' = \frac{1}{2}$.

EXERCISE: If the desired ratio is m/k, where m and k have no common factors, what is the smallest number of states per cell which will yield it? Pay careful attention to the fact that when we have a row of n actual cells rather than a continuous line, we cannot locate m/k but rather the cell whose coordinate is the integer closest to $m/k \cdot n$.

EXERCISE: Suppose we let the A pulse and B pulse continually propagate back and forth. Derive a formula for their jth intersection point.

Note that we can apply this technique in a multidimensional array as well; for instance, in a rectangle we could use the scheme of Fig. 7 in which we work in one coordinate direction at a time, first establishing $(\alpha, 0)$ and then "switching dimensions" to obtain (α, β).

Let us now see how this idea was exploited by Waksman [1966] in obtaining a time-optimal solution to the synchronization problem, that is, one which requires only time $2n - 2$ to synchronize a line of n cells, using an initiator located at one end. The solution is shown in Fig. 8.

At time 0, we initiate a number of pulses, propagating with velocities 1, $\frac{1}{3}, \frac{1}{7}, \ldots, 1/(2^k - 1)$. From our discussion of Fig. 6, we know that the returning 1-pulse will intersect the forward $1/(2^k - 1)$ pulse exactly $1/2^{k-1}$ of the way along the line. We let each intersection point be the initiation point of a new barrage of pulses traveling with velocities 1, $\frac{1}{3}, \frac{1}{7}$, and so on. It is then clear from inspection that the array is successively bisected until at time $2n - 2$, each cell is for the first time in the position not only of being an initiation point, but also of having *both* of its neighbors as initiation points at that time (the vertical dashed lines show the "memory" of being an initiation point). When this triplet occurs about a cell, it enters state F and thus all cells enter state F for the first time simultaneously at the minimal time of $t = 2n - 2$.

There are two points glossed over in the above paragraph which we should briefly discuss, while referring the reader to Waksman's paper for details of how all this logic can be packed into a 16-state cell. (Since it takes one bit to specify which end the first pulse came from, it is comforting to know that eight states will indeed suffice if we insist that the initiator must be at the left-hand end; see Balzer [1967].)

The first point is exemplified by the observation that in a finite array of cells, there may be no exact midpoint. For instance, in a string of $2l + 1$ cells, the 1 pulse initiated at time 0 will arrive in the middle cell, at position $l + 1$, at time $3l$ as will the $\frac{1}{3}$ pulse, and we can use their common arrival to determine this midpoint as initiation point; whereas if the string is of even length $2l$, then the 1 pulse will arrive at square $l + 1$ while the $\frac{1}{3}$ pulse is still in square l. A little thought will show that, to preserve symmetry of the two halves, cells l and $l + 1$ should *both* become initiators at this time. Thus the general convention in a cellular implementation of Fig. 8 is that a cell becomes an initiator either when two of our pulses propagating in opposite directions enter it or when it contains a pulse traveling toward a neighbor containing a pulse traveling in the opposite direction, whichever happens first.

It is the second point which shows Waksman's ingenuity, for the observant reader will have noted that his demand for pulses of velocities $1, \frac{1}{3}, \frac{1}{7}, \ldots ,$ $1/(2^k - 1)$ for every k up to $[\log_2 n] + 1$ where n is arbitrarily large does not seem consistent with the demand that cell size does *not* increase as n increases. Let us close our discussion of Waksman's solution, then, by showing how cells with a finite number of states can propagate pulses with all velocities up to $1/(2^k - 1)$ in array of length 2^{k+1}.

We send out an A pulse with unit velocity, which changes value from A_0 to A_1 as it passes from an even cell to an odd cell, and vice versa. Each cell in state A_1 will generate an R pulse which will propagate in the opposite direction to the A pulse. A P cell initiates a B_0 pulse each time it encounters an R pulse. There are two kinds of B pulses: each shifts forward only on intersecting an R signal, but B_1 annihilates the R signal, whereas B_0 lets it propagate. B pulses change kind each time they propagate.

That this scheme yields B pulses propagating with velocities $\frac{1}{3}, \frac{1}{7}, \frac{1}{15}, \ldots ,$ should be clear from Fig. 9.

EXERCISE: Prove that Fig. 9 works, and then fill in the details of Waksman's scheme.

Moore and Langdon [1967] showed that with a little extra machinery in each cell, initiation could start at any point, with an initiator k cells from the end yielding synchronization at the minimal time of $t = 2n - 2 - k$. We see from Fig. 10 that, for $k \leqslant n/2$, one can obtain at time $3k - 3$ the configuration that would not obtain with the initiator at the left-hand end until time $4k - 3$, thus saving k steps in the synchronization.

EXERCISE: Show that at most five states more are needed per cell for the Moore–Langdon solution than for the Waksman solution.

A more significant generalization is due to Rosenstiehl [1966] who showed that one may design an automaton of fixed size and structure, and with $d \geqslant 2$ input and output lines, such that *any* interconnection of the cells will syn-

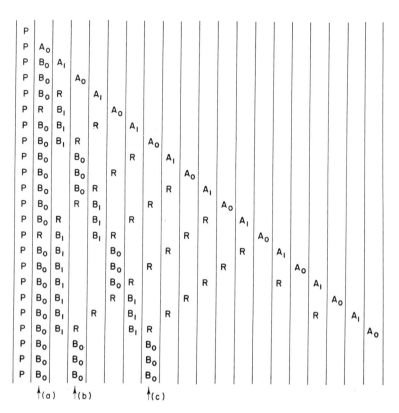

Fig. 9. Generate $1/(2^k - 1)$ pulses with finite machinery. (a) B pulse with velocity $\frac{1}{15}$; (b) B pulse with velocity $\frac{1}{7}$; (c) B pulse with velocity $\frac{1}{3}$.

chronize. The notion of "networds" which underlies his scheme can be used to determine position in arbitrary structures, though the reader should note that more efficient schemes exist when we restrict ourselves to special types of interconnection, such as the rectangular arrays shown in Fig. 7.

An automaton (Q, d, δ) *of degree d* is an automaton with input set Q^d, state set Q, and next-state function

$$\delta: Q \times Q^d \longrightarrow Q.$$

By a graph of degree at most d we mean a quadruple (X, U, I, T) where

    (i) X is a set of vertices;

    (ii) U is a set of edges;

    (iii) $I: U \rightarrow X$ is the initial node map;

    (iv) $T: U \rightarrow X$ is the terminal node map;

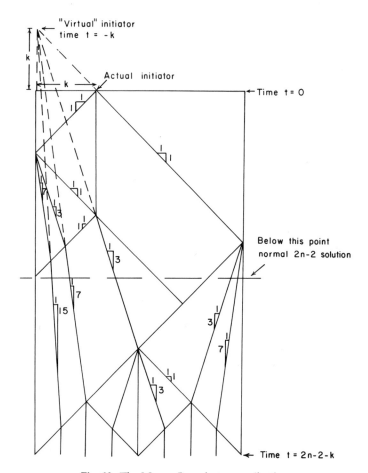

Fig. 10. The Moore–Langdon generalization.

and no node is initial or terminal for more than d edges; that is, $\omega(x) = \{u \mid O(u) = x\} \cup \{u \mid T(u) = x\}$ has at most d elements. *A network R based on the automaton $A = (Q, d, \delta)$ is specified by placing a copy of A at each node of a graph G.* To specify the interconnections, we take the graph $G = (X, U, I, T)$ of degree at most d, pick for each $x \in X$ a "connection map" that is one to one

$$\varphi_x: \; \omega(x) \longrightarrow \{1, 2, \ldots, d\},$$

which says that the ith input line of the automaton placed at x comes from $\varphi_x^{-1}(i)$ if that is defined, or has the fixed value 0 if not.

Figure 11a shows us a network based on an automaton of degree three:

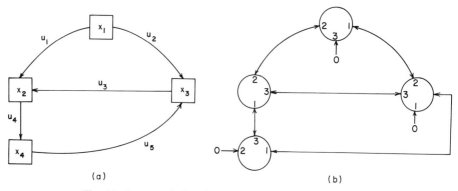

Fig. 11. A network based on an automaton of degree three.

It has graph $G = (X, U, I, T)$ where we have the table

U	I	T
u_1	x_1	x_2
u_2	x_1	x_3
u_3	x_3	x_2
u_4	x_2	x_4
u_5	x_4	x_3

We can then specify φ_x in the following table:

φ_x^{-1} x_1 $\quad$ i	x_1	x_2	x_3	x_4
1	u_2	u_4	φ	u_5
2	u_1	u_1	u_2	φ
3	φ	u_3	u_3	u_4

to yield the interconnection shown in Fig. 11b, in which all but one link is two way, and there are three null inputs.

Rosenstiehl's generalization of the synchronization problem may then be stated as follows:

PROBLEM: Find an automaton $A = (Q, d, \delta)$ with 0 state, and special states D, $\hat{M}$, and F such that *any* network R based on A has the following property: If at at time 1, $q_x(1) = \hat{M}$ for one x, and $q_x(1) = D$ for $\hat{x} \neq x$ then there is a time θ at which $q_x(\theta) = F$ for every automaton in the same component of R, where $q_x(t) \neq F$ for all $\hat{x}$ when $t < \theta$.

The usual problem is recaptured by setting $d = 2$, though the general statement allows the automata to be connected in a loop as well as a line, and with any cell initiating. In his paper Rosenstiehl gives a 13-state solution taking approximately $3n$ steps to synchronize a line of length n, but since we

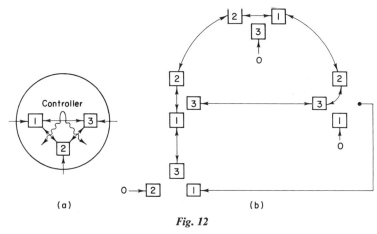

(a)   (b)

Fig. 12

have Waksman's solution before us, we shall not reproduce this simple but less efficient solution here.

Rosenstiehl solves the problem by reducing it to the standard straight line problem. The basic idea is to consider each automaton of degree d as decomposed into d subautomata, as shown in Fig. 12a, which can be interconnected by a central controller. Synchronization then proceeds in two states. In stage 1, the internal interconnections are set so that a "line" of subautomata is formed containing at least one from each automaton. Figure 12b shows how the network of Fig. 11 might look after stage 1. In stage 2, this "line" of subautomata synchronizes itself using the logic of the standard synchronization problem and we assume that the controller of each nodal automaton sends the whole automaton into the designated final stage simultaneously with the transition of one or more automata to its final state.

Clearly, the only point of any subtlety is to learn how the controller can use finite circuitry to establish the interconnections of stage 1 on the basis of local information, without making use of our ability to step back and study in which graph the component is presently interconnected. For our present purposes, it suffices to note that such finite-state control is possible, and to refer the reader to Rosenstiehl [1966, Section III] for the details, which are based on the idea of a *netword*—a string of edges which suitably traverses the graph.

B. Relative Position—A Slightly Embryological View

Let us now use these techniques to present an automaton-theoretic solution [Arbib, 1969a] to the French flag problem raised by Wolpert [1968] and which we discussed in Section II.

We shall indicate how arrays of one such type of cell may do the job in Figs. 13, 14, 15, and 16. We shall see that they have one great drawback from a biological modeling viewpoint, and then find to our pleasant surprise that the rectification of this deficiency yields a cellular behavior which is more biological than that of the original.

We are to imagine an array which can conduct five types of pulse, an A pulse, a B pulse, a head pulse, a tail pulse, and a body pulse. An A pulse will propagate with unit velocity in a given direction until it hits the end of the string of cells, whereupon it is reflected and moves with unit velocity in the opposite direction. We decree that if there are ever two A pulses traveling in the array, then upon their meeting the pulse traveling to the left is annihilated whereas the pulse traveling to the right is propagated as if it had not encountered the other pulse. B pulses are propagated with half the speed of A pulses and are created at a boundary each time an A pulse is reflected there (see t_0 in Fig. 13). If an A pulse overtakes a B pulse going in the same direction, then the two pulses do not interact but continue on their way at their specified velocities. If, however, the A pulse and the B pulse are going in opposite directions, then a fairly complex response is triggered.

If the A pulse is moving left and the B pulse is moving right, then the B pulse is annihilated whereas the A pulse will continue on its way. In addition to this effect, however, a tail pulse will be propagated with unit velocity towards the right-hand end of the string where it will be annihilated, and as it reaches each cell it will turn that cell into the tail state. At the same time a body pulse is triggered moving left at a velocity of $\frac{1}{3}$ (which leads to some problems in discrete systems which I shall overlook, using a continuum in all my diagrams). In general, this pulse will not propagate all the way to the left-hand end of the string but will be annihilated on meeting an A pulse.

If an A pulse moving right meets a B pulse moving left, then the B pulse will again be annihilated, and this time a head pulse will be propagated left at unit velocity turning each cell it encounters to the head state until it is annihilated upon reaching the left-hand end of the string.

To the reader who finds the switching logic of this scheme too patently irrelevant to our more biological framework, let us note that we can think of the A and B waves as only propagating once, corresponding to two cellular reactions, proceeding at different rates, each being initiated in a cell when its neighbor provides a critical level of "transmitter substance."

To appreciate how the rules for A and B cells work, look at Fig. 13 in which the array of cells is graphed in space from left to right while its evolution in time is graphed with time progressing as we move down the page. We see that at time 0 an A pulse and a B pulse are both initiated at the left-hand end and they propagate until they meet at time t_1, whereupon the B pulse is annihilated, but we see that the tail pulses and the body pulses work to cause cells to differentiate into the appropriate states. Note, however, that at time t_0 the A pulse was reflected from a boundary at which time a new B pulse was generated and so we see that at time t_2 the new B pulse meets our A pulse at which stage the cells in the front third of the array are told to convert to the head state. You will notice that the body pulse was annihilated by meeting an *A* pulse at just the same point in space and time at which the head pulses were triggered. It was to assure this coincidence that the propagation speed of $\frac{1}{3}$ was chosen for the body pulses. We now see that as time goes by, the *A* pulses and *B* pulses ricochet back and forth, interacting as they do so but the messages they send out do not change the configuration and so in fact at time t_3 the stable state with properly differentiated head, body, and tail is attained.

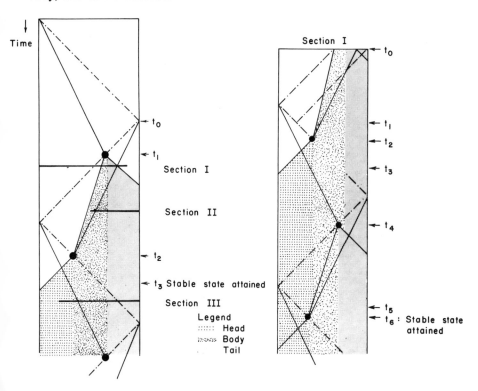

Fig. 13. [From Arbib, 1969a]   *Fig. 14.* [from Arbib, 1969a]

Let us now see what happens to our array when pieces are cut out of it. We have marked on Fig. 13 three sections which are to be removed. The evolution of Section I is indicated in Fig. 14, that of Section II in Fig. 15, and that of Section III in Fig. 16. We add one more condition. When a portion is cut from an array, an *A* pulse is triggered at the point or points of damage. We shall see how this works by tracing through the fate of Section I in Fig. 14 (the reader is invited to carry out a similar step-by-step analysis of Section II and Section III). We see that at time t_0 not only is there an *A* pulse and a *B* pulse traveling in the section but also there is a new *A*

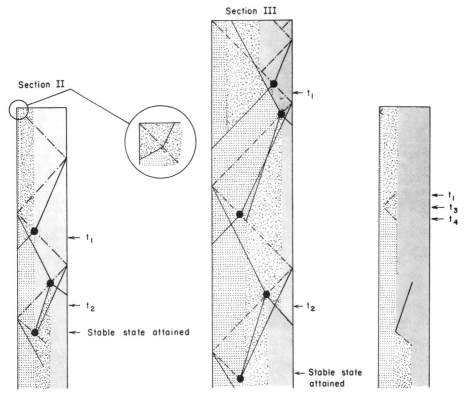

Fig. 15. ₁from Arbib, 1969a]   *Fig. 16.* [from Arbib, 1969a]   *Fig. 17.* [from Arbib, 1969a]

pulse initiated as a result of injury, but this A pulse only lasts until time t_1, when it is annihilated on encountering the first A pulse. The B pulse and the first A pulse meet at time t_2 just as they would have in the uninjured specimen, and so the head pulses propagate forward from time t_2 so that by time t_3 we do have an array in which the first portion is head, the second portion is body, and the third portion is tail. However, the proportions of the three components are not equal and so a stable state has not been attained. As we follow the evolution we see that at time t_4 the adjustments are made which ensure that exactly one-third of the array is in the tail state, but it is not until time t_6 that another encounter between an A pulse and a B pulse sets up the final equilibrium, in which head, body, and tail each occupy one-third of the array. Note that the propagation forward of a body pulse at time t_4 has no apparent effect until time t_5 when it first emerges from the front end of the string of cells that were already in the body state at t_4.

The reader will note one unpleasant feature of this evolution. Just after time t_4, a piece of body is caught between two pieces of tail. Similarly, in Fig.

15, we see head–body–head–tail at t_1 and head–tail–body–tail at t_2; while in Fig. 16 we see head–body–tail–head–tail at t_1.

I do not know whether embryologists have seen such effects, but it seemed to me worthwhile to look for a simple modification in my model which would avoid them.† We simply postulate that the states shown in my diagrams are not *actual* states, but *potential* states.

Let x be the location of a cell, $x - 1$ the location of the cell to the left of it, and $x + 1$ the location of the cell to the right of it. Let A_{xt} be the actual state of cell x at time t, P_{xt} the potential state of all x at time t. We then have the following rules:

1. If P_{xt} = head, then $A_{x,t+1}$ = head unless $A_{x-1,t}$ = body or tail.
2. If P_{xt} = body, then $A_{x,t+1}$ = body unless $A_{x-1,t}$ = tail or $A_{x+1,t}$ = head.
3. If P_{xt} = tail, then $A_{x,t+1}$ = tail unless $A_{x+1,t}$ = head or body.
4. In the other cases $A_{x,t+1} = A_{x,t}$.

Thus, the cells are polarized, and a tail cell can repress the cell behind it from turning into a head cell or a body cell, and so forth.

In Fig. 17 we have shown how the potential states of Section II, represented in Fig. 15, give rise to actual states without any "islands" when we make the repression assumption. For example, at time t_1 a pulse is sent forward changing cells to a *potential* head state, but until time t_3 each cell so changed finds a cell in front of it still in the actual body state. It is only at time t_3 that a cell can first change its actual state, and when it does so, the cell behind it can change one moment later, until at time t_4, all cells in *potential* body state have changed to *actual* body state.

I shall not explore this model further here. My purpose has simply been to show that the regenerative properties of the embryo are not incompatible with the automaton approach to development. With this we may now turn to the formal study of Turing machines and formal languages that we require to lay the basis for the further development of this approach.

IV. An Introduction to Turing Machines and Formal Languages‡

A. Turing Machines

This section surveys material in automata theory that may be of interest to theoretical embryologists, with some discussion of its "qualitative" value to the modeler. Most of the mathematics, though little of the discussion,

---

†Note, too, how in Section II, there is a period of time when no body cells are visible. Does such resorption occur in real preparations?

‡Much of this section is based on Michael A. Arbib, *Theories of Abstract Automata* © 1969. It is adapted here by permission of Prentice-Hall, Inc., Englewood Cliffs, New Jersey.

may be found in Arbib [1969b]. Through most of this section we emphasize, as does most of current automata theory, the ways in which information-processing structures (henceforth called automata) transform strings of symbols into other strings of symbols.

The usual notion of an automaton, though one we generalize when we emphasize theoretical embryology, is a device to which we may present a string of inputs. If and when the machine finishes computing on this string, the result will be a string of outputs. We say "if" because certain input strings may drive the computer into a "run-away" condition, for example, endless cycling through a loop, from which a halt is impossible without external intervention (which amounts to changing the input string). This case might correspond to associating with the machine a device which produces an infinite string of outputs for each input string.

Thus, if, given any set A, we use A^* to denote the set of all finite sequences of elements from A, we may say that automata theory in a very general form is the study of *partial* functions $F: X^* \longrightarrow Y^*$, that is, ways whereby *some* of the strings in X^* have assigned to them output strings from Y^*, it being understood that for other input strings θ the function $F(\theta)$ may not be defined at all. However, such a function becomes truly a part of "classical" automata theory only if we can relate it to a *finitely specifiable substrate* or if we are eager to prove that no such substrate exists for it.

There are two main types of process a function

$$F: \quad X^* \longrightarrow Y^*$$

which transforms *input sequences* (elements of X^*) into *output sequences* (elements of Y^*) can be thought to represent.

1. In an *on-line computation* we are to imagine a system with a set of possible inputs and possible outputs. F then tells us that if we start our system in some standard configuration and apply a sequence $x_1 x_2 \cdots x_n$ of inputs then the sequence of outputs emitted by the system over the *corresponding period of time* will be $F(x_1 x_2 \cdots x_n)$ (see Fig. 18).

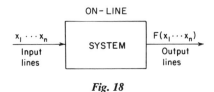

Fig. 18

Assuming that one output symbol is emitted for each input symbol, we may go from $F: X^* \longrightarrow Y^*$ to a function $f: X^* \longrightarrow Y$ defined by the equation

$$f(\theta) = \text{the last symbol of } F(\theta), \quad \text{for all} \quad \theta \in X^*.$$

We can then reconstruct F from f by the simple equation

$$F(x_1 x_2 \cdots x_{n-1} x_n)$$
$$= f(x_1)f(x_1 x_2) \cdots f(x_1 x_2 \cdots x_{n-1})f(x_1 x_2 \cdots x_{n-1} x_n).$$

The crucial point about an *on-line* computation is that the jth term of the output depends only on the first j input symbols; we call an F with this property *sequential*.

2. In an *off-line computation* we are to imagine a system which has some data structure into which a string of X^* may be read at the start (see Fig. 19).

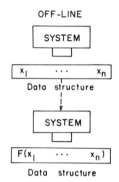

Fig. 19

The system then operates on this data and finally produces as output a string of Y^*. Now there may well be none of the sequential structure we associated above with on-line computations, as in the case of the function $F: X^* \longrightarrow X^*$ which *reverses* strings: $F(x_1 x_2 \cdots x_n) = x_n \cdots x_2 x_1$. Here the first output certainly does *not* depend only on the first input.

To get some feel for on-line computations, consider a device with two states. When it is in state q_0, it emits output 0, and when in state q_1 it emits output 1. If it receives input 1, it changes state, but if it receives input 0, it does not change state. In short we are to consider a system with the set $Q = \{q_0, q_1\}$ of states, the set $X = \{0, 1\}$ of inputs, the set $Y = \{0, 1\}$ of outputs, whose behavior is described by the state graph shown in Fig. 20.

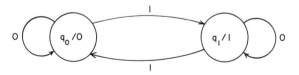

Fig. 20

If we start the system in state 0 and use it to process on-line a string θ the system will change state just as many times as there are 1s in the input string, and so its input–output behavior is given by the function,

$$f: X^* \longrightarrow Y: \quad \theta \mapsto \begin{cases} 1 & \text{if } \theta \text{ contains an odd number of 1's} \\ 0 & \text{if } \theta \text{ contains an even number of 1's.} \end{cases}$$

However, as our string-reversal example showed, not all computations $F: X^* \to Y^*$ are sequential, that is, not all can be built up from a function $f: X^* \to Y$, and so we must consider systems which do not function on-line. The classical model for an off-line computation is what is now called a Turing machine, a notion that was first set forth by Turing [1936] and Post [1936], as one possible formalization of the notion of an algorithm or effective procedure.

We may view a Turing machine as consisting of a finite tape divided into squares, each of which may bear a single symbol from some fixed set X, together with a control box which can move back and forth along the tape scanning a single square at a time (see Fig. 21). The control box contains a

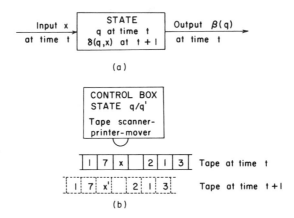

(a)

(b)

Fig. 21. (a) A causal system; (b) a Turing machine whose program contains $qxx'Rq'$.

finite program, each instruction of which, except for the halt instruction, consists of a process of testing the square under scan, and on the basis of the symbol observed, printing another, possibly the same, symbol on the square, moving the control box one square to left or right along the tape, and finally jumping to a new instruction of the program.

We often think of the control box of a Turing machine as a *finite* automaton, and represent its program by a list of quintuples of the form $qxx'Dq'$ to indicate that if the control box is in state q and the scanned symbol is x, then the machine does not stop, but instead prints x', moves in direction D (left, right, or not at all), and changes to state q'.

We start a computation by placing the control box executing its first instruction on the left-hand square of the tape bearing the initial data. If the control box comes to either end of the tape, we add new blank squares as desired. For a given input, the machine may halt eventually, or may go into various endless loops. Thus we may associate with each Turing machine Z a function $F_Z: X^* \to X^*$ from strings of symbols to strings of symbols, which is *partial* in that for some of the input values, no result may be defined.

Now suppose we have an algorithm (or "recipe" or "set of rules") for effectively producing one integer from another. Such a rule has always, in the past, been transcribable into a set of quintuples for a Turing machine; so that nowadays it is common to identify the intuitive notion of an effective computation with the formal notion of a computation which can be carried out by at least one Turing machine. (Observe that one cannot prove this equivalence, since one notion is informal and one is formal.) The statement, *"Any effective procedure can be implemented by an appropriate Turing machine,"* is often referred to as *Turing's hypothesis* or *Church's thesis*.

Given such a sweeping claim, the reader may well ask, "Is there anything a Turing machine *can't* do?" The answer is yes. Surprisingly the classic *unsolvable problem* (that is, one whose answer cannot be computed by a Turing machine) is one concerning a simple problem about Turing machines themselves, namely the *halting problem*.

If we think of Turing machines as actual physical devices, it is clear that the input symbols could be configurations of holes on punched tape, patterns of magnetization or handwritten characters, or the like, whereas the states could be that of a clockwork, of a piece of electronic apparatus, or of an ingenious hydraulic device. Such details are irrelevant to our study, and so if there are m inputs in X, we shall feel free to refer to them as $x_0, x_1, \ldots, x_{m-1}$ without feeling impelled to provide further specification; and if there are n states, we shall similarly find it useful to label them $q_0, q_1, \ldots, q_{n-1}$. Automata theory deals with abstract descriptions of machines, not their implementations. For each such abstract description there are many implementations, depending, for example, on whether we interpret x_i as a 0 or a 1, as a pattern of magnetization or a configuration of holes in a punched tape, but it should be clear that, from the information-processing viewpoint, there is a very real sense in which all these machines may be considered the *same* machine.

Since each Turing machine is described by a finite list of instructions, it is easy to show that we may *effectively* enumerate the Turing machines

$$Z_1, Z_2, Z_3, \ldots$$

so that, given n we may effectively find Z_n, and given the list of instructions for Z, we may effectively find the n for which $Z = Z_n$.

This implies that we can effectively enumerate all TM-computable (Turing-machine computable) functions as

$$f_1, f_2, f_3, \ldots$$

simply by setting $f_n = F_{Z_n}$. If we say that f_n is *total* if $f_n(\theta)$ is defined for all θ in X^*, we might ask: Is there an effective procedure for telling whether or not f_n is total; for example, does there exist a total TM-computable function h such that f_n is total if and only if $n = h(m)$ for some m (identifying a string with a suitable number that encodes it)? The answer is "no," for if such an h existed, we could define f by

$$f(n) = f_{h(n)}(n) + 1.$$

Then f would be total recursive, and so $f = f_{h(m)}$ for some m.

Then $f_{h(m)}(m) = f_{h(m)}(m) + 1$, a contradiction.

This is just one example of the many things we can prove to be *undecidable by any effective procedure.* To say that we cannot effectively tell that f_n is *total* is just the same as saying that we cannot tell effectively whether Z_n will stop computing no matter what tape it is started on. We may thus say that "*the halting problem for Turing machines is unsolvable.*"

We should note that the definition can be extended to include Turing machines in which the control box controls many heads on many different tapes, perhaps some of the tapes even being multidimensional, as indicated in Fig. 22.

The reader should appreciate how general a model of computers the generalized Turing machine is and thus how powerful the ordinary Turing machines are in capability, though not in efficiency. For instance, any present-day computer is really a generalized Turing machine: the input system corresponds to a read-only tape, the output system corresponds to a write-only tape, the tape units (although of only finite capacity) correspond to one-dimensional, one-head TM tapes, and addressable words of core-store correspond to TM tapes only one square long (the machine can read ater write but cannot move on these tapes).

In fact, it can be proved that the functions computable by these machines are all computable by the conventional one-head, one-tape, one-dimensional machine (see Arbib [1969b, Section 4.2] for the formal definitions, and a description of the simulation process). What we do gain by using more complicated machines is speed and efficiency. For instance, Barzdin [1965] has proved that no machine with one head can tell whether or not a word is symmetric, that is, is a palindrome, in less than a time that increases proportionately with the square of the length of the word; whereas it is easy to

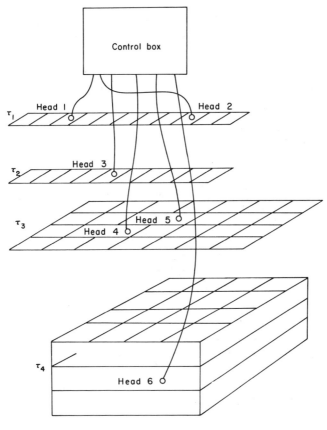

Fig. 22

see if we have two heads, we may move the heads to the opposite end of the tape and have them move in to the middle, reporting back to the control, so that when they meet, in a time that only goes up with the length of the tape, they will have been able to tell whether or not the string is a palindrome. Here is an extremely important question of automata theory: Given a problem, what is the computing structure best suited to it? For instance, I presume that one could rigorously prove that adding extra heads to the two-head machine could not make its recognition of palindromes any more efficient, and so the two-head structure is the one most suited for a device for recognising palindromes. We may note too that authors such as Wang [1957] and Minsky [1967] have shown that even simpler instruction sets than that of the Turing machine suffice to give us all effective procedures; the price one pays is even less efficiency than that of the Turing machine. We thus see the need to understand better the tradeoff between complexity of the "genetic" structures, and the complexity of the computations which take place within those structures. Some of these questions are broached in Chapter 3, Volume III of this treatise.

Let us indicate the power—and the inefficiency—of ordinary Turing machines by outlining how we might program one to handle the example of string reversal.

1. Start in initial state at left-hand end of tape, which bears a special marker followed by the string to be reversed

$$\overset{\downarrow}{*}\ x_1\ x_2\ \cdots\ x_{n-1}\ x_n.$$

2. Move right to the first blank. Go back one square. Note the symbol there, erase it, move one square right and print it:

$$*\ x_1\ x_2\ \cdots\ x_{n-1}\ b\overset{\downarrow}{x}_n$$

(where we use b to remind us of the presence of a blank square).

3. Move left until you first scan a nonblank symbol immediately after scanning a blank. If the symbol is $*$ go to 4. If the symbol is not $*$,

$$*\ x_1\ x_2\ \cdots\ \overset{\downarrow}{x}_j\ b\ \cdots\ bx_n\ x_{n-1}\ \cdots\ x_{j+1},$$

note the symbol there, erase it and move right until you first scan a blank immediately after scanning a nonblank symbol, and print the noted symbol

$$*\ x_1\ x_2\ \cdots\ x_{j-1}bb\ \cdots\ bx_n\ x_{n-1}\ \cdots\ x_{j+1}\ \overset{\downarrow}{x}_j$$

and repeat all the instructions in 3.

4. $$\qquad\qquad\overset{\downarrow}{*}\ bb\ \cdots\ bbx_n\ x_{n-1}\ \cdots\ x_2\ x_1;$$

erase $*$ and halt. The nonblank portion of the tape contains the reversed string.

An interesting point about this computation is that we only need about twice as much tape as that on which the data are printed originally. And so another question will be: If we do have a computation that is off-line, requiring us to present the total data in a structure on which the machine can operate, how much auxiliary storage will be required to complete the computation? Again there will often be a tradeoff between time and storage requirements, some of which are discussed by Arbib [1969b, Chapter 7].

One of the most interesting results in Turing's original paper [1936] was that there exists a *universal* Turing machine.

Theorem. *There exists a Turning machine U such that given the encoding $e(Z)$ of the program of a Turing machine Z, and a data string x, on its tape,*

it will proceed to simulate the computation of Z on x, halting if and only if
$F_Z(x)$ *is defined, in which case it will halt with* $e(Z), F_Z(x)$ *on its tape:*

$$F_U(e(Z), x) = e(Z), F_Z(x).$$

PROOF OUTLINE: The idea of the proof is that U, in simulating Z, is to have
its tape divided into two halves: the left-hand half contains the list of coded
instructions with a pointer indicating which state Z would be in at this stage
of the simulation, while the right-hand half contains the data string that Z
would have on its tape at this stage of the simulation, with a pointer indicating which square of its tape Z would be scanning. The strategy is for the
control box of U to read an instruction, use the next-state portion of it to
reposition the current-state marker (this involves some subtleties, since Z
may have far more states than U), and then move right until it finds the
current-symbol marker to carry out the designated tape manipulation
(problems arise here if new data is to be printed on a new square to the left of
Z's tape) before moving left to find the current-state marker and initiate the
cycle of simulation once more.

The reader may wish to try filling in the details himself before turning to
the elegant construction of Minsky [1967].

With this result before us, we can prove one of the most surprising theorems
about effective computations: that any enumeration of TM-computable
functions is redundant in any way that you can effectively specify.

The Recursion Theorem. *Let* $f_1, f_2, f_3, \ldots$ *be any effective enumeration of
the TM-computable functions, and let h be any total TM-computable function.
Then there exists a number* i_0 *such that*

$$f_{i_0} = f_{h(i_0)}.$$

PROOF: Consider the following functions:

$\alpha(z, x) = z, z, x$ This just involves copying the first argument
and so is effectively computable.

$\beta(z, y, x) = f_z(y), x$ This just involves applying the universal Turing
machine to the first two arguments, while preserv-
ing the third, and so is effectively computable.

$\gamma(w, x) = f_w(x)$ This is just the universal Turing machine program.

Thus

$$\gamma \circ \beta \circ \alpha(z, x) = f_{f_z(z)}(x)$$

is an effectively computable function of z and x, and so given z we may effectively find an index $g(z)$ for the function $\gamma \circ \beta \circ \alpha(z, .)$. We thus have for all
z and x

$$f_{g(z)}(x) = f_{f_z(z)}(x).$$

Let then v be the index of the total function $h \circ g$:

$$f_v(x) = h(g(x)).$$

Thus

$$f_{f_v(v)}(x) = f_{h(g(v))}(x).$$

But by definition of g

$$f_{f_v(v)}(x) = f_{g(v)}(x).$$

Thus, $g(v)$ is the desired index i_0 such that

$$f_{i_0} = f_{h(i_0)}.$$

An interesting application of this result is the following, already mentioned in Section I, which is due to Lee [1963].

Theorem. *There exists a self-describing Turing machine; that is, there exists an integer n_0 such that*

$$f_{n_0}(\Lambda) = n_0.$$

PROOF: Define h to be the function such that $h(n)$ equals an index of a machine such that $f_{h(n)}(\Lambda) = n$. Then let n_0 be the fixed point for h:

$$f_{n_0}(\Lambda) = f_{h(n_0)}(\Lambda) = n_0.$$

B. Biological String Processing

Before moving on to the study of formal languages, let us briefly look at some attempts to relate string processing to reproduction. Apter [1966, Chapter 5] described several programs for his computer, designed to capture certain aspects of biological programs, of which the following are typical:

1. The first program causes the computer to copy the program into many different sections of the store. The program is "self-reproducing."

2. The second program causes the computer to copy the program into many different sections of the store but with a parameter taking different values in each case. The program is "self-reproducing with differentiation."

3. The third program is like the second, save that it is "self-limiting" in that it halts after a specified number of self-reproductions.

4. The fourth program is a "general growth program," which has the virtue that if suitably appended to a finite list of instructions, it will cause the computer to copy the whole program into many different sections of the store.

Now these simple exercises in programming may yield some useful insights. What we wish to emphasize, however, is that these models are all far from biology. The program in each case is intended to model the genome of a cell. However, reproduction does not take place as in the biological case where the instructions are processed locally by the cytoplasmic machinery, but by the decoding of instructions by the whole "extracellular" machinery of the computer's control unit and arithmetic unit. Complicated organisms reproduce via zygotes of lesser complexity. It seems that interaction with a highly structured environment plays a crucial role in pushing up the level of complexity of the adult. How can we understand *the structure necessary in a substrate* to obtain the necessary complexity from the environment *within the time allowed for maturation*? In this regard, we should ponder the direct relationship between the time required for maturation, and the breadth of the ecological niche. This biological consideration contrasts with Apter's study in which the environment simply types a program into a computer.

Stahl [1965a, b; see also Stahl *et al.*, 1964] has used the present-day view of genetic processes as manipulation of macromolecules which may be considered as strings of bases, and so forth, to note that much of cellular processing may be represented by the logical manipulation of strings of symbols. To the extent that one believes that such manipulations are deterministic, it is clear from our above discussion that such processes may be carried out by Turing machines. Stahl undertook the task of programming Turing machines to execute molecular string manipulations, and of writing a computer program for juggling a large number of these molecular algorithms so as to keep track of the molecular strings in a dynamically operating simulation of a simple cell with 40 to 50 enzymes. It does not seem unwise to try to simulate more directly the type of automaton exemplified, for example, by a ribosome and the complex of chemical pathways *in vivo*. However, biologists unacquainted with algorithmic processes may find it a good pedagogical aid to chase through Stahl's Turing machine programs.

Stahl has programmed a mode of operation called "self-reproduction" in which the computer program simulates the string processing of a cell until concentrations reach a critical level, whereupon all concentrations are split in half and distributed among two new cells. My criticism of this model is that reproduction does not take place through local processing of instructions by the cytoplasmic machinery, but by the decoding of instructions by the whole "extracellular" machinery of the computer. However, Stahl [personal communication, 1966] states that his

> Turing automaton model of cellular reproductions involves no 'extracellular thinking'; the self-reproduction algorithm is incorporated entirely in the coding of the automata, which could in principle be operated on by any universal Turing machine. In practice, a special computer simulator and many heuristics are used, but these do not contribute to the essential logic of self-reproduction in a basic way.

Stahl [1966] provides details of this simulation.

Stahl [1965a] lists a number of problems which a cell may not be able to solve, such as predicting whether or not a new enzyme might cause the cell to become cancerous. However, we should emphasize that such a problem may *not* be "algorithmically unsolvable." In the theory of algorithms, we say a problem is *algorithmically unsolvable* if *no* effective procedure exists to solve the problem. Thus, there are many interesting problems raised by Stahl which a given machine (cell automaton) cannot solve; but one must not think that *no* automaton (possibly far more complex than a single cell) can solve such problems. (In deterministic operation, Stahl's automata preserve the length of their tapes. Thus, he considers a subclass of Turing machines to which many of the classic unsolvability results do not apply.) Thus, the correct automata theory for handling such problems is more likely to be a chapter of the new theory of complexity of computation [Blum, 1964; Hartmanis and Stearns, 1965; Arbib, 1966a; for a survey see Arbib, 1969b, Chapter 7] than the older theory of unsolvability exemplified by our discussion of the halting problem above.

C. Formal Languages

Turning now to formal languages, we should note that Turing originally developed his machines in order to derive theorems about what was and was not provable by finitely specified rules of inference in a formal system of deductive logic. Formal language theory may be seen to descend from an alternative approach taken by Emil L. Post.

Whereas Turing formalized effectiveness of rules of inference by demanding that a rule of inference R only be accepted for a formal logic if there exists a Turing machine $Z(R)$ such that

$$F_{Z(R)}(x_1, \ldots, x_n, y) = \begin{cases} 1 & \text{if} \quad R(y, x_1, \ldots, x_n) \text{ is true} \\ 0 & \text{if not,} \end{cases}$$

Post [1943] gave his finitistic restriction in terms of *productions*. A *k*-antecedent production is of the form

$$g_{11}u_1^{(1)}g_{12}u_2^{(1)} \cdots g_{1m_1}u_{m_1}^{(1)}g_{1(m_1+1)}$$
$$\vdots \qquad\qquad \vdots$$
$$g_{k1}u_1^{(k)}g_{k2}u_2^{(k)} \cdots g_{km_k}u_{m_k}^{(k)}g_{k(m_k+1)}$$

produce

$$g_1u_1g_2u_2 \cdots g_mu_mg_{m+1}.$$

In this display, the *g*'s represent specific strings including the null string, while the *u*'s represent the operational variables of the production, and, in the application of the production, may be identified with arbitrary strings. These variables need not be distinct; equalities among them constrain our substitutions. We then add the restriction that each operational variable in the conclusion of the production is present in at least one of the premises of the production, it having been understood that each premise, and the conclusion, has at least one operational variable. This production corresponds to the rule of inference R for which $R(\Theta_1, \ldots, \Theta_k, \Theta)$ is true if and only if there exist strings

$$h_j^{(i)}; \quad 1 \leqslant i \leqslant k, \quad 1 \leqslant j \leqslant m_i + 1$$

$$[\text{with} \quad h_j^{(i)} = h_{j'}^{(i')} \quad \text{if} \quad u_j^{(i)} = u_{j'}^{(i')}]$$

such that $\Theta_i = g_{i1} h_1^{(i)} g_{i2} h_2^{(i)} \cdots g_{i(m_i+1)}, 1 \leqslant i \leqslant k$ and $\Theta = g_1 h_1 g_2 h_2 \cdots g_m h_m g_{m+1}$, where h_j is the $h_i^{(i)}$ such that u_j is the variable $u_i^{(i)}$.

We may then use a set of productions to generate a formal language, that is, a set of strings on some alphabet X^*, as follows:

Definition. *A* Post generation system *is a quadruple $\mathscr{S} = (V, X, A, P)$ where*

(i) *V is a finite set;*

(ii) *$X \subset V$ (the* terminal *symbols);*

(iii) *A is a finite subset of V^* (the set of* axioms*);*

(iv) *P is a set of* productions *whose fixed strings lie in V^*.*

We say that a set B of strings *immediately generates* the set $\hat{B}$, and write $B \Rightarrow \hat{B}$, if for every string θ in $\hat{B}$, either θ belongs to B or θ can be obtained by applying a single production from P to strings already in B.

We say that θ is a *theorem* of $\mathscr{S}$ if there exists a sequence $B_1, B_2, \ldots, B_n$ of subsets of V^* such that

$$A = B_1 \Longrightarrow B_2 \Longrightarrow \cdots \Longrightarrow B_{n-1} \Longrightarrow B_n$$

and $\theta \in B_n$.

We shall say that a set is generated by $\mathscr{S}$, and call it $L(\mathscr{S})$, if it is the set of theorems of $\mathscr{S}$ which lie in X^*.

Given a set B, we write $B \underset{\mathscr{S}}{\vdash} x$ if x is a theorem of the system using the rules of inference of $\mathscr{S}$, and using B as the set of axioms. Thus, $L(\mathscr{S}) = \{x \in X^* \mid A \underset{\mathscr{S}}{\vdash} x\}$.

Definition. *We shall say a set of strings is* Post generable *if there exists a Post generation system $\mathscr{S}$ for which the given set equals $L(\mathscr{S})$.*

Definition. *A system in canonical form is said to be a* semi-Thue *system if each production is of the form*

$$u_1 g u_2 \longrightarrow u_1 g' u_2$$

for suitable strings g, g'.

Theorem. *Any computation which can be executed by a Turing machine can be carried out by a suitable Post canonical system, though it be only a semi-Thue system.*

The converse result, that even a general canonical system may be simulated by a Turing machine is harder; the reader may find the general strategy in Arbib [1969b, Section 6.2].

PROOF: Consider a Turing machine Z for which we assume Q (the state set of Z) and X (the input set of Z) are disjoint. We form a set of productions $\tau(Z)$ such that if we take hxh as our single axiom, then the only theorem lying in X^* will be $F_Z(x)$:

1. Certain productions of $\tau(Z)$ mimic the quintuples of Z, for example,

$$qxx'Lq' \quad \text{yields} \quad \begin{cases} u_1x_1qxu_2 \longrightarrow u_1q'x_1x'u_2, \\ u_1hqxu_2 \longrightarrow u_1hq'bx'u_2. \end{cases} \quad \text{all} \quad x_1 \in X$$

2. Second, there are productions which, for a string containing q_s (the halting state of Z), remove q_s and the h's.

Thus, the only strings on X^* obtainable are those we may find on the tape of Z at the end of a computation, since in no other case will the h's be removed.

Note that just as a Turing machine may never halt, so may a semi-Thue system produce an infinity of strings none of which contains only "terminal" symbols.

We now turn to the study of context-sensitive and context-free languages, which are the languages generated by restricted types of semi-Thue systems introduced by Chomsky [1959] as a formal approximation to natural languages (while noting that to gain linguistic insight it was necessary to refine that approximation by augmenting it with the notion of transformational grammar). We first present a slight modification of semi-Thue systems in a vocabulary appropriate to their interpretation as grammars. (We use Λ to denote the empty word, of length 0.)

Definition. *A* phrase-structure grammar *is a four-tuple*

$$G = (V, X, P, \sigma),$$

where

(i) V *is a finite nonempty set: the* total vocabulary.

(ii) $X \subset V$: *the set of* terminal symbols. *(We shall be interested in languages over X, and may think of $V - X$ as comprising the grammatical symbols.)*

(iii) *P is a finite set of ordered pairs (u, v) with u in $(V - X)^* - \Lambda$ and v*

in V^*. *We usually write* (u, v) *as* $u \rightarrow v$ *and call it a* production *or* rewriting
rule (*note that* v *may be empty*).

(iv) $\sigma \in V - X$: *the* initial symbol.

Given a grammar G, we write $y \Rightarrow z$, and say y *directly generates* z, if y
and z are words on V for which we can find u, u_1, u_2, and v such that (u, v)
is in P and $y = u_1uu_2$ and $z = u_1vu_2$.

We use $\Rightarrow$ * to denote the transitive closure of $\Rightarrow$; that is, if y and z are
in V^*, $y \Rightarrow^* z$ just in case $y = z$, or there is a sequence $z_1, z_2, \ldots, z_k$ in
V^* such that $y = z_1 \Rightarrow z_2 \Rightarrow z_3 \Rightarrow \cdots \Rightarrow z_{k-1} \Rightarrow z_k = z$. We call such a
sequence $z_1, \ldots, z_k$ a *derivation* or *generation* of z from y (by the rules of
the grammar G).

Definition. $L \subset X^*$ *is called a* phrase structure language *if there exists a*
phrase structure grammar $G = (V, X, P, \sigma)$ *such that*

$$L = L(G) = \{w \in X^* \mid \sigma \Rightarrow^* w\}.$$

The two best known classes of phrase structure languages are the context-
sensitive and context-free languages.

Definition. *A* context-sensitive grammar *is a phrase-structure grammar for*
which each production (u, v) *satisfies* $l(u) \leqslant l(v)$. L *is a* context-sensitive language *if* $L = L(G)$ *for some context-sensitive grammar* G.

Fact. No context-sensitive language contains Λ.

EXAMPLE: The set $\{a^n b^n c^n \mid n \geqslant 1\}$ is context sensitive. (Why is $\{a^n b^n c^n \mid n \geqslant 0\}$ *not* context sensitive?)

PROOF: Let us use $V - X = \{\sigma, \sigma_1, B, B'\}$ with initial symbol σ,

$$X = \{a, b, c\}$$

and productions

$$\left.\begin{array}{l} \sigma \longrightarrow a\sigma_1B' \\ \sigma \longrightarrow aB' \\ \sigma_1 \longrightarrow a\sigma_1B \\ \sigma_1 \longrightarrow aB \end{array}\right\} \text{produce strings of the form } a^nB^{n-1}B', \quad n \geqslant 1,$$

$$\left.\begin{array}{l} B' \longrightarrow bc \\ Bb \longrightarrow bB \\ Bc \longrightarrow bcc \end{array}\right\} \text{convert } a^nB^{n-1}B' \text{ to } a^nb^nc^n.$$

Sample derivations: $\sigma \Rightarrow aB' \Rightarrow abc$.

$\sigma \Rightarrow a\sigma_1B' \Rightarrow aa\sigma_1BB' \Rightarrow aaaBBB' \Rightarrow aaaBBbc \Rightarrow aaaBbBc$
$\Rightarrow aaabBBc \Rightarrow aaabBbcc \Rightarrow aaabbBcc \Rightarrow aaabbbccc$.

To make clear the name, "context sensitive," consider a grammar which is context sensitive in the strict sense that every production is of the form $\varphi u \psi \longrightarrow \varphi v \psi$, where $\varphi \in V^*$, $\psi \in V^*$, $u \in V - X$, and $v \in V^+$; that is, each production replaces a single letter by a nonnull string, but the replacement may depend on the context. Let us see that *any* context-sensitive production can be replaced by a series of strictly context-sensitive productions.

Given $x_1 \cdots x_n \longrightarrow x_1' \cdots x_m'$ with $m \geqslant n$, we introduce n new nonterminal symbols $A_n, \ldots, A_1$ and replace the above production by $2n$ strictly context-sensitive productions:

$$x_{n-1}x_n \longrightarrow x_{n-1}A_n x'_{n+1} \cdots x_m',$$
$$x_{j-1}x_j A_{j+1} \longrightarrow x_{j-1}A_j A_{j+1}, \qquad 1 < j < n,$$
$$A_{j-1}A_j \longrightarrow A_{j-1}x_j', \qquad 1 < j \leqslant n.$$

Theorem. *For every context-sensitive language there exists a decision procedure to test membership in it.*

PROOF OUTLINE: Given v we use the length condition to see that there are only finitely many derivations which yield distinct strings of length $\leqslant l(v)$ and we check all of these effectively to see if v is itself derivable.

The notion of context in this discussion refers to the context of a single *letter. This motivates the following definition.*

Definition. *A* context-free (CF) grammar *is a phrase structure grammar in which each production $u \longrightarrow v$ has the property $l(u) = 1$; that is, each rule has the form $u \longrightarrow v$ with $u \in V - X$, $v \in V^*$.*

L is a context-free language (CF language; CFL) *if $L = L(G)$ for some CF grammar G.*

EXAMPLE: If $X = \{a, b\}$, $V = \{a, b, \sigma\}$, $P = \{\sigma \longrightarrow a\sigma b, \ \sigma \longrightarrow ab\}$, then $L(G) = \{a^n b^n \mid n \geqslant 1\}$, which is thus a CFL.

Unfortunately, it is not usually possible to obtain so explicit a description of a language defined implicitly by a CF grammar.

Fact. Λ may be in a CFL, but L context free implies $L - \{\Lambda\}$ context sensitive.

EXAMPLE: Let $G = (V, X, P, \sigma)$ with $X = \{a, b, c, d\}$, $V = \{\sigma, \alpha, \beta, \gamma, a, b, c, d\}$ and $P = \{\sigma \longrightarrow a\alpha, \alpha \longrightarrow \beta\alpha, \alpha \longrightarrow \beta, \beta \longrightarrow dc, \beta \longrightarrow bb\gamma, \gamma \longrightarrow cc\}$. Then the derivation $\sigma \Rightarrow a\alpha \Rightarrow a\beta\alpha \Rightarrow a\beta\beta \Rightarrow a\beta dc \Rightarrow abb\gamma dc \Rightarrow abbccdc$ may be "graphed" by a tree known as the *derivation tree* in which the word generated is given by the terminal modes, read from left to right as in Fig. 23, a structure reminiscent of a parsing tree.

By looking at derivations in terms of these trees, we may obtain the following result known as the *pumping lemma* for context-free languages.

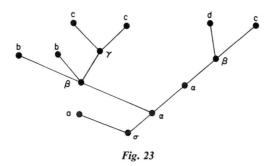

Fig. 23

Theorem [Bar-Hillel, Perles, and Shamir, 1961]. *Let $G = (V, X, P, \sigma)$ be a CF grammar. It is possible to determine two integers p and q such that every sentence z of $L(G)$ with $l(z) > p$ can be decomposed into the form $z = xuwvy$, where $u \neq \Lambda$ or $v \neq \Lambda$, $l(uwv) \leqslant q$ and $z_k = xu^k wv^k y \in L(G)$ for $k = 1, 2, \ldots$.*

PROOF: We may assume without loss of generality $l(x) \geqslant 2$ whenever some $\xi \to x$ is in P. Let n be the number of symbols in $V - X$, and let p be the length of the longest string that can be derived with a tree of height at most n. Then every string z of length greater than p must have some symbol W of $V - X$ occurring at least twice in a branch of its derivation tree, say $W \Rightarrow^* u'Wv' \Rightarrow^* uwv$ where $u' \Rightarrow^* u$, $W \Rightarrow^* w$ and $v' \Rightarrow^* v$. By our initial assumption, u or v is nonnull. Thus, $z = xuwvy$ and clearly $W \Rightarrow^* (u')^k W(v')^k \Rightarrow^* u^k wv^k$. So $z_k = xu^k wv^k y \in L(G)$. The bound q is obtained by noting that we may take $W \Rightarrow^* u'Wv' \Rightarrow^* uwv$ to have total branch length $\leqslant n + 1$.

We can use this theorem to verify that $\{a^n b^n c^n \mid n \geqslant 1\}$, which we have seen to be context sensitive, is not a CFL. Show that $\{ww^R \mid w \in X^*\}$ is a CFL, whereas $\{ww \mid w \in X^*\}$ is context sensitive but not a CFL.

Fact. *The class of CFL's is not closed under intersection.*

PROOF: $L_1 = \{a^i b^j c^j \mid i, j \geqslant 1\}$ and $L_2 = \{a^i b^i c^j \mid i, j \geqslant 1\}$ are both CFLs, but $L_1 \cap L_2$ is *not* a CFL.

We say a subset R of X^* is *regular* if it is *accepted* by some finite-state automaton, that is, if there exists a next-state function $\delta: Q \times X \to Q$, with Q finite, a given *initial* state $q_0 \in Q$, and designated set of *final* states $F \subset Q$ such that R is the set of strings sending the machine from q_0 to a state in F:

$$R = \{x \in X^* \mid \sigma(q_0, x) \in F\}.$$

We may also characterize the regular sets as CFL's of a particular kind.

Definition. *A CF grammar* $G = (V, X, P, \sigma)$ *is said to be* right linear *if each production in P is of the form* $\xi \longrightarrow u$ *or* $\xi \longrightarrow u\alpha$ *where u is in* X^* *and* α *is in* $V - X$.

Theorem. *A set* $R \subset X^*$ *is regular if and only if there exists some right-linear grammar G such that* $R = L(G)$.

PROOF: Let $M = (Q, X, \delta, p_0, F)$ be a finite acceptor for R (that is, $R = \{x \in X^* \mid \delta(q_0, x) \in F\}$), and let $G = (V, X, P, \sigma)$ where $V = X \cup Q \cup \{\sigma\}$ and let P contain

    (i)   a production $\sigma \longrightarrow q_0$;

    (ii) and for each $q, q' \in Q$ and $x \in X$ such that $\delta(q, x) = q$ a production $q \longrightarrow xq'$;

    (iii) and for each q in F a production $q \longrightarrow \Lambda$.

Then $L(G) = R$. A somewhat modified reversal of this construction yields the converse.

D. Tessellation Automata

Let us now give a few results which tie all this in with the arrays of finite automata which occupy our attention in Sections III and V. We recall from Section III that an *automaton of degree d* is simply a finite automaton with state set Q for which the next-state function takes the form $\delta: Q \times Q^d \to Q$, as if the input was obtained by observing d automata with the same state set. We also saw how, given any graph with at most d edges per vertex, we could insert a copy of a given automaton of degree d at each vertex to obtain a network. There we considered synchronization in finite, but otherwise arbitrary, graphs of degree d. Now we wish to introduce the study of graphs that are infinite, but extremely regular. We consider graphs the vertices of which are the points with integer coordinates in some n-dimensional Euclidean space, that is, we consider sets $\mathbb{Z}^n = \{(z_1, \ldots, z_n) \mid z_j \in \mathbb{Z}, 1 \leqslant j \leqslant n\}$, $\mathbb{Z}$ being the set of all integers. ($\mathbb{Z}^n$ may be replaced by any Abelian group, but the easy generalization need not concern us here.) Then the edges are so disposed that any cell has the same edges, as far as *relative* position goes, that it would have anywhere in the graph.

Definition. *A* tessellation automation *is given by specifying*

    (i) *A graph G with set of vertices* $\mathbb{Z}^n$ *for some n, and a finite subset* $\langle g_1, \ldots, g_d \rangle$ *of* $\mathbb{Z}^n$ *called the* template *such that an edge leads from* z_1 *to* z_2 *in G if and only if* $z_1 - z_2$ *is in the template.*

    (ii) *At each vertex of G there is a copy of a given automaton of degree d; with the jth input to the automaton at z being supplied by the state of the automaton at* $z + g_j$.

We require that the automaton have a quiescent state q_0 *such that a quiescent cell will remain quiescent if all its neighbors are quiescent:* $\delta(q_0, q_0, \ldots, q_0) = q_0$.

Thus, a linear array in which each cell receives inputs from its neighbors has underlying graph $\mathbb{Z}$ with template $\langle -1, 1 \rangle$, while a two-dimensional array in which a cell receives inputs from the adjacent cells along the axis directions has $\mathbb{Z}^2$ with template $\langle (-1, 0), (0, 1), (1, 0), (0, -1) \rangle$.

It is usual to graph a tessellation with cells occupying squares rather than points; in Fig. 24 we have sketched a fragment of a two-dimensional array, and also shown the template which is such that if the crosshatched square is placed over any cell, then the other cells of the template will be over the cells from which that cell receives its input.

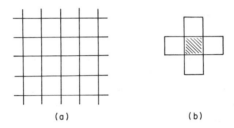

(a)   (b)

Fig. 24. (a) Fragment of array; (b) template.

Let us check our understanding of these ideas by observing the following trivial fact:

Fact. Given any Turing machine Z, there exists an automaton M_Z of degree 2 which, if inserted at the nodes of the tessellation $\{\mathbb{Z}, \langle -1, 1 \rangle\}$, will simulate Z.

PROOF: We let M_Z have two halves: the top half corresponds to Z's control head, either in some state or unactivated, while the bottom half corresponds to a square of Z's tape, possibly blank. Then if at any time Z is in state q scanning x_j of the string $x_1 \cdots x_n$ of symbols, the tessellation will have all but n squares blank, and those will have $x_1, \ldots, x_n$ on their "tape halves" while all but the jth will have unactivated state halves, the jth having q in its state half, as shown in Fig. 25.

It is left as an exercise to the reader to write out how a quintuple $qxx'mq'$ is to be reflected in the next-state function $\delta \colon \hat{Q} \times \hat{Q}^2 \to \hat{Q}$ where $\hat{Q} = (Q \cup \{b\}) \times X$ is the state set of M_2.

[*Hint:* A cell does not change state unless it or its immediate neighbor has an activated "state half."]

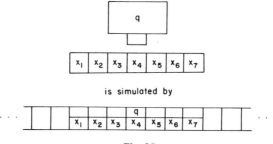

Fig. 25

Corollary. *Given a tessellation and an initial pattern it is undecidable whether or not a "steady state" (that is, no further change of state in any cell) will be reached.*

PROOF: Just consider the embedding of a Turing machine with an unsolvable halting problem.

Just as we saw that any Turing machine may be simulated by a semi-Thue system, so may a one-dimensional tessellation be simulated by a suitable grammar. The study of multidimensional tessellations thus should spur a study of multidimensional grammars, a study initiated in the context of theoretical embryology by Arbib and Laing [1972], following the lead given by Lindenmayer [1968a, b, 1971] in using conventional linear grammars to study plant growth.

V.   Automata Which Construct as Well as Compute†

A. Self-Reproducing Automata

Von Neumann [1951] noted that when machines built other machines, there was a degradation in complexity (an assembly line is more complicated than what it produces), whereas the offspring of an animal seems generally to be at least as complex as the parent, with complexity increasing in the long range of evolution. Von Neumann asked whether there was an immutable difference here, or whether one could design "self-reproducing machines" which could produce other machines without a degradation in complexity.

Such a question suggests that we study automata theory, not from that aspect emphasized in Section IV, namely processing of input sequences to yield outputs, but rather a more constructional approach in which we study

†Much of this material appeared in Chapter 10 of Michael A. Arbib, *Theories of Abstract Automata* © 1969. Adapted by permission of Prentice-Hall, Inc., Englewood Cliffs, New Jersey.

how information can be read out and processed to control the growth and change in structure of an automaton.

When we have an adaptive network which is changing certain parameters with time, it may *not* be useful to make the distinction between growth processes and parameter adjustment processes. If we design a network to carry out a computation, there are at least two ways to build it. One is to connect each component to all other components, and then adjust the coupling coefficients through computation. One then might claim that information-processing did not change structure, even though in some cases it would lead to a zero coupling coefficient which would be equivalent to no connection at all (see Section III, Chapter 3, Volume III of this treatise). Alternatively, one might use a process which starts with very few connections. Then if we compute that one component should, in fact, influence another which it has not before, we may use this to "grow" a connection between the two components. Thus, as we come to understand in more detail how genes help to control embryological development, we may hope to be able to understand, by extrapolation of those processes, something of the mechanisms that underlie learning and memory.

To return to von Neumann's question, let us see how we might describe, within the context of Turing machines, a complicated machine which can produce something as complicated as itself. In the next section we shall analyze some of the many ways in which the reproduction of such an automaton differs from biological reproduction. However, for now we shall study "pure" automata theory with little explicit concern for its applicability.

Let us first observe that the question: "Is there a machine which, if set loose in a component-rich environment, will form components into a copy of itself?" can be made trivial, as in the "domino example" in Fig. 26.

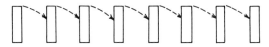

Fig. 26. Trivial self-reproduction. A domino on edge is the basic component. We stand dominoes in a chain, as shown, and let the automaton we want to reproduce be a *falling* domino. A falling domino knocks down its neighbor, and thus "reproduces." *Falling* is propagated down the chain.

Motivated by the fact that in human reproduction a single cell, containing a large but nonetheless finite number of ongoing chemical reactions, can develop into an aggregate with much greater capabilities, let us consider the following artificial, but at least automata-theoretic, question: "How, starting from one fixed kind of finite automaton as basic cell and given any Turing machine, can we design an automaton able to simulate that Turing machine, which also can reproduce itself?" If the question is phrased in this way, we can avoid the domino objection.

To complete the formalization of this (admittedly nonembryological) problem, consider an infinite "chess board," with each square either empty or containing a single component. Each component can be in one of various states, and we think of an organism as represented by a group of cells, collected together somewhere in the plane (see Fig. 27). We are thus talking of regions in a tessellation automaton.

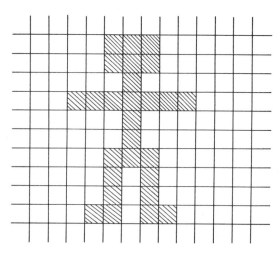

Fig. 27. An "organism" of active (hatched) cells embedded in a chessboard array (see Fig. 24).

A mathematical simplification: We have said that any square of the board may be empty or contain some component, say of type C_j ($1 \leqslant j \leqslant N$), in some state, say q_i. We may lump these $N + 1$ alternatives into one super component C, which has one more state than the total number of states of the N components. For mathematical purposes, it is easier to think of there being a copy of one fixed component in every cell, so that rather than study the kinematics of components moving around in the plane we look at a more tractable process of how an array of identical components C, consisting initially of an activated array with the remaining cells in the passive state, passes information to compute and to "construct" new configurations.

Von Neumann [1966] was able to show that with a 29-state cell "super-component," he could set up a simulation of a complex Turing-type machine which, besides being able to carry out computations on its tape, would also be able to "reproduce" itself. The 29 states could be seen as several states corresponding to an OR gate, several states corresponding to an AND gate, several states corresponding to different types of transmission line, and so forth. Von Neumann's proof was not completed at the time of his death, but the manuscript he left was edited by Arthur Burks, and has since been put out as a book called *The Theory of Self-Reproducing Automata* [von Neumann, 1966]. The proof is over 100 pages long. (The price we pay for simple components is a complex program. To take an analogy from computer programming, it is like trying to program in machine language, rather than in an appropriate assembly language. In other terminology, or in biological terms, we might say that it is like trying to understand a complicated organism directly in terms of macromolecules, rather than via the intermediary of cellular structure.)

Turing's result that there exists a universal computing machine suggested to von Neumann that there might be a universal construction machine A, which, when furnished with the suitable description I_N of any appropriate automaton N, will construct a copy of N (see Fig. 28). In what follows, all automata for whose construction we use A will share with A the property of having a place where an instruction I can be inserted. We thus may talk of "inserting a given instruction I into a given automaton."

If the automaton A has description I_A inserted into it, it will proceed to construct a copy of A. However, A is *not* self-reproducing, for A with appended description I_A produces A without I_A; it is as if a cell had split in two with only one of the daughter cells containing the genetic message. Adding a description of I_A to I_A does not help; now $A + I_A + I_{I_A}$ produces $A + I_A$ and we seem to be in danger of an infinite regress. Such a consideration suggested to von Neumann that the correct strategy might involve "duplication of the genetic material." He thus introduced an automaton B that can make a copy of any instruction I with which it is furnished, I being an aggregate of elementary parts, and B just being a "copier" (see Fig. 29). Next C will insert the copy of I into the automaton constructed by A. Finally, C will separate this construction from the system $A + B + C$ and "turn it loose" as an independent entity.

Let us then denote the total aggregate $A + B + C$ by D. In order to function, the aggregate D must have an instruction I inserted into A. Let I_D be the description of D, and let E be D with I_D inserted into A. Then E *is* self-reproductive and no vicious circle is involved, since D exists before we have to define the instructions of I_D.

We thus see that once we can prove the existence of a universal constructor for automata constructed of a given set of components, the logic required to proceed to a self-reproducing automaton is very simple, though there is something somewhat whimsical in the idea of a universal constructor, as if a mother could have offspring of any species, depending only on the father. While this may be appropriate to Greek myths, it does seem inappropriate

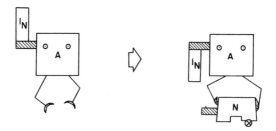

Fig. 28. Fanciful description of universal constructor *A*.

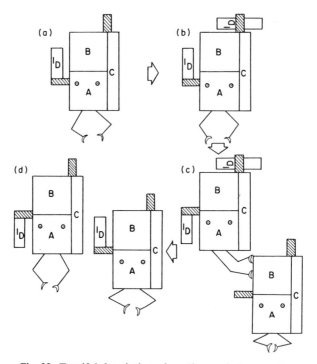

Fig. 29. Fanciful description of a self-reproducing machine.

to biological modeling—unless we think of experiments in which cytoplasm plays host to a transplanted nucleus. It would seem that animals of the same species are such that (the cytoplasm is compatible for readout of the code of any member and) the "construction program" is structured in the same way for all members at the level which treats the subroutines as single constructions. Members of the species then seem to differ in the details of these subroutines. With the added mechanism of dominance, we see why sexual recombination always yields valid programs within a species, but may yield meaningless programs between species when two different program structures are paired in what is computationally a random fashion. This also suggests why the notion of a universal constructor seems so whimsical *biologically*, though perhaps cytoplasm may be a universal constructor for the combined genetic message of mother and father; however, I suspect that cytoplasm itself has more evolutionary memory than we normally ascribe to it.

Our concern now is to examine the difficulties involved in actually providing a universal constructor. Von Neumann did not do this in his original paper [1951] and the task involves over a hundred pages of his book [1966]. The problem is essentially this. A Turing machine is only required to carry out logical manipulations on its tape, sensing symbols, moving the tape, printing symbols and carrying out elementary logical operations. A universal *computer* only has to carry out the same operations, but a universal *constructor* must also be able to recognize components, move them around, manipulate them, join them together. Thus, presumably, constructors of Turing machines require more components than do Turing machines themselves. We are immediately confronted with the possibility of another infinite regress. Given a set of components $\mathscr{C}_1$, to construct machines which build all the automata made of components from $\mathscr{C}_1$, we may need a bigger set of components $\mathscr{C}_2$. To build all machines constructed of components from $\mathscr{C}_2$, we may need machines put together from a bigger set of components $\mathscr{C}_3$. The question is: "Is there a fixed point? Can we find a set of components $\mathscr{C}$ such that all automata built from components of $\mathscr{C}$ can be constructed by automata built from the same set $\mathscr{C}$?" I have called this [Arbib, 1967b] *the fixed-point problem for components*. This is the fundamental problem in the theory of self-reproducing automata. Once we have found a set of components $\mathscr{C}$ in which for each automaton A, there can be found an automaton $c(A)$ which constructs A, it turns out to be a fairly routine matter to prove the existence of a universal constructor. We then know from von Neumann that it is a simple matter to prove *the construction fixed-point theorem*, namely that there exists a self-reproducing machine U which can construct a copy of U. There have been several procedures following von Neumann's to exhibit a set of components that satisfy the component fixed-point theorem. Von Neumann [1966] used 29-state components and gave an elaborate construction taking about 100 pages. Thatcher [1965] used the same 29-state components as von Neumann, but gave a more elegant construction of perhaps half the length. Codd [1965], with remarkable ingenuity and interaction with a computer, showed that a construction similar to von Neumann's could go through using components with only eight states. Arbib [1966b,

1967a] showed that the construction could be done with great simplicity, in a matter of eight pages, if one allowed the use of much more complicated components. My rationalization for this use of complex components was that if one wishes to understand complex organisms, one should adopt a hierarchical approach, seeing how the organism is built up from cells, rather than from macromolecules.

Myhill [1964] (see Arbib, [1969b, Section 10.5] for an alternative exposition) has given an axiomatic theory of self-reproduction, but the axioms are so formulated that they cannot be directly applied to different sets of components, but rather allow one to generate theorems about self-reproduction when one already has theorems about universal constructors. Intriguingly, Myhill shows that results in recursive function theory can lead to the startling (though biologically inapplicable) conclusion that infinite programs can yield unbounded improvements in successive generations of offspring without requiring any randomness in the mutations.

Rather than go into any details of my construction I shall just briefly present two pictures which give some idea of the basic notions involved. We are to imagine a CT machine (Construction and Turing machine) which under the control of a program in its logic box can read and write on a one-dimensional tape in just the way a Turing machine does, and which can write but not read on a two-dimensional tape. The idea is that the two-dimensional tape is to be thought of as a construction area, and the writing of a symbol is to be thought of as equivalent to the placing of a component. Our task is to find a set of components from which we can build tape, logic box, and

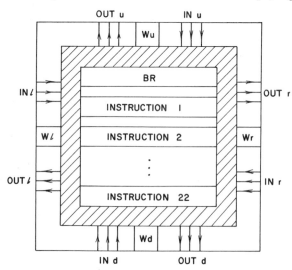

Fig. 30. The basic module.

construction area. Such a component as shown in Fig. 30 is a finite-state module that can contain up to 22 instructions from a rather limited instruction set. We are to think of a two-dimensional plane in which these cells are embedded in the manner of Figs. 24 and 27. An automaton is then represented as an activated configuration of these cells. I might mention that the little boxes marked W are *weld registers* which serve to "weld" squares together so that a number of squares may be "welded" into a one-dimensional tape in such a way that when any one square of that tape is instructed to move, all cells will move in the indicated direction. The assumption of such *a weld operation greatly simplifies our programming.*

In Fig. 31 we see an overall plan of an embedded CT machine. We see that the logic box has been broken into two pieces, a one-dimensional tape which contains the program and two cells which form a control head. The idea is that on activation by the control head squares of the program tape may either be used to guide the control head in manipulating the computation tape in a Turing machine fashion or else may be used to place selected components in the constructing area and move welded blocks of components around. Arbib [1969b, Chapter 10] shows that in fact an instruction code can be specified for our basic modules so that it is not only possible to embed arbitrary Turing machines in the array of those components in the indicated fashion but in fact to program these machines so that they can construct other such machines in the construction area and to go on from there to show that there exists a universal constructor made of these components. It is then a standard procedure, following von Neumann's argument, to present an actual self-reproducing machine.

Smith [1968] has shown that Lee's results on self-description of Turing machines (see Section IV) may be used to construct self-reproducing configurations in a manner far simpler than that just outlined. We have already seen that for any Turing machine Z we can define a one-dimensional cellular space $\tilde{Z}$ which simulates Z in real time, with the state of each cell encoding

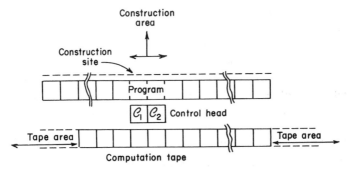

Fig. 31. Overall plan of embedded CT machine.

a state-symbol pair of Z, having introduced a new "null state." For such a space, we say that Z is wired into $\tilde{Z}$, or that $\tilde{Z}$ has Z wired in.

Let the wired-in computer be a universal Turing machine U. We use the notation $x \xrightarrow{P} y$ to mean that Turing machine program P acting on initial tape x halts with y on the tape as its final result, so that we have

$$f(P'), x' \xrightarrow{U} y',$$

where f is the encoding function required by U and $x' \xrightarrow{P'} y'$. We require a program P such that

$$x \underset{\uparrow}{\xrightarrow{P}} (f(P, x), y, f(P, x)) \underset{\uparrow}{\xrightarrow{P}} \cdots,$$

where y is the result of some computation on x.

Lemma [Smith, 1968; see Lee, 1963]. *For an arbitrary one-to-one recursive function f from (program, tape) pairs to tapes and for an arbitrary total recursive function g such that $g(x) = y$, there exists a self-describing machine with program P such that*

$$x \underset{\uparrow}{\xrightarrow{P}} (f(P, x), y, f(P, x)).$$

PROOF: Define the function h from programs to programs such that $h(Q)$ is the first program which reads arbitrary input tape x, encodes program Q and tape x by given function f to get $f(Q, x)$, prints $f(Q, x)$, computes $g(x)$ to get y, prints y, prints $f(Q, x)$ again [either by reencoding $f(Q, x)$ or by copying the result of the first encoding], and finally moves the head to the leftmost symbol in the rightmost encoding $f(Q, x)$. That is,

$$x \underset{\uparrow}{\xrightarrow{h(Q)}} (f(Q, x), y, (f(Q, x)).$$

Clearly h can be chosen total recursive. Thus, by the recursion theorem, there exists P which is a computational fixed point of h such that

$$x \underset{\uparrow}{\xrightarrow{P}} (f(P, x), y, f(P, x)).$$

Thus, in cellular space $\tilde{U}$ with U wired in, the following situation can hold:

$$f(P, x) \xrightarrow{U} (f(P, x), y, f(P, x))$$

at some time $T > 0$. If we decree that $U_{(,)}$ is U so modified that on completing such a computation it will backtrack to the rightmost comma, and start anew without reading or writing to the left of that comma, we will obtain

$$f(P, x) \xrightarrow{U_{(,)}} (f(P, x), y, f(P, x), y, f(P, x))$$

at some later time $T' > T$ and so forth. Hence, we have shown the following:

Theorem. *Let $\tilde{U}_{(,)}$ be a computation-universal cellular space with a modified universal Turing machine $U_{(,)}$ wired in. Then there exists a configuration* *c in Z_U which is self-reproducing and computes any given total recursive function g.*

This result puts us on guard against too hasty acceptance of the identity between a biological reality and any mathematical formalization. Whereas Smith's formalization makes us uneasy because self-reproduction becomes too simple for us to believe that we have captured the true complexity of the biological situation, Rosen [1960] has presented another formalism in which self-reproduction is impossible. Essentially, he notes that if a system is specified by a list of all the input–output pairs that could arise in its function, then a self-reproducing automaton would occur as one of its outputs; but the list describing the automaton cannot be given without specifying all the outputs, and at least one of the outputs (namely the "offspring") cannot be specified until we have completed the description of the original system. This paradox invalidates self-reproduction *in Rosen's system.*

The reason such a paradox does not arise to destroy von Neumann's, and our, construction of a self-reproducing machine is that we only considered machines that can be described in terms of finite arrays of components. We then manipulated these *finite* descriptions, and then proved our fixed-point theorems to avoid any infinite regress. Zeno's paradox arose because of the attempt to equate a finite length of $1\frac{1}{9}$ with its infinite description in terms of the series $1 + \frac{1}{10} + \frac{1}{100} + \frac{1}{1000} + \cdots$. The reader may find it useful to reread our proof of the recursion theorem in Section IV, and note the essential use we made of the availability of finite programs to describe our functions; had we to use an infinite set of pairs to describe the function, Rosen's paradox would vitiate this result too.

With this reminder that the domain of application of any formalism has its limitations, let us pay more attention to biology.

B. Modified Modules for Greater Biological Relevance

We may summarize von Neumann's result as saying: "In a tessellation we may replace every cell by copies of a single finite automaton (one state of which is passive, and from which an automaton can only be shifted by having nonpassive neighbors) such that finite arrays may be so actuated that they can not only compute as universal Turing machines, but also construct replicas of themselves (with these same properties) elsewhere in the tessellation."

This meets R. D. Hotchkiss' definition [see Gerard, 1958, p. 109] of life as *the repetitive production of ordered heterogeneity*, and thus challenges us to refine that definition, avoiding the simplicities of "common sense" discourse in our theoretical biology. Buckingham [personal communication, September 6, 1968] has suggested an even simpler system which meets this criterion, and which may be rephrased in terms of our domino example of Section V.A by considering a space consisting of n, rather than one, rows of dominoes, and then noting that 2^n different patterns of falling dominoes may be reproduced. What our model adds is that the units of the reproducing entity are *not* independent, but rather can *behave* together in effecting complex computations. [Biological systems further possess some ability of flexible adaptation; that is, the cells can *change* together. They endure (exhibiting growth, learning, and homeostasis) and they multiply.]

However, in a space (such as ours of Section V.A) in which one can countenance the welded "move" instruction, why not just allow a welded "move and divide" instruction, which allows a two-dimensional array to continue function at one value of the third ordinate, while producing a functioning copy at the adjacent value of the third ordinate? This "Xerox" approach to reproduction seems to be cheating. But why? Is it because it would not work for a three-dimensional array in three dimensions, or is there another reason? Note that DNA *may* be considered to reproduce by this method, so perhaps this points to a hierarchy of levels of reproduction with higher levels associated with increased *complexity* but decreased *fidelity*.

Our self-reproducing automata do share with living systems the property that initial information serves to regulate their *growth* and *change in structure*, but our construction ignores the role of external factors which bias development and affect evolution.

For example, it does seem that an organism is more highly ordered than its environment and maintains a boundary—be it the membrane of the cell, the shell or outer covering of an egg, or the skin of an adult—which demarcates a discontinuity of order. Our self-reproducing automata have no such boundary, and there is nothing in the theory which would predict that much of the growth of biological reproduction takes place within the parent's boundary. We could "program in" such a feature, but then we would not be explaining anything. This suggests that a true theory of reproduction should have reproduction deduced as a property of systems trying to maintain order in a less ordered environment, rather than an *ad hoc* postulation of reproduction as the design criterion for the constructs of our theory. Again, as we emphasized in Section I, genetic commands are not absolute commands (grow a leg 32 cm long) but rather serve to adjust settings in a hierarchical control system, so that the growing system is in a delicate balance with its environment, and *we must search for a theory in which the command hierarchies of development can be seen to be necessary consequences of as well as prerequisites for those required for the maintenance of the adult.*

Mammalian growth starts from a single egg, which, fertilized and properly nourished, can grow by repeated cell division and pattern formation into an adult. Instead, our automaton is formed as if a mother had started off by building laboriously a little toe, and then a foot to attach that on, and then putting the other toes on, and then working up the leg, and so on, until finally, as she put the skin in place over the skull, she was able to give her child to the world. How then can we transform this automata theory of a completely serial process, transforming one component at a time, to obtain a highly parallel process where all cells of a collection are possible growth points, so interacting that each cell, even though it only interacts with nearby cells, will make an appropriate contribution to the organism? Here is perhaps the crucial problem for theoretical embryology:

How can a single cell in a continually active tissue, where all cells may be growing, dividing, differentiating, or dying, use only information from nearby cells to so grow as to contribute properly to the overall form of the organism?

It is in getting some feel for the ways in which such local interactions govern a global process that we start moving toward a better understanding of the biological system. Section V.A showed how a multicellular organism could construct a copy of itself, but did not answer the question, "How can a complex multicellular automaton grow from a *single* cell?" In a sense, it is only the cell which reproduces "itself" in the two stages of growth and division. Multicellular organisms produce zygotes, "models" which independently grow and develop to be similar to the parents. A human zygote grows into a human, not into a replica of one of the parents. This leads to a whole series of questions—which here I can merely raise—of the "identity" of an automaton. How may we handle a whole complex of environmental interactions, so that we may see an organism as a "node in a structure of dynamically interacting hierarchies?" What does it mean to say that two automata belong to the same "species?" Embryological reproduction gives rise to offspring with similar structure and this implies similarity of function. What are measures of structural similarity and functional similarity, such that the first implies the second? Are there interesting classes of automata for which we may carry out *functional* decompositions into a species-dependent automaton, and an individuality-expressing automaton? Can we study reliability of reproduction for such systems, where the genetic information determines the species-dependent automaton with high probability, random influences making their appearance chiefly in modifying the individuality-expressing automaton?

Basic to the problems of replication (perhaps a better term than self-reproduction in view of our above remarks) are:

1. *Cell Reproduction.* This we have touched on. Our construction rests on the assumption that we can produce new cells at will, our only problem being to ensure that they contain the proper instructions.

2. *Organism Replication.* Given cell reproduction, how do we replicate an organism? This is the topic we treat and it makes sense to subsume a lot of hard work in 1 by using complicated cells. The living cell, with its synthetic machinery involving hundreds of metabolic pathways, can rival any operation of our module, as well as being under the control of DNA molecules, we believe, with far more bits of information than our module stores. So it may be relevant that our model was easier to "grow" with larger cells than von Neumann's 29-state elements. In fact, we might hypothesize (see the discussion of Simon [1962, Section 4.4]) that "sophisticated" organisms can evolve (by whatever mechanism) only when there are complicated reproducing cells available.

Contrasting our model with organism reproduction, we note that

(1) our program was embedded in a string of cells, whereas the biological program is a string stored in *each* cell;

(2) we use a complete specification, whereas "nature uses" an incomplete specification;

(3) we did not use anything like the full power of our model (that is, the operation was sequential instead of parallel);

(4) we constructed a passive configuration—we set up all the cells with their internal program, and only then did we activate the machine by telling the control head to execute its first instruction. Contrast the living, growing embryo. Our construction *relied* on the passivity of the components and demanded that any subassembly would stay fixed and inactive until the whole structure was complete. The biological development depends on active interaction and induction between subassemblies.

The question of how reproducing cells evolved in the first place is somewhat outside the scope of the present paper, but should be borne in mind. Codd [1965] considers tessellations with even simpler components than von Neumann's. A pure automata problem is to embed our module in Codd's model, where one of our cells is simulated as an aggregate of Codd's cells with appropriate change of time scale [Holland, personal communication, 1966]. Perhaps we can approach the cellular evolution problem by imagining a subtessellation with components comparable to the macromolecules of biology, and consider reproduction of our modules as aggregates of these pseudomacromolecules. Our constructions would then treat arrays of arrays.

In any case, our distinction between cell reproduction and organism replication suggests that we need a careful classification (or at least relating) of different types of processes of reproduction, in which even DNA replication and Xerography find a place.

Let us now give a modified module model in which each cell contains the whole control string, and examine some of its advantages and disad-

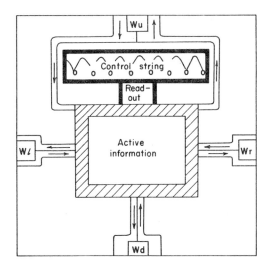

Fig. 32. Mark II module.

vantages. The Mark II module is shown in Fig. 32. We have kept the tessellation structure, sidestepping the morphogenesis of individual cells. The control string is segmented in words which correspond to the possible internal instructions of the original module. The whole control string corresponds to the whole program in our original model. Only a small portion of the control string can be read by an individual cell. *Every cell in an organism has the same control string.* Individual cells differ only in the portion of the control string which can currently be read out. *The change in activation of portions of the control string is our analog of differentiation. The increase in the number of cells in a comoving set is our analog of growth.*

Of course, even these notions are too simplistic. *Differentiation,* biologically, is accompanied by structural change, so that cells may differ despite activation of the same control string due to differing histories of activation due to their differing positions within the cell mass. Similarly, biological cells grow individually as well as increase in number.

A cell computes under the control of its activated control string portion and of its inputs. Thus a cell which functions one way as part of a comoving set may well behave completely differently when removed from the aggregate, a phenomenon familiar in embryological experiments.

Besides computation on inputs to produce outputs, a cell can divide (this corresponds to the activity in the construction site in the original model) or die (this corresponds to returning a cell to the quiescent state). (To obtain sexuality one simply requires a means for two cells to redistribute the instruction of their control strings. We shall be somewhat more insightful about this in Section VI.) Thus, we are really considering the quiescent cell of Mark I (that is, the cell of Fig. 30) as a noncell in the Mark II model, and thus our tessellation is to be thought of as empty save where there are comoving sets. These masses grow only when individual cells divide.

The "active information" region corresponds to the bit register and 22 instruction registers of the Mark I module. It has access to the activated portion of the control string.

When a cell divides, the original cell may be considered as preserved at its original site, while a replica has been produced at a neighboring site. This replica will have the same control string as the original cell, but may have "differentiated" in that the activated region of the control string may be adjacent to, rather than identical with, that of the original cell.

Given an automaton A in the Mark I model, we may replace it by a Mark II automaton simply by replacing each cell by a Mark II cell. We then grow A from a single cell by encapsulating the program of the constructor of A within a single cell $\hat{c}(A)$. This single cell grows into the Mark II version of $c(A)$ simply by "reading out" the control string into a linear string of cells, the last two (leftmost) of which then produce a control head. This proceeds to build a copy of A.

This procedure has several fascinating concomitants. First, the problem of self-reproduction becomes very simple in this model. We simply require that $\hat{c}(A)$ initially produces a copy of itself (a single cell) that is rendered dormant. Then $\hat{c}(A)$ produces $c(A)$. In turn, $c(A)$ builds A, but before turning A loose, attaches to it the "germ cell," that is, the copy of $\hat{c}(A)$. Then A reproduces by releasing this copy of $\hat{c}(A)$ into the tessellation at "maturity."

Second, in this model, the control string is not that of the reproducing automaton A itself, but rather that of the constructor of A.

At this stage, I would not push the latter observation as a clue to the operation of real biological systems. Rather, its purpose is to shock the biologist, and the theoretician, into a reappraisal of how devious genetic coding may be. Our progress in decoding the fashion in which DNA directs the production of enzymes has been so impressive of late that we tend to forget how little we know about the way this relates to overall cell function, let alone the morphogenesis of the multicellular organism. I should also emphasize that we have considered the basic units of our control string as explicit instructions with respect to cellular computation or reproduction. This must be contrasted with the exceedingly indirect commands in the control string (DNA) of a real cell, which serve to bias overall metabolic activity rather than explicitly modify specific bits of data storage.

Let me reiterate that this model shares with the Mark I model an essentially serial mode of operation. We shall have a far greater feel for the demands we must make on a genetic coding system when we have programmed the Mark II system in a parallel growth mode, in which *every cell of an organism may be considered a growth point.*

In "designing" an organism, we must use subroutines to build "components," that is, tissues with a basically repetitious structure. Higher-level routines then serve to arrange these components in appropriate topographical relationships. This is one way in which we reduce the length of the program required to specify the growth of an organism. Another factor which helps economize the growth program is that *an organism will only grow normally in an appropriate highly structured environment.* Thus, the growth program may rely on "self-organizing mechanisms" to select information out of the environment, thus economizing on the information required in the zygote. That is why both nature and nurture are important.

Let us note then that our Mark II model admirably meets the five "axioms" for development presented by Apter [1966] that establish the desiderata for the models presented both in his book and in this chapter. Our own comments (the "nonquotes") will be placed in square brackets in the rest of this section only.

1. *The unit of development is most conveniently the cell.* Each cell contains one set of genetic instructions, and the nature of the resulting organism or regions or tissue would seem to depend on the nature of these instructions in the cells composing it.
[An instruction only has meaning inasmuch as it is an instruction *to* something. The same string of symbols can mean different things to different computers. In biological terms, information is contained not merely in the genetic "instruction," but also in the cytoplasmic "processors."]

2. *All cells in the organism are genotypically identical.* (There is some evidence [Kroeger, 1963] that in some organisms the genotypes might not be identical, but on the whole the assumption which we have given is one with which few biologists would disagree.)

3. *An organism develops through the self-reproduction of the cells composing it.* Thus one of the outputs of our "black box" should be an identical black box. (Note that there is no formal difference as far as results are concerned between the situation in which a new individual is formed identical to its predecessor which remains in existence, and the situation in which two new individuals are formed which are each identical to each other and to their predecessor which has, however, disappeared).

4. Intercommunication between cells is a prerequisite for coherent development.

5. *An organism controls the important aspects of its own development.* Food and energy deprivation will have certain effects on the size of the organism and its ability to develop normally. But in general the animal controls its own development; that is, it is closed to information but open to energy. [Axioms 1–5 from Apter, 1966, pp. 44–46]

[I do not agree with this. It is true of the models I shall present here that information from the environment is not considered in the development of the automaton, but I regard this as a limitation to be removed, not as the realization of an axiom of development. Wiesel and Hubel

[1963a, b] have shown that *patterned* visual input is crucial for the development of the cat's visual system; and the work on imprinting reviewed, for example, in Thorpe [1957], seems to show the importance of external environment for proper functional development. However, we may agree with Apter that this importance is less in development *per se* than in a learning situation.]

VI.   Evolution and Entropy†

Among the many other questions which our discussion of self-reproducing automata raises, area: "Whence come the components out of which our automata are made?" and "Given that such automata exist, how might one imagine them to evolve?" It is not our purpose in this section to answer these questions—would that we could—but rather to suggest some interesting avenues toward their solution.

Pattee [1968] has suggested that the crucial problem of the origin of life may be reduced to the problem of finding a molecule which is both stable enough to survive but yet not so stable as not to evolve. In our automaton-theoretic approach we tend to neglect these crucial problems of reliability. However, we may note in passing that Cowan and Winograd have erected a theory of reliability in neural nets which proceeds by the judicious application of redundancy, and it seems to me quite conceivable that a similar redundancy scheme might be used to replace the function of one cell in our iterative array by the function of a number of cells in the new array, it further being the case that each cell in the new array shares the properties of several cells of the old array. However, the question still remains of what is the simplest component with which one may hope to build up interesting self-reproducing automata. We pointed out that if the only system one wishes to reproduce is a falling domino, then the problem of self-reproduction becomes very trivial. If, however, one's criterion of interesting self-reproduction is that the organism involved can have computational power akin to that of a universal Turing machine, then the problem becomes nontrivial. It is known that if one embeds one's automaton in a tessellation in which the state of each cell is affected by only the states of its four immediate neighbors, then eight states will suffice to yield computation universality [Codd, 1965]. Codd has shown further that two states are insufficient, but no minimal number of states for such an array has yet been shown. We should note that even an eight-state cell is rather complex in that each transition, depend-

†From "Self-Reproducing Automata—Some Implications for Theoretical Biology" by Michael A. Arbib, in *Towards a Theoretical Biology. 2. Sketches* (C. H. Waddington, ed.), pp. 219–225. 1969 Edinburgh University Press; and Some Comments on Self-Reproducing Systems, in *Systems and Computer Science* (J. F. Hart and S. Takasu, eds.), pp. 51–52. 1967 University of Toronto Press.

ing as it does on the state of the cell and its four neighbors, requires the specification of actions for each of 32,768 possible circumstances. Unfortunately, it is not at all clear how this sort of discussion can give us any insight into the minimal complexity required of the macromolecule that can enter into a process of hereditary evolution. This is especially so, since we cannot be sure that the present DNA system occured at any stage at all near the origin of life. Just as a virus, which appears to be intermediate in complexity between simple molecular systems and the cell, could not have evolved until after cells had evolved to provide an environment for their reproduction, so it may well be possible that DNA could only evolve after various other subsystems, such as those now available in the cytoplasm, were available to help it in its replication. At the present stage it still requires cytoplasm or Kornberg for the replication of active DNA. While on this topic of possible precursors for present replicating systems we would draw the reader's attention to the papers of Pattee [1968] and Cairns-Smith [1968].

Turning now to the broader questions of evolution, once we have already obtained a suitable supply of existent automata from which to evolve, we see that a radically different point of view must be imposed on that of our basic automaton-theoretic approach. We have already mentioned, in Section V.B, the transition to the Mark II module model in which the whole program of the organism is contained in each cell, differentiation serving to activate different portions of the program. To study evolution, one must proceed to the Mark III model in which these internal programs are themselves modifiable (mutable) programs. It will be a matter of some interest to see how necessary stochastic factors are in modeling this situation. We must then study not only what is the simplest program for a self-reproducing organism but also how the program may be complicated to yield "fitter" programs. We need a framework in which we can *compute* the parameters which are usually at best *fitted* in population genetics (the only attempt in the literature of which I know at erecting such a framework is in the work of John Holland; see, for example, Holland [1962]). In such a study we move to a higher level at which we neglect the details of the dynamics of cell division and work instead perhaps with some measures of mitosis intervals, and stable lifetimes of cells. In fact, it is not at all clear to me whether study at this level will depend upon its automaton-theoretic base or will in face become indistinguishable from Fisher-type statistical studies of natural selection. Having set up such basic parameters we might then study free movement in the tesselation and consider for various aggregates what is their reproduction time and stable life time. One would hope to prove that the initial aggregates arise by a transition from symbiosis to functional reproducing organism [compare Cairns-Smith, 1968]. One might expect that aggregates would be better suited to utilize material. One should also be able to prove theorems that parallel operation would be favored, as would information and material transmission channels, to increase the use of materials encountered by the cell. (Crick [1966] has observed that if one removed all the DNA from a single human being and stretched it end to end, it would span the whole solar system—surely a very strong argument for parallel processing in any organism.) We might then ask for theorems which contrast the short generation time of individual cells with the long generation time of aggregates. Simon [1962] presented a rather relevant parable of two watchmakers, one of whom had his watches arranged in a hierarchical organization so that whenever his work was interrupted the only damage done was that his work was put back to its component subsystems, whereas another watchmaker was in the unfortunate position that whenever he put down a watch it dismantled right down to the very smallest components. Clearly, the first watchmaker could get much, much more done than the second watchmaker. With such parables in mind we may hope to proceed to a really rigorous theory of evolution in which we can show the selective advantage of systems which are not only highly parallel but also hierarchical in structure.

Let us turn now to the relation between the origin and evolution of life on the one hand, and the laws of thermodynamics on the other, and see wherein a statistical mechanics may prove useful.† Attempts so far to deduce thermodynamics from mechanics require an *additional hypothesis*, namely one of random phase, and I suspect that living systems are ones for which random phase hypotheses do not apply. There may be subsets of phase space, albeit of very small measure, where thermodynamics do not hold true, and such regions may contain the trajectories of populations of living systems. More strongly, we may even question whether thermodynamic parameters are well defined in such regions. Thus, it is only as a sop to a limited statistical mechanics that in what follows, I shall follow the standard line of reconciliation between evolution (increasing order) and the second law of thermodynamics (decreasing order) by saying that "in the *total* system, the entropy must increase."

I hope that developments in statistical mechanics, such as those discussed by Prigogine [1967], will eventually show how this simplistic statement may be refined.

Thermodynamic approaches to problems of organized complexity, integration, adaptation, or evolution, fail to expose the mechanisms underlying these phenomena, and at best deal only with the direction in which a situation will develop and not with any quantitative aspects. In the same vein, we viewed evolution and progressive adaptation as directed processes. But,

†The rest of this section owes much to discussions with Warren McCulloch in May of 1962.

fine conceptually though this may be, it cannot tell us what will evolve, or what properties of the organism are going to emerge. A self-organizing system is characterized by change, whether it be growing or adapting. Thus, we need to find a mechanism for the description of such a system that is essentially dynamic in nature. Further, it must be noted that the bulk of present thermodynamics and statistical mechanics is designed to deal with reversible systems which are essentially homogeneous. Self-organizing systems, on the contrary, are highly structured and have a definite "positive arrow in time," and thus great care must be exercized in applying thermodynamic notions.

Be that as it may, let us now consider in more detail what we want our dynamic mechanism to be. It seems clear that whether we wish to analyze biological systems or synthesize adaptive machines, we are going to need a theoretical study of automata. An automaton is a structure; as such it is essentially characterized by its form, but the real expression of this form is in the behavior the automaton realizes, including its "information-handling" behavior. We must explain how automata endure. (Endurance has three aspects: growth, learning, and homeostasis. Homeostasis must repair what would otherwise be permanent damage. The correction of purely temporary errors is a problem of information handling rather than self-perpetuation.) Hopefully, we then may deduce under what conditions reproduction is necessary for endurance of the species (whatever *that* means).

Although we gained some insight from the study of deterministic systems, it is clear that the eventual form of our theory of biological automata must be statistical, since they are continually subject to noise, both environmental and internal. The noise interferes with both enduring and multiplying and if uncompensated would cause the automaton to disintegrate rapidly, and would prevent it from begetting its like; also it interferes with the information handling, and if uncompensated would cause drastic computational errors.

Clearly these aspects are not independent. We have already mentioned the work of Cowan and Winograd on combatting computation errors in automata, by applying information and coding theory to the problem. A major task for our theory is to learn how to apply coding theory to the problem of enduring and multiplying in the presence of noise. We already have suggested that automata theory may help refine theoretical biology by allowing us to completely describe systems that share more and more of the properties we ascribe to living systems, so that we may avoid oversimplifications which are almost inevitable at the level of verbal discourse. Let us now note that our strategy may be a fruitful one, whether or not one holds the reductionist view that all vital processes can be reduced to physicochemical terms, that is, whether one believes that one is converging to an understanding of the nature of life, or sharpening an inevitable distinction between living things and man-made machines. I personally believe reduction is possible, but cannot share the confidence expressed by Crick [1966] in its imminence, since I believe that we are at the very beginning, and require immense break throughs in the mathematical theory of automata and other complex systems, before we can hope really to understand the hierarchy of processes that extends from molecule to man. It seems to me that the notion of DNA $\longrightarrow$ RNA $\longrightarrow$ enzyme transduction is of as vital importance to understanding life as the conversion of decimal numbers to binary notation is to understanding digital computers. However, just as computers may be built to operate in a nonbinary mode, so may there be life without DNA; and we are no more entitled to say we understand embryology when we understand DNA than we are entitled to say we understand computation when we understand radix two. Perhaps our automaton models may break us of too parochial a view of the goals of theoretical biology.

It is, then, perhaps appropriate here to mention an argument by Wigner [1961] that self-reproduction is virtually impossible in a quantum mechanical system, and show wherein we believe Wigner's argument fails [Arbib, 1968]. We may summarize Wigner's argument in the following points 1–7:

1. Let there be a quantum mechanical state v which is the living state. (Wigner later replaces this by a whole subspace, which changes neither his conclusion nor our refutation.)

2. The state at time t_0 is given by

$$\Phi = v \times w,$$

which is the initially noninteracting combination of the organism and environment.

3. If and when self-reproduction has taken place (say at time t_1), we have two organisms present in a different environment. The state is then

$$\Psi = v \times v \times r,$$

where r describes the new environment and the relative position of the two organisms.

4. *Assumption 1.* Choosing coordinates so that

$$\Phi_{\kappa\lambda\mu} = v_\kappa w_{\lambda\mu}, \qquad \Psi_{\kappa\lambda\mu} = v_\kappa v_\lambda r_\mu,$$

we assume there are only finitely many dimensions, κ and λ taking on N values, say, while μ runs through R values.

5. The change of state from t_0 to t_1 is given by a collision matrix S:

$$v_\kappa v_\lambda r_\mu = \sum_{\kappa',\lambda',\mu'} S_{\kappa\lambda\mu;\,\kappa'\lambda'\mu'} v_{\kappa'} w_{\lambda'\mu'}. \tag{$*$}$$

6. *Assumption 2* There are no special relationships between the components of S.

7. Since (∗) must be valid for any κ, λ, and μ, we have N^2R complex equations.

There are several identities here, but since N^2 is enormous, this fact will be disregarded. There are N unknown v components; r has R unknowns, and w has RN unknowns. Thus, there are $N + R + NR$ unknowns. These are very much fewer than the number of equations, and so Wigner states that "it would be a miracle" if the equations could be satisfied.

Wigner [1961] comments that, "If one tries to confront the above with von Neumann's specific construction [of a self-reproducing automaton], one finds the confrontation impossible because the von Neumann model can assume only a discrete set of states whereas our model is continuous." This argument seems specious since (a) continuity was not used in Wigner's argument; and (b) the discreteness of the von Neumann model can be obtained to a sufficient degree of accuracy using quantum mechanical variables. The argument is not saved by the comment, "The inapplicability of his [von Neumann's] model to biological applications was realized by von Neumann," since Wigner's argument does not depend on the bioticity of v.

We believe that our self-reproducing automaton is, in fact, a counter-example to Wigner's claim, and that the weak link (as, in fact, Wigner has conceded) lies in Assumption 2. Self-reproduction only occurs when v is highly structured, and w is sufficiently structured to ensure rich interaction between the organism and environment. This rich interaction is represented by many relationships between the components of S, which vitiate Assumption 2, and annihilate the "need for a miracle" of 7 above.

It is amusing to see that Wigner's argument may be applied to show that there cannot exist a pair of stable objects. Let v describe a chair, and let $w = v \times r_0$ describe a second chair, and the remainder of the environment, at t_0. Let $\Psi = v \times v \times r_1$, where the two v's describe the same two chairs, and r_1 the environment at t_1. Pursuing Wigner's argument, we would have to conclude that "it would be a miracle" if the two chairs survived from t_0 to t_1. Once again, Assumption 2 is at fault: in the reproduction example it was the development of organism–environment correlations that vitiated the choice of random S in our stability example, whereas, in Assumption 2, it is the "interchair" correlations that vitiate the assumption.

Of course, Wigner's argument should caution us not to expect life to have evolved in "one large step" from the primeval slime, and Pattee [1968] has pointed out that Wigner's argument is well taken in that the problems of transition from genotype to phenotype are inextricably bound up with the yet unresolved problems of the quantum theory of measurement. However, this does not counter the claims of the reductionist, unless one is extreme enough to insist that the continued existence of pairs of chairs defies a physical explanation. But it *is* a very profound problem to understand how verbally classifiable objects can emerge from the whirl of the wave function. Presumable the law of large numbers and the cooperative effect of gravitational forces can solve the problem of stability of a chair. If we can handle the problem of molecular stability, then we may have the one small step whose compounding explains the origin of life.

The probability of making a living creature, then, is vanishingly small, when just given a box of "components," but this is not the way it came about. The great trick of the living system is that it slows down the entropic degradation. The huge structural complexity of a human being is coded into the chromosomes selected for the job the organism is to do. The genetic instructions interact with the environment to produce the organism. (Note that the environment of a cell includes adjacent cells; for example, a zygote usually gives rise to an organism, but if the two halves are separated, two organisms will often be formed.) This organism is, as a result of the interaction, adaptive both in structure and in function.

We may define *adaptation* as the property of that which has been kicked once which enables it to change in such a way that, when kicked again, it undergoes less internal change. Nature uses this process for adjustment to the environment. Note that this adaptation can only occur if the organism has the requisite variety from which to choose; this is the important point that Ashby has made time and again. Whether we build a machine or investigate a biological system, we must have components capable of entering into sufficient relationships to each other to gain flexibility and reliability. (We note that bisexual reproduction also serves to increase the variety.) Greater diversity in function might also handle these problems. But variety is useless unless it can proceed from a sufficiently rich structure. It is this combination of natural structure and requisite variety which yields adaptability.

The necessity of the basic structure can be illustrated by an example [McCulloch, personal communication, 1962] from neurosurgery. If certain parts of the brain are removed, the patient becomes "stimulus bound"; he only acts in the present. He has a poverty of associational structure; his speech becomes less ordered in the large, his sentences shorter. He lacks adaptability. Now, his brain is still very complex and abounds in variety but a vital part of its structure is missing. The warning is clear: In building a self-organizing system, we must ensure that we transfer enough information in forming the machine to allow it the complexity of structure requisite for the adaptive tasks we set it.

We need to be able to say precise things about the structures and their functions required to subserve a certain basic range of adaptability. And not just to subserve, but also to take active advantage of it. As we discuss in Section II, Chapter 3, Volume III of this treatise, it does not seem useful to describe an adaptive system in the old automaton-theoretic way which enumerates all its possible states. To cut down the number of states under consideration we should consider the system to be nonstationary so that one state space is that of the basic system, while another keeps track of the transition function. Again, a hierarchical description helps, in which, not only do we carve the system into subsystems, somewhat simplistically, but realize that the activity of one system is often relevant to the rest of the system only as "caricatured" in a high-level language, rather than in the full details of state variables required to keep track of its internal functioning. Note, too, the misleading nature of working directly (rather than in terms of a dynamic sequence of subspaces) in terms of the infinite-dimensional spaces required to capture all possible states of a growing system.

We should thus reiterate at this point the importance of hierarchical structuring. The unity of an organism overrides the unities which make up the lesser parts. In large groups of organisms, the regularities are greater than those for the individual (this is reminiscent of the behavior of fluctuations in the statistical mechanics of gases).

One would like a theory that is adequate to treat, and distinguish between, on the one hand, "genesis, growth, adaptation, and learning," all of which are in terms of a single device, and, on the other hand, effects due to the interaction of organisms, which thus develop "language" in the sense of Occam's conventional terms and Pavlov's second signaling system. A human derives energy and information from food, perception, and verbal communication. (It is interesting to note that the method of gaining information by verbal inputs is itself a product of learning; we have a regenerative loop of adaptation.) The food is composed of useful proteins, sugars, and so forth, and thus contains much necessary energy and structure; perception and verbal inputs provide him with information. He grows and learns. His output is informed when he communicates and reproduces, but is otherwise degraded. Thus, his internal entropy is generally decreased, (despite environmental and internal noise), but, again, the net entropy of the universe may increase. In any case, we are reminded of the importance of a richly structured environment in securing the development of an organism.

What we recognize as order is something describable in our language in a small number of words. We strongly suspect that we have had the wrong way of looking at living systems. We are beginning to see proper order. If we exclude the case of a crystal, we find that when we evolve something, there will be an apparent disorder simply because we have not figured out how to look at it. Just as the very concept of evolution itself brought a great deal of order to biology, so our further studies should reveal to us even further order in the resultant systems.

When we analyze biological systems, we are trying to uncover a rich order. In the optimum code for a noise structure each message may be encoded in many ways. All possible words will be used as often as possible and the result will look like noise. Thus, *apparent* disorder may well be due to adaptation. It does not follow that because animals are complex, the genetic information to specify them must be extensive; apparent complexity, if viewed appropriately, may turn out to be ordered and relatively simple.

Our goal is to develop an automata theory that will be helpful anywhere, as in embryology, in which we have that integration of diverse processes that yields the progressive emergence of increasing functional stability in complex systems. For we cannot end on this note without stressing that most of the "explanations" above are, rather, implicit observations that certain contributory mechanisms must be accounted for by our theory. Even if one accepts that the total entropy (whatever that means) of the universe must increase so that the organism eventually degrades (dies), we are still far from explaining the exceptional stability of "entropic reversal" in a population (or, dare I say, ecosystem?) of organisms.

Acknowledgments

Much of the material in this exposition was published previously in three of my articles [1967a, 1967b, 1969a] and I should like to thank the respective editors, J. F. Danielli, J. Hart, and C. H. Waddington, and publishers, Academic Press, University of Toronto Press, and University of Edinburgh Press, for their kind and much appreciated permission to use this material here, somewhat reworked to meet the present purpose.

Dr. Lewis Wolpert and the editor and publishers of the Journal of Theoretical Biology have kindly agreed to the reproduction of the figures from Wolpert [1969], which appear in Section II. Again, a generous amount of material appearing here in Sections IV and V has appeared in similar or identical form in my book *Theories of Abstract Automata*, copyright 1969, and is here reprinted by kind permission of Prentice-Hall Incorporated, Englewood Cliffs, New Jersey. The reader is referred to this source for further details on much of the material treated in the two mentioned sections.

Preparation of the more novel portions of this work was supported in part by PHS Research Grant Number 1 RO1 NS09102-01 COM from the National Institute of Neurological Diseases and Stroke.

References

Apter, M. J. [1966]. "Cybernetics and Development." Pergamon, Oxford.
Arbib, M. A. [1966a]. Speed-up theorems and incompleteness theorems, *in* "Automata Theory" (E. R. Caianiello, ed.), pp. 6–24. Academic Press, New York.
Arbib, M. A. [1966b]. A simple self-reproducing universal automaton, *Inform. Control* **9**, 177–189.

Arbib, M. A. [1967a]. Automata theory and development, I, *J. Theor. Biol.* **14**, 131–156.
Arbib, M. A. [1967b]. Some comments on self-reproducing systems, *in* "Systems and Computer Science" (J. F. Hart and S. Takasu, eds.), pp. 42–59. Univ. of Toronto Press, Toronto.
Arbib, M. A. [1969a]. Self-reproducing automata—some implications for theoretical biology, *in* "Towards a Theoretical Biology. 2. Sketches" (C. H. Waddington, ed.), pp. 204–226. Edinburgh Univ. Press, Edinburgh.
Arbib, M. A. [1969b]. "Theories of Abstract Automata." Prentice-Hall, Englewood Cliffs, New Jersey.
Arbib, M. A., and Laing, R. A. [1972]. Automata theory and development. II. Morphogenesis of simple artificial organisms, *J. Theor. Biol.* (in press).
Balzer, R. [1967]. An 8-state minimal time solution to the firing squad synchronization problem, *Inform. Control* **10**, No. 1, 22–42.
Bar-Hillel, Y., Perles, M., and Shamir, E. [1961]. On formal properties of simple phrase structure grammars, *Z. Phonetik Sprachwiss. Comm. Forsch.* **15**, 143–172.
Barzdin, Y. M. [1965]. Complexity of recognition of symmetry in Turing machines, *Problemy Kibernetiki* **15**.
Blum, M. [1964]. Measures on the computation speed of partial recursive functions, *Quart. Progr. Rep. 72*, Res. Lab. Electronics, MIT, 237–253.
Bullough, S. [1967]. "The Evolution of Differentiation." Academic Press, New York.
Cairns-Smith, A. G. [1968]. An approach to a blue-print for a primitive organism, *in* "Towards a Theoretical Biology. 1. Prolegomena" (C. H. Waddington, ed.), pp. 57–66. Edinburgh Univ. Press, Edinburgh.
Chomsky, N. [1959]. On certain formal properties of grammars, *Inform. Control* **2**, 137–167.
Codd, E. F. [1965]. Propagation, computation and construction in 2-dimensional cellular spaces, *Tech. Publ. Univ. of Michigan*; reprinted as "Cellular Automata." Academic Press, New York, 1968.
Crick, F. H. C. [1966]. "Of Molecules and Men." Univ. of Washington Press, Seattle.
Gerard, R. W. (ed.) [1958]. Concepts of biology, *Behavioral Sci.* **3**, No. 2 (April).
Goodwin, B. C. [1963]. "Temporal Organization in Cells: A Dynamic Theory of Cellular Control Processes." Academic Press, New York.
Goodwin, B. C., and Cohen M. H. [1969]. A phase-shift model for the spatial and temporal organization of developing systems, *J. Theor. Biol.* **25**, 49–107.
Hartmanis, J., and Stearns, R. E. [1965]. On the computational complexity of algorithms, *Trans. Amer. Math. Soc.* **117**, 285–306.
Holland, J. H. [1962]. Outline for a logical theory of adaptive systems, *J. Assoc. Comput. Mach.* **9**, 297–314.
Lee, C. Y. [1963]. A Turing machine which prints its own code script, *Proc. Symp. Math. Theory of Automata Symp. Ser.* **12**, 155–164.
Lindenmayer, A. [1968a]. Mathematical models for cellular interactions in development. I. Filaments with one-sided inputs, *J. Theor. Biol.* **18**, 280–299.
Lindenmayer, A. [1968b]. Mathematical models for cellular interactions in development. II. Simple and branching filaments with two-sided inputs, *J. Theor. Biol.* **18**, 300–315.
Lindenmayer, A. [1971]. Developmental systems without cellular interactions, their languages and grammars, *J. Theor. Biol.* **30**, 455–484.
Michie, D., and Longuet-Higgins, C. [1966]. Party game model of biological replication, *Nature* **212**, 10–12; Reprinted in "Towards a Theoretical Biology. 1. Prolegomena" (C. H. Waddington, ed.). Edinburgh Univ. Press, Edinburgh, 1968, with a comment by H. H. Pattee.
Minsky, M. [1967]. "Computation: Finite and Infinite Machines." Prentice-Hall, Englewood Cliffs, New Jersey.

Moore, E. F. [1964]. The firing squad synchronization problem, *in* "Sequential Machines" (E. F. Moore, ed.), pp. 213–214. Addison-Wesley, Reading, Massachusetts.

Moore, F. R., and Langdon, G. [1967]. A generalized firing squad problem, *Syracuse Univ. Res. Inst. Tech. Rep.* RADC-TR-67-521.

Myhill, J. [1964]. The abstract theory of self-reproduction, *in* "Views on General Systems Theory" (M. D. Mesarovic, ed.), pp. 106–118. Wiley, New York.

Pattee, H. H. [1968]. The physical basis of coding and reliability in Biological evolution, *in* "Towards a Theoretical Biology. 1. Prolegomena" (C. H. Waddington, ed.), pp. 67–93. Edinburgh Univ. Press, Edinburgh.

Post E. L. [1936]. Finite combinatory processes—Formulation I, *J. Symbolic Logic* **1**, 103–105.

Post, E. L. [1943]. Formal reductions of the general combinatorial decision problem, *Amer. J. Math.* **65**, 197–268.

Prigogine, I. [1967]. Temps, structure et entropie, *Bull. Cl. Sci. Acad. Roy. Belg. Ser. 5* **53**, 4, 273–287.

Rosen, R. [1960]. On a logical paradox implicit in the Notion of a self-reproducing automaton, *Bull. Math. Biophys.* **28**, 149–151.

Rosenstiehl, P. [1966]. Existence d'automates finis capables de s'accorder bien qu'arbitrairement connectés et nombreux, *ICC Bull.* **5**, 245–261.

Saunders, J. W., Jr. [1966]. Death in embryonic systems, *Science* **154**, 604–612.

Simon, H. A. [1962]. The architecture of complexity, *Proc. Amer. Phil. Soc.* **106**, 6, 467–482.

Smith, A. R. [1968]. Simple computation-universal cellular spaces and self-reproduction, *Rec. Ann. Symp. Switching and Automata, 9th* 269–277.

Stahl, W. R. [1965a]. Algorithmically unsolvable problems for a cell automaton, *J. Theor. Biol.* **8**, 371–394.

Stahl, W. R. [1965b]. Self-Reproducing automata, *Perspect. Biol. Med.* **8**, 373–393.

Stahl, W. R. [1966]. A model of self-reproduction based on string processing finite automata, *Natural Automata Useful Simulations Proc. Symp. Fundam. Biol. Models* 43–72.

Stahl, W. R., Coffin, R. W., and Goheen, H. E. [1964]. Simulation of biological cells by systems composed of string-processing finite automata, *Proc. Spring Joint Comput. Conf.* 89–102.

Thatcher, J. [1963]. The construction of a self-describing Turing machine, *Proc. Symp. Math. Theory of Automata, April 1962*, New York 165–171.

Thatcher, J. [1965]. Universality in the von Neumann cellular model, *Univ. Michigan Tech. Rep.* 132–186; also in "Essays on Cellular Automata" (A. W. Burks, ed.). Univ. of Illinois Press, Urbana, 1970.

Thorpe, W. H. [1957]. "Learning and Instinct in Animals." Cambridge Univ. Press, London and New York.

Trinkaus, J. P. [1969]. "Cells into Organs: The Forces that Shape the Embryo." Prentice-Hall, Englewood Cliffs, New Jersey.

Turing, A. M. [1936]. On computable numbers, with an application to the Entscheidungsproblem," *Proc. London Math. Soc. Ser. 2* 42, 230–265; Errata **43**, 544–546 (1936–1937).

von Neumann, J. [1951]. The general and logical theory of automata, *in* "Cerebral Mechanisms in Behavior: The Hixon Symposium," pp. 1–32. Wiley, New York.

von Neumann, J. [1966]. "Theory of Self-Reproducing Automata" (edited and completed by A. W. Burks). Univ. of Illinois Press, Urbana.

Waksman, A. [1966]. An optimum solution to the firing squad synchronization problem, *Inform. Control* **9**, 66–78.

Wang, H. [1957]. A variant to Turing's theory of computing machines, *J. Assoc. Comput. Mach.* **4**, 63–92.

Wiesel, T. H. and Hubel, D. H. [1963a]. Effects of visual deprivation on morphology and physiology of cells in cat's lateral geniculate body, *J. Neurophysiol.* **26**, 978.

Wiesel, T. H., and Hubel D. H., [1963b]. Single-cell responses in striate cortex of kittens deprived of vision in one eye, *J. Neurophysiol.* **26**, 1003.

Wigner, E. P. [1961]. The probability of the existence of a self-reproducing unit, *in* "The Logic of Personal Knowledge: Essays Presented to Michael Polanyi," pp. 231–238. Free Press, Glencoe, Illinois.

Wolpert, L. [1968]. The French flag problem: A contribution to the discussion on pattern development and regulation, *in* "Towards a Theoretical Biology. 1. Prolegomena" (C. H. Waddington, ed.), pp. 125–133. Edinburgh Univ. Press, Edinburgh.

Wolpert, L. [1969]. Positional information and the spatial pattern of cellular differentiation. *J. Theor. Biol.* **25**, 1–47.

Chapter 4

SOME RELATIONAL CELL MODELS: THE METABOLISM-REPAIR SYSTEMS

Robert Rosen

Center for Theoretical Biology
State University of New York at Buffalo
Amherst, New York

I. Introduction

This chapter will be devoted to the exploration of the properties of a class of models of the cell, which are aimed toward an understanding of the same phenomena of regulation, homeostasis, and epigenesis as were those described in the past few chapters. However, at first sight, the models to be discussed will seem to have no relation at all to those we have already discussed because the basic principles which underlie the construction of the two kinds of models are quite different. Complicated systems, such as those found in biology, generally allow themselves to be described in many different kinds of ways, each with its own emphases and its own domain of applicability. A comparison of the models developed in this chapter with those the reader has already seen will illustrate the enormous range of conceptually distinct theoretical approaches possible when dealing with complex systems like cells.

In this introduction we will briefly describe the different philosophies which motivate the two kinds of cell models, leaving until the last section (after the reader has familiarized himself with the properties of the new models) how the two can be related to each other.

In the cell models we have already seen, the point of departure was entirely structural and biochemical in tone. It was assumed explicitly that the basic quantities involved in the study of regulation were biochemical in nature, and that the basic physical processes underlying regulatory and epigenetic phenomena in cells are chemical reactions and diffusion. As much use as possible was made of the kinds of interactions which biochemical experimentation on cellular systems have told us might be important in such regulation, particularly with regard to activation (induction) and inhibition (repression). As we saw, this gave us a large class of *plausible* models, wherein we could demonstrate the capability of such systems to exhibit a wide variety of properties suggestive of those which actually occur in cells.

However, it must be noted that the main emphasis in all of these models is of a *structural* character. We emphasize the materials of which a particular system is made, and seek to understand how these materials change in the course of time. This is, of course, a thoroughly legitimate and most important way to proceed, and it has the advantage of being more or less closely tied with the kinds of experiments that we can actually perform on cellular systems. However, it must also be noted that such models, as a general rule, ignore completely the *functional organization* of the systems from which they are abstracted. Indeed, it is often vaguely said that this functional behavior is at a "higher level" than the structural studies in question, and one of the articles of faith accepted in this area is that *the functional organization* of biological systems will in fact follow from sufficiently comprehensive studies of a purely structural character. This is the hypothesis of *reductionism*,† which dominates biology today.

---

†Oddly, perhaps the first relational theory in biology was Mendel's attempt to explain his results on inheritance in peas. For his arguments were essentially of a nonstructural, functional character; he did not proceed from the structure of the gene, nor from the physicochemical forces involved in the transmission of information from gene to phenotype, but rather from a set of simple numerical relations, which could be satisfied by a large class of physically diverse systems. Population genetics is still dominated by the functional idea of "fitness," which is very hard to express in structural terms. The apotheosis of the relational character of the laws of genetics is found in the works of Woodger [1952]. However, the first to explicitly recognize the meaning and status of relational models was Rashevsky, who in a series of papers beginning in 1954 constructed a class of relational models based on powerful principle of comparison between diverse systems. The point of departure for these developments was a class of cell models quite different from those to be developed below; the reader is referred to the original papers [Rashevsky, 1954–1955a, b, c, d, 1956a, b, c, 1957] for a detailed development. The culmination of this work, from the standpoint of general principles, can be found in Chapter 2C of Volume III.

Indeed, the first step in any structural study of a biological system, whether the study is experimental or theoretical, is to forget about, or even to destroy, the higher-level functional organization, leaving behind a purely physicochemical system, which is then studied entirely by physicochemical (that is, nonbiological) means. For instance, the molecular biologist studies "fractions" obtained from cells by a variety of physical processes; these fractions are simplified model cells or abstract cells just as surely as the mathematical models to which these volumes are devoted. In the process of abstraction or simplification, whether it be theoretical or experimental, there is a loss of information. And it is quite obvious from the means whereby such fractions are produced, that the information lost concerns the higher-level organizational and functional properties of the system from which the fraction was taken. In other words, the first step in conducting any structural study of a biological system is to abstract away the organizational properties of the system, leaving behind a purely structural residue to be studied entirely in structural (that is, physicochemical) terms.

The class of cell models described below are constructed according to a principle which is almost the exact opposite of that used in the structural studies we have considered thus far. This is restricted in part by the observation that systems of the utmost structural diversity, with scarcely a molecule in common, are nevertheless recognizable as cells. This indicates that the essential features of cellular organization can be manifested by a profusion of systems of quite different structure. What we shall do, in effect, is to begin by *abstracting away the structure* (that is, the physics and chemistry) of the system, leaving behind only the functional organization, which then can be characterized and studied abstractly. Clearly this kind of abstraction can only be made in theoretical terms; but as we shall see, when we consider the organization of the system as the fundamental object of study, divorced from any particular *realization* in terms of specific physical and chemical structures, we can obtain new insights and results of a generality not easily possible with purely structural studies. Of course, we must pay for this; we do so by losing the easy interpretability of our results in terms of the structural observations of specific systems which have constituted the bulk of our biological knowledge. Therefore it is important to establish relationships which will allow us to pass back from the models of functional organization to be developed below to the kinds of structural models we have already considered. Just how this may be done will be considered in the final section.

Our procedure will be reminiscent of that employed in the study of primary genetic processes in Chapter 3, Volume I of this treatise. There too we began by ignoring the specific structural basis of the genetic process, but made only a very general functional assumption (namely, that microphysical processes were involved in this process). So here too, we shall begin by making a very simple and general assumption about the organization of cells, independent of any specific structural realization. The cell models themselves will be defined in Section III, and their properties developed therein and in Section IV. In the next section we shall define some basic preliminary notions.

Since the models to be developed do not proceed from an assumption of specific structural properties, but rather from a set of relationships assumed to hold between functionally defined elements of unspecified structure, such models are frequently called *relational*; the use of such models in the study of biological systems is often called *relational biology*.†

II. General Input–Output Systems

A. Components and Component Networks‡

Our point of departure in the construction of the class of cell models called metabolism–repair systems is the idea of an input–output system. We will begin with the simplest possible type of such a system, which we will call a *black box* or a *component*.

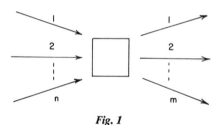

Fig. 1

Generally speaking, a black box is a structure which can accept a certain family of *inputs* from its environment, and convert these into a corresponding family of *outputs* which are discharged back into the environment. In the most general situation, we visualize the black box as provided with a number of *input lines*, each corresponding to a distinct and necessary class of environmental inputs, and a (generally different) number of *output lines*, which represent the distinct output classes which can be discharged back into the environment. This type of structure is shown in Fig. 1.

How do we represent such a structure mathematically? Let $A_1, A_2, \ldots,$ A_n denote the sets of individual inputs that can be accepted along the input

†For a more detailed discussion of reductionism, and some of the difficulties (of a relational character) faced by a purely reductionist approach to biology see Rosen [1968].

‡The basic ideas for the material in this and the next section may be found in Rosen [1958a] and further developed in Rosen [1958b, 1959]. The definitions we use probably should be compared with other definitions of general systems [see especially von Bertalanffy and Rapoport [1959–1968] or Mesarović [1964].

lines 1, 2, . . . , n, respectively, and let $B_1, \ldots, B_m$ denote the sets of input entities that can be released into the environment along the output lines, 1, 2, . . . , m, respectively. Then the behavior of the black box can be regarded as a set-theoretic mapping of the form

$$F: A_1 \times A_2 \times \cdots \times A_n \longrightarrow B_1 \times B_2 \times \cdots B_m.$$

In our subsequent work we shall consider the case for which $m = 1$, that is, the case in which the black box produces only a single class of output elements.

Given a family of such black boxes, we can begin to string them together into networks or arrays, where (some of the) output elements produced by one of the black boxes can be accepted as input elements by another box. Thus, typically, we will have situations of the forms depicted in Figs. 2–4. These represent what is commonly called the *block diagram* of the array. Intuitively, we want to characterize such arrays, which we shall call general *input–output systems*, in such a way that the following intuitively evident conditions are satisfied:

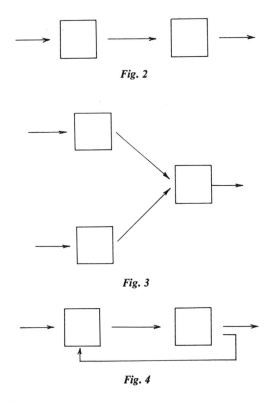

Fig. 2

Fig. 3

Fig. 4

1. Every component shall have at least one input and one output.

2. If a component requires more than one input line (that is, if the domain of the mapping corresponding to the component is a Cartesian product), then each factor of this Cartesian product is generated within the system.

3. The system can be "functionally isolated" from the external world.

If we think about these conditions for a moment, we find that we can give a formal definition for such arrays of components satisfying the above conditions.

Definition. An abstract input–output system M is a pair $(\mathscr{A}, \mathscr{F})$ where $\mathscr{A} = \{A_i\}$ is a family of sets and $\mathscr{F} = \{f_j\}$ is a family of mappings such that

1. If $f \in \mathscr{F}$, then there exist sets $A, A' \in \mathscr{A}$ such that the domain of f is contained in A, and the range of f is contained in A'.

2. If $A_1, \ldots, A_n \in \mathscr{A}$, then the Cartesian product $A_1 \times A_2 \times \cdots \times A_n \in \mathscr{A}$.

3. Every set $A \in \mathscr{A}$ contains either the domain or the range of a mapping $f \in \mathscr{F}$, or else is a factor of the domain of such a mapping.

Because of the relation of this abstract definition of input–output system to the conventional block diagram, we shall sometimes call such a system an *abstract block diagram.*

A bit more terminology will also be helpful. A set $A \in \mathscr{A}$ which does not contain the range of a mapping $f \in \mathscr{F}$ will be called an *environmental input* to the system; likewise a set A that does not contain the domain of a mapping f will be called an *environmental output.* If $A_1, A_2, \ldots, A_n$ represent the environmental inputs to a system M, then a particular choice of elements $a_i \in A_i, i = 1, \ldots, n$ will be called an *environment* of the system.

It is important to note that there is nothing unique about the block diagram or abstract block diagram of a particular input–output system. Given such a system, we can dissect it into components in many different ways, each of which will have its own topology of interconnection. Thus the block diagrams are meaningful only in terms of an initial choice of components. Indeed, we can regard the entire input–output system as a single black box or component in the following way: If $A_1, \ldots, A_n$ represent the environmental inputs to the system M, and $B_1, \ldots, B_m$ represent the environmental outputs, then we can define a *single* mapping F, constructible canonically out of the mappings $f \in \mathscr{F}$, such that

$$F: A_1 \times \cdots + A_n \longrightarrow B_1 \times \cdots \times B_m,$$

which represents the input–output behavior of the entire system. The block diagrams represent a synthetic and not an analytic tool; we shall see examples of this in the next section.

B. Some Examples of Input–Output Systems

There are in general two classes of input–output systems which have received extensive study. These classes are the *discrete-time systems*, in which time is measured as a discrete set of instant $t_0, t_1, t_2, \ldots$, and the *continuous-time systems*, in which, as the name indicates, time moves continuously. Although conceptually these two classes of systems are very similar, they differ greatly in technical terms; the study of discrete-time systems comprises the theory of automata, while the study of continuous-time systems comprises essentially the theory of dynamical systems in Euclidean space. It will be helpful to see how the formalism of the preceding section can be represented (or *realized*) within these two classes of system.

In the discrete case, a typical input–output system is a finite automaton, or a finite-state machine.† We are given two finite discrete sets A, B, which represent the input elements and output elements, respectively. (Note that A and B may be themselves Cartesian products of other sets). We are also given an auxiliary set S of "internal states" of the system. Finally, we are given two mappings, the *next-state map* $\delta: S \times A \longrightarrow S$, and the output map $\lambda: S \times A \longrightarrow B$. We suppose that at the initial instant t_0 of time, the system is presented with the input element $a_0 \in A$, and is in the internal state s_0. Then at the next instant of time t_1 the internal state of the system becomes $s_1 = \delta(s_0, a_0)$, and the system produces the output $b_1 = \lambda(s_0, a_0)$. If at time t_1 the input element $a_1 \in A$ is presented (we of course allow a_0 and a_1 to be the same) then at time t_2 the internal state of the system becomes $\delta(s_1, a_1)$ and the output $\lambda(s_1, a_1)$ is released into the environment. Thus, if we know the sequence of inputs which have been provided to the box and its initial state, we can compute the corresponding output sequence.

We can express this kind of discrete automaton in the formalism of Section II.A by defining new sets

$$A' = S \times A, \qquad B' = S \times B$$

and a new mapping

$$F: S \times A \longrightarrow S \times B,$$

where $F(s, a) = (\delta(s, a), \lambda(s, a))$. In this kind of representation the set of internal states, so prominent in the theory of automata, disappears from explicit view. There is a considerable advantage in this, in that it allows us to describe time-dependent automata (that is, the state set, or next-state or output mappings, of which are themselves functions of time) in the same formalism as ordinary automata.

---

†A good development of the various aspects of automata theory, with an eye towards biological applications, is given by Arbib [1966a]. See also Chapter 3 of this volume.

Let us illustrate our remarks regarding block diagrams and abstract block diagrams in terms of automata. First, given a collection of automata, we can obviously iterate them to construct more complex automata in the fashion indicated in the preceding section. On the other hand, it is well known that we can choose a particular set of elementary automata, out of which all others can be constructed in a canonical fashion. For instance, the McCulloch–Pitts formal neurons† represent a set of such elementary components; an arbitrary automaton can be represented (usually in many ways) as a block diagram whose components are McCulloch–Pitts neurons. If we choose some other set of basic components (for example, Sheffer stroke organs) the block diagrams of the same automata would appear quite differently.

Let us now turn to the continuous time case. Here the theory is not developed in anything like the generality of the discrete-time case, being in practice limited to the study of components which are *linear*. Linearity exploits a specific kind of algebraic (and topological) structure on the input and output sets, specially chosen to be compatible with the corresponding structures possessed by the continuum of instants (that is, the time axis). We shall not go into great detail about the theory of linear systems, but shall merely review enough to explain how the theory of linear components can be subsumed into the general framework we have developed.

Let us then suppose we have a linear component, whose set of inputs is the set of all real-valued mappings of the time axis R into a Euclidean n-dimensional set; thus any input to the component may be regarded as an n-tuple of real-valued functions of time. Likewise we may consider the set of outputs to consist of all m-tuples of real-valued functions of time; for simplicity we shall take $m = 1$ (the single-output component).

It is well known‡ that, in the most general terms, given any input $x(t)$ to the linear component, the corresponding output $y(t)$ at time t can be expressed in the form

$$y(t) = \int_0^t H(t, \alpha)x(\alpha)\, d\alpha + y(0) = f(x(t)),$$

where $H(t, \alpha)$ is usually called the *weighting function* of the system. It is characteristic of the component, and its Laplace transform is the well-known *transfer function* of the component. In these terms, the behavior of the component may be represented as a linear functional f on the set of inputs to the component. For fuller details, see Chapter 5 of this volume or Chapter 1 of Volume III.

---

†Arbib [1966a] develops the relationship between networks of McCulloch–Pitts neurons and abstract finite automata. It also provides numerous good references to the voluminous literature on this subject which, as the reader will readily appreciate, belongs primarily to relational biology.

‡A good reference is Zadeh and Desoer [1963].

With this kind of component, we retain the input and output sets of the finite sequential machine, but there is no state set visible. The integral representation plays the role of the next-state and output function combined. However, under certain conditions, we can build a set of states out of the integral representation of the action of a linear component, and obtain a picture of linear component activity which is much more like that of the finite sequential machine.

In particular, if $H(t, \alpha)$ is of an appropriate form, then this function can be regarded as the Green's function† of a corresponding linear differential equation of the form

$$d^n y/dt^n + \beta_1(d^{n-1}y/dt^{n-1}) + \cdots + \beta_n = x(t),$$

where every solution of the integral equation [that is, determination of the output $y(t)$ for a given input $x(t)$] is a solution of the differential equation, and *vice versa*. This nth order differential equation can itself be expressed as a system of n first-order simultaneous differential equations in a variety of ways; the most obvious is to write

$$y = y_1, \qquad dy/dt = y_2, \ldots, d^n y/dt_n = y_n,$$

so that we have the system

$$dy_1/dt = y_2,$$
$$dy_2/dt = y_3, \ldots, dy_{n-1}/dt = y_n,$$
$$dy_n/dt = -\beta_1 y_{n-1} - \beta_2 y_{n-2} - \cdots - \beta_n + x(t).$$

This represents the equations of motion of a dynamical system with state variables $y_1, \ldots, y_n$. These equations of motion specify the manner in which the state variables change in time when the system receives an input or "forcing" $x(t)$. In this representation we break up the input–output problem for a linear component in a manner analogous to that described for the finite sequential machine; we obtain the effect of the inputs on a set of states, and compute the output from the effect of the inputs on the state set. The dynamical equations then become the exact analog of the next-state function of the sequential machine, telling us in effect the "next" (infinitesimal) state corresponding to the value of the input function at an instant and the values of the state variables at that instant. The output can then be regarded as a function of the state variables alone; indeed in the representation we have used, the output is already one of the state variables, and the output function is a projection on the appropriate state variable.

---

†A good reference to this topic, and the relation between differential and intergral equations is general, is Courant and Hilbert [1953]. The theory is mainly due to Hilbert and can be found in numerous texts.

C. Dependency Structure in Block Diagrams†

Whenever we define a system constructively or synthetically, as we have done in defining block diagrams of general input–output systems in terms of a specific set of components, almost the only means we have available for the study of the properties of the system is to pull individual components out of the system and determine the effect which this has on overall system behavior. To fix ideas, let us look at a particular block diagram (Fig. 5) and see what happens to its behavior when we remove different components from the system.

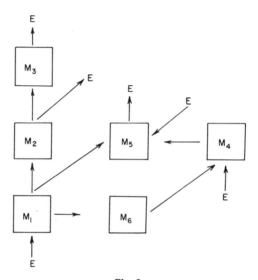

Fig. 5

One obvious measure of the importance of a particular component to the overall operation of the entire system is the number of components which can be reached from that component along a directed path in the block diagram. These represent the components which will cease to function (through absence of input) when that particular component is removed from the system. Thus, in the above diagram, if we remove the component M_3, say, no other component is affected because no other component receives inputs from M_3. On the other hand, if I remove the component M_2, the component M_3 is effectively removed from the system as well, since it receives an input directly from M_2. Likewise, if I inhibit the component M_6, the component M_4 (which receives an input directly from M_6, and the component M_5,

†Rosen [1958a, b, 1959], von Bertalanffy and Rapoport [1959–1968], and Mesarović [1964] are appropriate here.

which receives an input from M_4 (and thus indirectly from M_6) are effectively removed from the system also.

The totality of components which must be considered inhibited in this way when a particular component is removed from the system is called the *dependent set* of the component. If we know the block diagram of the system, we can read off from it the dependent set of each component. However, the converse proposition is not true; knowing the dependent set of each component in a system does not in general allow one to reconstruct uniquely the block diagram of the system. Indeed, the algorithmic construction of block diagrams compatible with a particular "dependency structure" seems to be an open problem, and one of some interest for general system theory.

A particularly interesting kind of component in this connection is the component M_1 in the above diagram. The block diagram is such that *every* component of the system can be reached from M_1 along a directed path; that is, every component in the system receives an input directly or indirectly dependent on the activity of M_1. Thus, the removal of M_1 from the system will inhibit all activities of the system, or stated another way, the dependent set of M_1 contains all the components of the system. Components with this property are called *central* components.

D. The Roles of Time in General Input–Output Systems

In the general representation of input–output systems developed in Section II.A, time plays no role. We provide the system with an input and the corresponding output appears as the value of the function corresponding to the system. This happens, as it were, instantaneously, regardless of how big the system is, that is, regardless of how many components it may have.

This is in distinction to the representation of input–output systems either as sequential machines or as linear components. In those representations time enters into the definition in an apparently essential way. However, even in those cases, it is possible to remove explicit reference to time. For instance, in the theory of finite sequential machines, a particular sequence of inputs is just a mapping from the set of instants into the input alphabet, with corresponding mappings from the set of instants into the sets of states and outputs. Thus, the totality of possible input sequences is simply a monoid or semigroup, and likewise for the corresponding state and output sequences. We can develop the theory of automata completely within the framework of the abstract theory of monoids,† and thereby entirely avoid the sequential aspect through which time manifests itself. We can do a similar thing in the theory of linear components or of dynamical systems. Conversely, given a

---

†A particularly brutal development of this type is found in the paper by Eilenberg and Wright [1967]. Many other references could be given.

theory of automata, say, developed completely in abstract terms, we can recover the sequential aspect (in many ways, in fact) by simply representing the monoids in question as built out of free monoids on a finite set of generators, and then build back the corresponding next-state and output mappings from the abstract mappings which characterize the automaton.

Thus although our general theory of input–output systems lacks a temporal element, we can always recover one by adding back sufficient structure to give us a temporal representation. The particular aspect of time which will be important for us here is that of the *lag* involved between the presentation of an input to a system and the appearance of the corresponding output. Let us consider a simple system of two components in series, whose block diagram is

$$\longrightarrow A\ \boxed{M_1}\ \xrightarrow{\ B\ }\ \boxed{M_2}\ \longrightarrow C$$

and whose corresponding abstract block diagram is

$$A \xrightarrow{f_1} B \xrightarrow{f_2} C.$$

There are two kinds of entities involved in these block diagrams; the inputs and outputs themselves (represented by sets in the abstract block diagram) and the components (represented by mappings). Each of these may give rise to a lag. The lags arising from the action of the components will be called an *operation lag*, and intuitively corresponds to the time taken in processing a particular input. The lags arising from the connections between components will be called *transport lags*, and correspond intuitively to the time required for the output of one component to be moved or transported to where they can be utilized as inputs to those components which require them. Thus in the above diagrams there will be two operation lags corresponding to the two components, and a single transport lag, corresponding to the connection between the components. In general, these lags must be added as further elements of structure of the system. In its most general form, the lag structure is a mapping from the collection $\mathscr{F}$ of sets and mappings defining a general input–output system (that is, from the abstract block diagram considered as a collection of sets and mappings) to the nonnegative real numbers.

In McCulloch–Pitts networks, for example, the transport lags are all taken to be zero, while the operation lags are all taken to be unity. The zero transport lags have been regarded as a serious deficiency in that theory, since real neural conduction times (the analogs of our transport lags) are in fact not zero. However, it is customary to introduce nonzero transport lags by adding *delay units*, neurons that compute the identity function of their inputs, and whose operation lags thus can be interpreted as transport lags.

There is one other bit of temporal structure that may be added to our abstract block diagrams of general input–output systems. In neural networks, for instance, we know that our neurons are mortal; that is, they have finite *lifetimes*. This means that each component of such a system, after a characteristic time elapses, will simply cease to function, and the system will behave as if this component had been removed from the system. This kind of mortality is characteristic particularly of biological input–output systems, and if we wish to represent this property within our formalism we must assign to each component (that is, mapping) a number that represents its lifetime.

The introduction of time lags and finite lifetimes into the general theory of input–output systems makes certain restrictions necessary in the general theory of such systems. For instance, if the lag structure is such that the lifetime of some component is exceeded before its inputs can reach it, the system cannot function properly. Moreover, the specification of the lifetimes of the components of an input–output system means that, as we watch the behavior of our system in time, components will begin to fail as their lifetimes are exceeded, and the overall activity of the system will deteriorate in a characteristic way until all activity ceases.

III. Theory of a Single Metabolism–Repair System

A. The Basic Idea of Metabolism–Repair Systems†

If we regard a metabolizing cell as a general input–output system, receiving inputs from its environment and processing them in a characteristic way to produce environmental outputs, we are struck at once by the fact that the lifetime of a cell far exceeds the functional lifetimes of any of its parts. The question we must put ourselves, then, is the following: How is it possible for the overall lifetime of an input–output system to be greater than that of any of its components?

If we think for a moment about the properties of biological cells, we see at once that these cells are continually *repairing* themselves. Indeed, to simply look at a typical cell under the microscope is to be struck by the fact that the cell is compartmented into two obviously different regions, which we may roughly call *nucleus* and *cytoplasm*. On closer investigation, we discover that the cytoplasmic part of the cell is mainly concerned with what we customarily call the metabolic activity of the cell, while one of the basic nuclear functions is concerned with repair. This repair takes the form of a continual synthesis of basic units of metabolic processing (enzymes), utilizing as inputs materials *provided by the metabolic activity itself.*

†Basic references are Rosen [1958a, b, 1959], von Bertalanffy and Rapoport [1958–1969], and Mesarović [1964].

We shall now show how this kind of behavior can be represented within the theory of general input–output systems. Systems of the kind we have described, which can be regarded as partitioned into subsystems concerned with metabolism and repair, will be called (M, R) systems. The basic motivation for the construction and study of these systems should be kept in mind as we proceed with the more formal developments.

B. The Conversion of Input–Output Systems to Metabolism–Repair Systems

It will be simplest to proceed initially by means of a specific example. Consider the input–output system whose block diagram is shown in Fig. 6. This is a typical input–output system represented as an array of interconnected components, and the dependency structure follows immediately from the pattern of connection of these components. And if these components have finite lifetimes, the entire system will cease to function as soon as the lifetimes of all of its components has been exceeded.

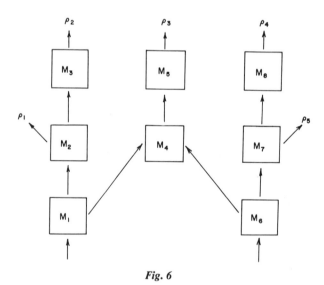

Fig. 6

In order to keep the system functioning beyond the lifetimes of its components, it is necessary to *replace* components before their lifetimes are exceeded. It is suggested, from analogy with the situation in biological cells, that this be done by appending to the system a number of new components, one for each of the original components, with the following properties:

1. If R_i is the component corresponding to the original component M_i, then among the outputs of R_i are *copies* of the component M_i.

2. Each R_i receives at least one input which is an environmental output of the original system.

Thus, for instance, we can embed our original input–output system into an (M, R) system as shown in Fig. 7. A different assignment of environmental outputs of the original system to the new components R_i would lead us to another (M, R) system containing the original system.

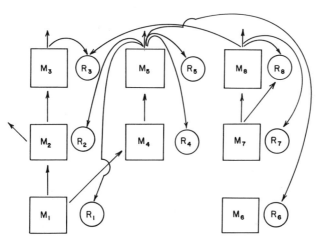

Fig. 7

Let us see what this construction amounts to in terms of the abstract block diagrams which are the natural habitat of our (M, R) systems. In the abstract, each component M_i of our system is represented by a corresponding mapping f_i, mapping a set A_i into a set B_i. Let us denote the statement

$$f_i: A_i \longrightarrow B_i$$

by the more compact statement $f_i \in H(A_i, B_i)$, which should be read, "f_i belongs to the set of mappings which map the set A_i into the set B_i." This terminology also has a much deeper significance, which will become clear shortly.

Our general rule in passing from block diagrams to abstract block diagrams is that components are replaced by mappings, and inputs and outputs by the sets which comprise the domains and ranges of these mappings. This rule applies equally well in the transition from block diagrams of (M, R) systems to abstract block diagrams of (M, R) systems. Let us then see what kind of mappings must be appended to a general abstract block diagram to convert it into the abstract block diagram of an (M, R) system; that is, how we are to represent the components R_i as mappings. Our conditions are that the domain of such a mapping must be a set corresponding to an environmental output of the system, or in general a Cartesian product of such sets. The range of such a mapping must contain a copy of the corresponding component; that is, must contain the mapping f_i. Therefore, it is natural to represent the component R_i by the mapping

$$\Phi_{f_i}: \Pi C_i \longrightarrow H(A_i, B_i),$$

or, in our more compact notation, we can say that

$$\Phi_{f_i} \in H(\Pi C_i, H(A_i, B_i)),$$

where the C_i are environmental outputs of the original system. Thus in general, an abstract block diagram of an (M, R) system consists of the following data:

  1. the abstract block diagram $\mathscr{F}$ of an arbitrary input–output system;
  2. a set of mappings Φ_{f_i} in one-to-one correspondence with the mappings f_i in $\mathscr{F}$;
  3. the sets $H(A_i, B_i)$ which contain the ranges of the new mappings Φ_{f_i}.

C. The Dependency Structure of Metabolism–Repair Systems

To motivate the discussion of the dependency structure in (M, R) systems, and to see how it differs from the corresponding structure in general input–output systems, look at the diagrams in the preceding section. In the first diagram, suppose that the component M_1 is removed from the system. Then the components M_2, M_3, M_4, and M_5 do not receive their inputs, and cease to function, while the components M_6, M_7, and M_8 continue to operate unaffected. On the other hand, if we remove the same component M_1 from the corresponding (M, R) system, the components M_2, M_3, M_4, M_5 will be inhibited as before. But now notice that the component R_6 receives an environmental input produced by M_5. This means that if M_1 is removed, R_6 will no longer receive its input, and hence can no longer manufacture copies of M_6. Thus, as soon as the lifetime of M_6 is exceeded, it, and consequently M_7 and M_8, will be removed from the system, and the entire system will cease to function. Thus, in the (M, R) system, M_1 is in effect a central component, whereas it fails to be a central component in the original input–output system. This kind of cascading, or enlargement of dependency structure, is typical of (M, R) systems.

Another typical, and in fact crucial, property of (M, R) systems, is illustrated by removing the component M_2 from the (M, R) system. This will immediately inhibit the component M_3 and the two environmental outputs that M_3 produces. However, the reader should check that the component M_5 is in no way affected by the removal of M_2 from the system. Moreover, M_5 produces an environmental output that is the sole environmental input to the new component R_2. Thus, the removal of M_2 will be *repaired* by the system; copies of M_2 will continue to be made in the event that M_2 is removed or damaged. Thus, we see that certain components of an (M, R) system, such as M_2, have the capacity of being replaced by the system in the event of their destruction or inhibition, while other components, such as M_1 or M_4, do not have this property. Components like M_2 will be called *reestablishable* components, while components like M_1 will be called *nonreestablishable*.

We now come to one of the most interesting properties of (M, R) systems.

Theorem. Every (*M*, *R*) system must contain a nonreestablishable component.†

PROOF: Choose a reestablishable component, relabeling it M_1 if necessary. By hypothesis its corresponding repair component R_1 is not dependent on M_1. But since R_1 must by hypothesis receive inputs from some component of the system, there must be a component M_2 the output of which is an input to R_1. If M_2 is nonreestablishable we are done; otherwise R_2 receives an input from another component M_3 distinct from M_1, M_2. The corresponding repair component R_3 must, if M_3 is reestablishable, receive an input from a component M_4 distinct from M_1, M_2, M_3. Proceeding in this fashion we must ultimately reach a nonreestablishable component, since the total number of components is finite.

An immediate corollary we can draw from this theorem is the following: *If an (M, R) system contains a single nonreestablishable component, then that component is central* [in the extended sense of (M, R) systems]. Otherwise we could remove the single nonreestablishable component and be left with an (M, R) system containing only reestablishable components, which contradicts our theorem.

The biological significance of this theorem is the following: If we regard a cell as an (M, R) system, then by virtue simply of the fact of the functional interrelationships of the metabolic and repair structures of the cell, there must be metabolic components whose loss or injury is not repairable by the system, even though the repair components are intact. *Which* components will fail to be repairable depends on the specific relationships that exist between the metabolic and repair components; the same component may be reestablishable in one (M, R) system but may fail to be reestablishable in another (M, R) system containing the same metabolic components interrelated in exactly the same way. But there must always be one component that is nonreestablishable.

Moreover, there is a certain subtle interplay between reestablishability and centrality that is worthy of mention. It seems evident that an (M, R)

†This theorem was first proved by Rosen [1958a] with an unnecessarily strong restriction. It was proved in general by Foster [1966]. We follow Foster's proof in the text.

system will be most effectively constructed if the relation between the metabolic and repair components is such that the largest possible number of components are reestablishable. But this means, according to our corollary, that the nonreestablishable components will be central, or close to it. Thus, a system with a large number of reestablishable components will be able to survive many types of injury essentially unaffected, but certain kinds of injury will cripple the entire system. It may thus be better to allow a larger number of nonreestablishable components, but such that the damage caused by removal of these components is not so serious. We thus have questions of selective advantage entering into our theory at the outset.†

The preceding discussion proceeded independent of any time-lag structure on the (M, R) system. In the presence of time lags the discussion must be modified somewhat, but the basic aspects remain. To see this, consider the simplest (M, R) system, whose block diagram is of the form

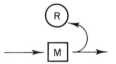

and whose abstract block diagram is of the form

$$A \xrightarrow{f} B \xrightarrow{\Phi_f} H(A, B).$$

A system like this is obviously nonreestablishable, in the sense we have used the term above. However, if we introduce finite lags, the component M may be replaced by its corresponding repair component R before irreversible damage is done. Thus, if the combined operation lag of R and transport lag from M to R is T, and if at $t = 0$, M produces an output and is immediately removed, then T units later R will produce a copy of M, and the presence of finite lags allow M to be built back into the system even though M is graph-theoretically nonreestablishable. However, our theorem in this case implies that if the component M is not simply instantaneously removed from the system, but is inhibited or repressed for a time greater than or equal to T, then the irreversible damage will be done. This finite suppression of the component M is the time-lag analog of our previous result.

D. Replication of Repair Components‡

We have introduced no special mechanism whereby the repair components R_i in an (M, R) system may be repaired. Moreover, if we assume that these components have finite lifetimes, we have really gained very little, for the

---

†This point is developed in some detail in the paper by Martinez [1964].

‡This replication scheme first appeared in Rosen [1959]. Some further properties of this replication scheme which are of biological significance will be discussed in Section IV.C.

activity of an (M, R) system, like that of a general input–output system, must terminate when the lifetimes of the R components are exceeded. The obvious idea of iterating the construction we have used, and introducing repair components for the R_i, leads to an infinite regress and is useless.

In biological cells the problem is solved by making the repair components R_i (self-) *replicating*. In this way, new copies of the R_i are continually being produced, and it is not necessary to assume that the, R_i are immortal or to fall into an infinite regress of repairers to repairers. A metabolic theory with any claims to completeness thus must account for the replicability of the R_i in a natural fashion. One of the most interesting aspects of the (M, R) formalism is that a variety of mechanisms for replicating the R_i come along essentially free of charge, in that the basic ingredients for this replication are already at hand within the structure of the (M, R) system itself.

To fix ideas, let us consider the simplest (M, R) system,

$$A \xrightarrow{\ f\ } B \xrightarrow{\ \Phi_f\ } H(A, B),$$

and let us proceed purely formally.

Quite generally, if X, Y are arbitrary sets, we can define for each element $x \in X$ a mapping

$$\hat{x}: H(X, Y) \longrightarrow Y$$

by writing $\hat{x}(f) = f(x)$ for every $f \in H(X, Y)$. We thus obtain an embedding of the *set* X into a *set of mappings*, namely $H(H(X, Y), Y)$. If it should happen that the mapping $\hat{x}$ corresponding to an element $x \in X$ is *invertible*, that is, if

$$\hat{x}(f_1) = \hat{x}(f_2) \qquad \text{implies} \quad f_1 = f_2,$$

then there is an inverse mapping $\hat{x}^{-1}: Y \to H(X, Y)$, with the property that $\hat{x}^{-1}\hat{x}(f) = f$ for all $f \in H(X, Y)$.

This argument holds for arbitrary sets X, Y; indeed the embedding of the set X into the set of mappings $H(H(X, Y), Y)$ is the abstract version of the familiar embedding of a vector space into its second dual space. In particular, let us put

$$X = B, \qquad Y = H(A, B).$$

We thus obtain, for each element $b \in B$ (that is, each output of the metabolic component) for which the associated mapping $\hat{b}$ is invertible, a mapping $\beta_b = \hat{b}^{-1}$ with the property that

$$\beta_b: H(A, B) \longrightarrow H(B, H(A, B)).$$

But the reader will observe that this mapping β_b actually replicates the mapping Φ_f; in other words, if $b = f(a)$ for some $a \in A$ and if $\Phi_f(f(a)) = f$ [that is, Φ_f makes copies of the component f when supplied with the input $f(a)$], then

$$\beta_b(f) = \Phi_f.$$

As stated before, all of the ingredients required for this "replication map" β_b are already present within the system. The condition that the mapping $\hat{b}$ be invertible is a most interesting condition, which can best be explored in a more general context; we shall do this in the next section. This result is the only one of its kind in the literature, that is the only one in which metabolic activity can, under precise and natural conditions, be made to imply replicability. Indeed, rather more than this is true. We may observe that the system

$$ B \xrightarrow{\Phi_f} H(A, B) \xrightarrow{\beta_f} H(B, H(A, B)) $$

is itself an (M, R) system, but one whose metabolic component was itself the genetic component for the system with which we started. From this it follows that there is an *equivalence between metabolic and genetic activities* in (M, R) systems, a rather unexpected result. The study of this equivalence raises certain technical problems, but should prove most illuminating.

IV.   Global Theory of Metabolism–Repair Systems

A. The Theory of Categories†

We have seen in the preceding sections that an appropriate language with which to describe the input–output behavior of very general systems was in terms of set theoretic mappings. As we saw, every component of such a system could be represented by a mapping the domain of which was the set of inputs to the component and the range of which was its set of outputs. A general input–output system could be regarded as a family of such mappings together with their domains and ranges, and an (M, R) system as a particular kind of input–output system. In many cases the use of this language is absolutely imperative, as, for example, in the case of the replication of repair components in (M, R) systems; this argument could not begin to be formulated without the mathematical apparatus of sets and mappings between them.

Thus, it seems that the natural habitat of our theory of cellular systems is a mathematical world populated by sets of various kinds and appropriate families of mappings between these sets. Quite coincidentally, it happens that such a mathematical world has been in existence at least since 1945; it first saw the light of day in the rarefied realms of algebraic topology, and after a slow start has during the last 10 or 15 years become one of the major forces in modern mathematics. This world is called the theory of categories,†

†The first presentation of the theory of categories which was at all definitive was by Eilenberg and MacLane [1945]. Although originally developed to treat certain problems in algebraic topology (giving rise to a new branch of mathematics called homological algebra)

and is due originally to Eilenberg and MacLane [1945]. We shall briefly review the basic landscape of the theory of categories in the form most useful for the further development of our biological models.

Although the theory of categories has reached an extreme level of abstraction in recent years, which has even begun to alarm many mathematicians, the basic ideas are quite simple and may be enumerated easily.

1. If $f: M \rightarrow N$ is a mapping, and $g: N \rightarrow P$ is a mapping, then there exists a mapping $h: M \rightarrow P$ defined by $h(m) = g(f(m))$ for every m in M. We write $h = gf$, the *composite* of f and g.

2. The composition of mappings just defined is associative; given the diagram of mappings

$$A \xrightarrow{f} B \xrightarrow{g} C \xrightarrow{h} D,$$

then $h(gf) = (hg)f$; thus, we can assign an unambiguous meaning to the three-termed product hgf.

3. If A is any set, there is a mapping i_A (the identity mapping of A onto itself) with the following properties:

(i) if $f: A \rightarrow B$ is any mapping, then $fi_A = f$;
(ii) if $g: C \rightarrow A$ is any mapping, then $i_A g = g$.

We now proceed to give the formal definition of a category (nowadays categories of the type we define are called *small* categories), which the reader will observe says nothing more than that the three properties enumerated above are satisfied.

Definition. A *category* $\mathscr{A}$ is a family or class of *objects*, $A, B, C, \ldots$ To each ordered pair (A, B) of objects in the category we assign a set, which we designate as $H(A, B)$. An element of $H(A, B)$ will be called a *mapping* of A onto B. The following axioms hold:

Axiom 1. If $f \in H(A, B)$ and $g \in H(B, C)$, then there is a unique map, denoted by gf, in $H(A, C)$; gf is the *composite* of f and g.

Axiom 2. If $f \in H(A, B)$, $g \in H(B, C)$, $h \in H(C, D)$, then $h(gf) = (hg)f$.

---

it has become, after a slow start, one of the major unifying influences in mathematics. It does not seem to be generally appreciated, though, that the theory of categories is a natural tool in any science which involves the use of model systems or abstractions of any type. Indeed, this is precisely what algebraic topology is all about; the study of a topological or geometric object in terms of a sequence of abstract algebraic models or images of it. Thus, algebraic topology occupies the same position, within mathematics itself, as does the building of mathematical models to understand physical or biological systems outside of mathematics. Seen in this light, the importance of category theory for mathematical biology may not be so surprising. A modern text is Mitchell's volume [1965].

Axiom 3. To each object A we associate a unique mapping $i_A \in H(A, A)$ such that

   (i)   for any object B, and any $f \in H(A, B)$, $fi_A = f$;
   (ii) for any object C, and any $g \in H(C, A)$, $i_A g = g$.

It is important to recognize that the sets $H(A, B)$ are an essential aspect of the structure of a category. Thus, two categories can contain exactly the same class of objects, and yet be two entirely different categories. Further, these sets $H(A, B)$ are at our disposal; at one extreme we can, for instance, choose them all to be empty (except for the identities required in all sets of the form $H(A, A)$); at the other extreme we can fill these sets $H(A, B)$ with all possible set-theoretic mappings between the sets A, B (where the objects A, B are themselves interpreted as sets), or we can take anything in between.

The most important kinds of categories for pure mathematical studies have been the categories of all mathematical objects of a particular type (the category of topological spaces, the category of groups, the category of commutative groups, and so forth), the sets $H(A, B)$ in each case being chosen as the full set of mappings that preserve the structure in question. Thus, in the category of topological spaces, $H(A, B)$ consists of all continuous mappings between A and B; in the category of groups, $H(A, B)$ consists of all group homomorphisms between A and B, and so on.

The next important idea in the theory of categories is that of a *functor*. A functor is a kind of mapping between two categories, which preserves the category-theoretic structure. It thus plays a role in the theory of categories analogous to the role played by group homomorphisms in the theory of groups, or by continuous mappings in topology. As with any structure-preserving mapping, it allows us to *compare* the structures imposed on its domain and its range.

More precisely, let $\mathscr{A}$ and $\mathscr{B}$ be categories. Let T be a rule which associates to each object $A \in \mathscr{A}$ another object, denoted by $T(A)$, in $\mathscr{B}$. Moreover if if $f \in H(A, B)$ is a mapping in $\mathscr{A}$, we associate with f a mapping $T(f)$ in $\mathscr{B}$. Functors come in two distinct types, depending on the relation of $T(f)$ to the images under T of the domain A and range B of f in $\mathscr{A}$:

   (i) if the diagram

$$A \xrightarrow{f} B$$

transforms under T into a diagram of the form

$$T(A) \xrightarrow{T(f)} T(B)$$

then the functor is called *covariant*;
   (ii) if the diagram

$$A \xrightarrow{f} B$$

transforms under T into a diagram of the form,

$$T(B) \xrightarrow{T(f)} T(A)$$

(in which the direction of the arrow is reversed) the functor is called *contravariant*.

In order to be structure preserving, a functor must also preserve compositions and identities:

1. If f and g are mappings in $\mathscr{A}$ such that the composite gf is defined, then $T(gf) = T(g)T(f)$ if the functor is covariant; $T(gf) = T(f)T(g)$ if the functor is contravariant.

2. If i_A is an identity in $\mathscr{A}$, then $T(i_A) = i_{T(A)}$, whether the functor is covariant or contravariant.

EXERCISE: To clarify the nature of the concepts involved, the reader should prove statements (a) and (b).

(a) The correspondence that assigns to every set X the set $T(X)$ of all its subsets, and to every mapping $f: X \longrightarrow Y$ the mapping $T(f): T(X) \longrightarrow T(Y)$ defined as follows: if A is a subset of X, then $T(f)(A) = f(A)$, is a covariant functor.

(b) The correspondence T defined as in (a), except that if $f: X \longrightarrow Y$, we now define $T(f): T(Y) \longrightarrow T(X)$ as follows: If B is a subset of Y, then $T(f)(B) = f^{-1}(B)$, is a contravariant functor.

It is no problem to extend this definition of functor to functors of two, three, or more variables. For example, if A and B are sets, we can define a new set

$$P(A, B) = A \times B$$

(the Cartesian product). If we are given a pair of mappings $f: A \longrightarrow A'$, $g: B \longrightarrow B'$, we can define a mapping $P(f, g): P(A, B) \longrightarrow P(A', B')$ as follows: if $x \in A$, $y \in B$, then $P(f, g)(x, y) = (f(x), g(y))$. The reader should have no difficulty establishing that this "Cartesian product functor" is covariant in both its arguments A and B.

A functor of two variables need not be covariant (or contravariant) in both of its arguments. For instance, if A and B are sets, define $H(A, B)$ as the set of all mappings of A into B. If X', B' are any other sets, and if $f: A' \longrightarrow A$, $g: B \longrightarrow B'$ are arbitrary mappings with the indicated domains and ranges, then we can define a mapping

$$H(f, g): H(A, B) \longrightarrow H(A', B')$$

by writing, for any mapping θ in $H(A, B)$

$$H(f, g)(\theta) = g\theta f.$$

This "hom functor" is contravariant in its first argument and covariant in its second.

The Cartesian product and hom functors are particularly important, because we can use them to build up other functors of several arguments; for instance the functors

$$H(A, H(B, C)), \qquad H(A \times B, C), \dots$$

There is one final concept required before we leave the general theory of categories. Just as we needed the idea of *functor* to preserve and compare structures between different categories, we need another kind of mapping to compare functors. We can see the need for this, for example, from the Cartesian product functor, which we defined as $P(A, B)$. We just as well could have defined the functor $P(B, A)$, and if we do so, there is nothing to tell us that these two functors, which look different, are really the same. Thus, we introduce the idea of a *natural transformation* between functors, a mapping between functors defined as follows: Let $S: \mathscr{A} \to \mathscr{B}$, $T: \mathscr{A} \to \mathscr{B}$ be functors. Then these functors (supposed covariant for simplicity) transform a diagram of the form

$$A \xrightarrow{f} A'$$

in $\mathscr{A}$ into two different diagrams in B; namely

$$S(A) \xrightarrow{S(f)} S(A'), \qquad T(A) \xrightarrow{T(f)} T(A').$$

A natural transformation is a way of comparing these two different diagrams in B; namely, it is a rule which associates with every object A in $\mathscr{A}$ a *mapping* denoted by $t(A)$, where $t(A): S(A) \to T(A)$, in such a way that the diagram

$$\begin{array}{ccc} S(A) & \xrightarrow{S(f)} & S(A') \\ {\scriptstyle t(A)}\downarrow & & \downarrow{\scriptstyle t(A')} \\ T(A) & \xrightarrow[T(f)]{} & T(A') \end{array}$$

commutes. [By a commutative diagram, we mean that any two composite mappings connecting the same domain and range in the diagram are equal. Thus, in the above diagram, we can map $S(A)$ into $T(A')$ by two different composite mappings; to say that the diagram commutes means that these two composites are the same for every pair A, A' of objects in $\mathscr{A}$.] If the natural transformation $t: S \to T$ has the property that, for every object $A \in \mathscr{A}$ the mapping $t(A)$ is an equivalence (that is, one to one and onto), then the natural transformation is called a *natural equivalence*. Two naturally equivalent functors are abstractly identical, however different they may appear in initial definition. Thus, the two Cartesian product functors $P(A, B)$, $P(B, A)$ are naturally equivalent. Somewhat more surprising is the fact that the two functors

$$H(A, H(B, C)), \qquad H(A \times B, C)$$

mentioned above are also naturally equivalent (the proof is somewhat tedious, though not difficult). Indeed, the above natural equivalence, which indicates a deep relation between the Cartesian product and sets of mappings, is the motivation for the important theory of *adjoint functors*,[†] which may have important biological applications. It will be noticed, for instance, that the repair maps Φ_f defined in Section III.B belong to sets of the form $H(A, H(B, C))$.

B. Metabolism–Repair Systems in Arbitrary Categories[‡]

In this section, we address ourselves to the following question (and its cognates):

Query. Given a category $\mathfrak{S}_0$, *under what circumstances can an arbitrary map* $f \in \mathfrak{S}_0$ [*or more accurately,* $f \in H(A, B)$ *where* $A, B \in \mathfrak{S}_0$] *be embedded in an* (M, R) *system within the category?*

Let us see what this means. Let $\alpha \in H(X, Y)$, $X, Y \in \mathfrak{S}$, be given. Then α can be either a "repair" map (that is, $X = \Pi B_j$, $Y = H(A, B)$), or a "metabolic" map in an (M, R) system.

1. Suppose α is a "metabolic" map $f: A \to B$. If $H(B, H(A, B)) \neq \varnothing$ in $\mathfrak{S}_0$, then any map $\Phi_f \in H(B, H(A, B))$ will, together with f, form the (M, R) system we may designate by $\{f, \Phi_f\}$; that is,

$$A \xrightarrow{f} B \xrightarrow{\Phi_f} H(A, B).$$

2. Suppose $H(B, H(A, B) = \varnothing$ in $\mathfrak{S}_0$, but there is a set $H(X_1, H(A, B))$ $\neq \varnothing$ in $\mathfrak{S}_0$. Let $\Phi_f \in H(X_f, H(A, B))$. Then $\{f, \Phi_f\}$ is not now an (M, R) system (since Φ_f does not now accept an output of the "metabolic" sub-system). We can hope to make $\{f, \Phi_f\}$ into an (M, R) system by adjoining a "metabolic" map whose range is X_1; say $g_1 \in H(Z_1, X_1)$. This requires that $H(Z_1, X_1) \neq \varnothing$ in $\mathfrak{S}_0$ for some $Z_1 \in \mathfrak{S}_0$.

If we can find such a g_1, adjoin it to the system $\{f, \Phi_f\}$. The result is not an (M, R) system, for we need a "genetic" component Φ_g to replace g. We can

[†]Adjoint functors were first defined by Kan [1958].

[‡]The material in this section first appeared in Rosen [1962]. Further developments may be found in Rosen [1963a]. In this paper it is shown that if a category $\mathfrak{S}_0$ satisfies a number of simple conditions, then any sufficiently large finite family of mappings drawn from that category must contain an (M, R) system. This result has obvious implications for the origin of life; the proof proceeds by applying graph-theoretic techniques to diagrams of mappings belonging to the category.

do this if one of the sets

$$H(B, H(Z_1, X_1)), \qquad H(X_1, H(Z_1, X_1)), \qquad H(B \times X_1, H(Z_1, X_1))$$

is not empty in $\mathfrak{S}_0$, in the obvious manner.

If the above sets are all empty, but there is some set X_2 such that

$$H(X_2, H(Z_1, X_1)) \neq \varnothing,$$

we can pick $\Phi_{g_1} \in H(X_2, H(Z_1, X_1))$ and form the system $\{f, g_1, \Phi_f, \Phi_{g_1}\}$. This is still not an (M, R) system, since X_2 is not a "metabolic" output set. But we can seek a mapping $g_2 \in H(Z_2, X_2)$ for some set Z_2 such that $H(Z_2, X_2) \neq \varnothing$ in $\mathfrak{S}_0$. We then must find a map Φ_{g_2}. Once again we get an (M, R) system if one of the sets

$$H(B, H(Z_2, X_2)), \qquad H(X_1, H(Z_2, X_2)),$$
$$H(X_2, H(Z_2, X_2)), \qquad H(\Pi, H(Z_2, X_2))$$

is not empty, where Π is $B \times X_1$, $B \times X_2$, $X_1 \times X_2$, $B \times X_1 \times X_2$.

Otherwise we proceed again as before: seek X_3 such that $H(X_3, H(Z_2, X_2)) \neq \varnothing$, $g_3 \in H(Z_3, X_3)$ when $Z_3 \in \mathfrak{S}_0$ is such that $H(Z_3, X_3) \neq \varnothing$, seek Φ_{g_3}, and so forth. Thus, we see immediately that the embeddability of a mapping α as a metabolic map of an (M, R) system means that the above process must terminate; roughly this means that there shall not be too many empty sets $H(X, Y)$ in $\mathfrak{S}_0$.

Let us give an indication of what kind of category $\mathfrak{S}$ allows these luxuries.

Definition. A sequence $A_1, A_2, A_3, \ldots$ of objects in a category will be called *normal* if no pair A_i, A_{i+1} occurs infinitely often in the sequence.

The category $\mathfrak{S}_0$ will be called *normal* if, for any normal sequence of objects of $\mathfrak{S}_0$, the associated sequence

$$H(A_1, A_2), H(A_2, A_3), H(A_3, A_4), \ldots$$

contains only a finite number of empty sets.

Lemma. If $\mathfrak{S}_0$ is normal, and possesses infinitely many objects, then to each object $X \in \mathfrak{S}_0 \; \exists \; Z \in \mathfrak{S}_0$ such that $H(Z, X) \neq \varnothing$.

PROOF: Let $Z_1, Z_2, Z_3, \ldots$ be a sequence of pairwise distinct object in $\mathfrak{S}_0$. Consider the sequence

$$Z_1, X, Z_2, X, Z_3, X, \ldots.$$

This is clearly a normal sequence, and the result follows from the normality of $\mathfrak{S}_0$.

Theorem. If $\mathfrak{S}_0$ is normal, then any map $\alpha \in H(X, Y)$, $X, Y \in \mathfrak{S}_0$, can be embedded as a metabolic map of an (M, R) system.

PROOF: Suppose not. Then $\exists\ f \in H(X_0, Y_0)$, which cannot be so embedded. Hence, in particular

$$H(Y_0, H(X_0, Y_0)) = \varnothing$$

and, pursuing the previous argument,

$$H(Y_1, H(X_1, Y_1)) = \varnothing,$$
$$H(Y_2, H(X_2, Y_2)) = \varnothing,$$
$$H(Y_n, H(X_n, Y_n)) = \varnothing, \ldots.$$

Consider now the sequence

$$Y_0, H(X_0, Y_0), Y_1, H(X_1, Y_1), Y_2, H(X_2, Y_2), \ldots.$$

This is a normal sequence, and the theorem follows from the resulting contradiction.

Corollary. Any finite family of mappings in a normal category can be embedded into an (M, R) system. The converse assertion, that any category $\mathfrak{S}_0$ such that any map $f \in H(A, B)$ can be embedded into an (M, R) system is normal, is *false*. However, various simple partial converses may be stated.

Theorem. If every α is embeddable, and if A, B are objects in $\mathfrak{S}_0$ such that $H(X, H(A, B)) = \varnothing$ for all $X \in \mathfrak{S}_0$, then $H(A, B) = \varnothing$.

Theorem. If every α is embeddable, and if $\exists\ X \in \mathfrak{S}_0$ such that $H(Z, X) = \varnothing$ for all $Z \in \mathfrak{S}_0$, then $H(X, H(X, H(A, B))) = \varnothing$ for every pair A, B $\in \mathfrak{S}_0$.

REMARK 1: These theorems mean that there are no "Garden of Eden" objects or mappings in the theory of (M, R) systems.

REMARK 2: In a normal category, and map can be either a "metabolic" or a "genetic" map in an (M, R) system. It is easy to see that by making slight restrictions on $\mathfrak{S}_0$, we can preserve the result, except that every map $\alpha \in \mathfrak{S}_0$ could be *either* a "metabolic" map *or* a "genetic" map, but not both.

REMARK 3: Our theorem gives us information about *how large an (M, R) system is required*, on the average, to embed an arbitrary map $\alpha \in \mathfrak{S}_0$. We may get *quantization*.

REMARK 4: If $\mathfrak{S}_0$ is normal and $F: \mathfrak{S}_0 \to \mathscr{A}$ is a functor, then $F(\mathfrak{S}_0)$ is a normal subcategory of $\mathscr{A}$.

REMARK 5: Different categories $\mathfrak{S}_0$ give rise to different classes of (M, R) systems, that is, to different *abstract biologies*. Indeed, the class of (M, R) systems definable in a given category may be taken as an index of the structure of the category.

C. The Behavior of Abstract Metabolism–Repair Systems in Changing Environments†

Let us begin by the study of particular examples. Let us consider the simplest kind of (M, R) system, which we can write as $\{f, \Phi_f\}$ where $f \in H(A, B)$ and $\Phi_f \in H(B, H(A, B))$. That is, $\mathscr{A} = \{A, B\}$, $\mathscr{F} = \{f\}$, $\mathscr{F}' = \{\Phi_f\}$, and $\mathscr{A}' = \{B, H(A, B)\}$. Each $a \in A$ specifies an *environment* of the system.

The situations of stable operation of the system are those environments $a \in A$ such that

$$\Phi_f(f(a)) = f. \tag{1}$$

Thus, the "metabolic" structure is "stable" in this environment. Now let us suppose we change the environment, by letting a vary in A. Let $a' \neq a \in A$ be given. The property (1) holds if and only if either

$$f(a) = f(a') \qquad \text{or} \qquad \Phi_f(f(a')) = f.$$

Next, suppose a' is such that

$$\Phi_f(f(a)) = f' \neq f.$$

This means that the metabolic machinery has been completely changed for this system. We still have an (M, R) system, except that now $\mathscr{F} = \{f'\}$. What are the possibilities?

(a)
$$f'(a') = f(a),$$

or more generally,

(b)
$$\Phi_f(f'(a')) = f.$$

In these cases, the metabolic structure of the system undergoes periodic changes in time. Or, we could have

(c)
$$\Phi_f(f'(a')) = f'.$$

In this case the environment a' causes the system to make a permanent transition to a new metabolic form.

Finally, it may happen that

(d)
$$\Phi_f(f'(a')) = f'' \neq f, f'.$$

We may iterate the possibilities as before. In general, it will be seen that the effect of environmental change on the (M, R) system is to cause it to "hunt" through $H(A, B)$, changing its "metabolic" structure through a sequence of maps,

$$f, f', f'', \ldots, f^{(n)}, \ldots.$$

†See Rosen [1961, 1963b]. For the effect of environmental alteration on the replication of repair components, see Rosen [1966a, 1967].

This process will stop if $\exists$ n such that

(a) $$\Phi_f(f^{(n_0)}(a')) = f^{(n_0)},$$

or more generally

(b) $$\Phi_f(f^{(n_0)}(a')) = f^{(n_0-k)}, \qquad k = 1, \ldots, n_0 - 1.$$

The case (a) means that the system becomes *stable* in the new environment. The case (b) means that the system goes through *periodic* changes in "metabolic" structure. If no such n_0 exists, the system is *unstable* and *aperiodic* in its metabolic structure.

REMARK: This last possibility requires $H(A, B)$ to be *infinite*.

Throughout the above discussion it has been tacitly supposed that the repair mapping Φ_f remains the same throughout the various environmental alterations we have described. However, we may inquire whether a change in the environment of the system might not change the repair mapping of the system, via the replication scheme we described in Section III.D. Indeed, it would be a drawback to discover that simple environmental alterations could produce changes in the repair mappings; in real cells this would amount to a change in the genome itself, due in a deterministic fashion to alterations in cellular environments.

Thus, let us once again alter the environment by replacing $a \in A$ by $a' \in A$. Then at this stage the induced replication map is changed from $\beta_{f(a)}$ to $\beta_{f(a')}$, where

$$\beta_{f(a)} = \widehat{f(a)}^{-1}, \qquad \beta_{f(a')} = \widehat{f(a')}^{-1}$$

and

$$\widehat{f(a)}(\Phi_f) \equiv \Phi_f(f(a)), \qquad \widehat{f(a')}(\Phi_f) \equiv \Phi_f(f(a'))$$

(*assuming* that the inverse maps exist). But now applying $\beta_{f(a)}, \beta_{f(a')}$, respectively to these last two relations, we find that

$$\beta_{f(a)}(\Phi_f(f(a))) = \beta_{f(a')}(\Phi_f(f(a'))) = \Phi_f,$$

which shows that the new replication map $\beta_{f(a')}$, *if it exists*, replicates the existing repair component Φ_f perfectly.

The same result holds if we replace the original metabolic component f by some other mapping h in $H(A, B)$, either by environmental alteration or otherwise. For again we have by definition

$$\widehat{h(a')}(\Phi_f) = \Phi_f(h(a'))$$

and applying the mapping $h(a')^{-1}$, *if it exists*, to both sides of this relation, we find that

$$\beta_{h(a')}(\Phi_f(h(a'))) = \Phi_f.$$

Thus, the induced replication maps under all conditions replicate the original repair component of the system; in other words, these repair components are insulated against "Lamarckian" alterations caused solely by changes in metabolic environment. This behavior arises solely from the functional organization of the system as an (M, R) system and not because of any physical properties of particular realizations of the system.

There is, according to the above analysis, only one way remaining in which alterations of environment can affect replication, and that is for the required replication map to fail to exist. Let us see whether such a failure can actually come about; this analysis will be seen to throw interesting light on the entire category from which the (M, R) system is constructed.

For a replication map to exist, we recall that the induced mapping b must be one to one, that is,

$$\hat{b}(\Phi_1) = \hat{b}(\Phi_2) \qquad \text{implies} \qquad \Phi_1 = \Phi_2.$$

But by definition of the induced maps, this means precisely that

$$\Phi_1(b) = \Phi_2(b) \qquad \text{implies} \qquad \Phi_1 = \Phi_2$$

whenever Φ_1, Φ_2 are mappings in the set $H(B, H(A, B))$ in our category. But this means that *two repair maps in our category which agree on any element of their domains must be identical.* This result may be regarded as the abstract version of the celebrated one-gene–one-enzyme hypothesis.

But it also follows from this analysis that, if the replication map exists for *any* environment of our system, then it exists for *all* environments. And this existence is seen to depend ultimately on the size of the set $H(B, H(A, B))$ in the category. Specifically, if this set of mappings is too large, then there will have to be mappings in the set which agree on some element of their common domain, and in this case no replication is possible. Thus the requirement of replication in (M, R) systems (at least according to the scheme we have proposed) places an *upper bound* on the sizes of the sets of mappings in the category. This makes an interesting comparison with one of the results of Section, IV.B, which showed that for (M, R) systems to exist at all within the category, the sets of mappings in the category could not get too small; that is, they were *bounded below.* Thus, the existence of replicating (M, R) systems can occur only in a highly restricted class of categories, and can be regarded as an index of the structure of such categories.† Some further results along these lines may be found in Rosen [1966a, 1967].

D. The Problem of Control and Reversibility in Metabolism–Repair Systems

We have seen in Section IV.C that the metabolic components, which correspond roughly to the *phenotype* of an (M, R) system, can be altered in various ways according to the characteristics of the environment in which the

---

†This result is also of interest for the existence of cellular types of organizations in different physicochemical milieus from those of earth, that is, for extraterrestrial life.

system is embedded. On the other hand, the repair structure, which corresponds roughly to the *genotype* of the (M, R) system, cannot be so affected. If we turn these arguments around, however, we begin to touch on one of the most basic questions of biology, namely: Given an (M, R) system with a particular metabolic structure ("phenotype"), is it possible, by means of an appropriate sequence of environmental alterations, to bring this system to have a *preassigned* metabolic structure (the *control problem*)? As a special case of this, we may ask whether a change in metabolic structure, arising as a result of previous environmental alterations, can be *reversed* by an appropriate sequence of further environmental alterations (the *reversibility problem*). Questions of this kind touch on some of the most important aspects of biology and medicine; for instance on problems of differentiation and development, cell–virus interactions, and carcinogenesis. As might be expected, the answers to these questions depend very heavily on the structure of the category in which we are working.

This problem may be pursued entirely in the abstract, in the fashion we have employed throughout the previous analysis. However, to show the ramifications and difficulty of this problem, we shall cast it into two rather more specific contexts.

Metabolism–Repair Systems as Sequential Machines[†]

We shall consider only the example of the simplest (M, R) system,

$$A \xrightarrow{\ f\ } B \xrightarrow{\ \Phi_f\ } H(A, B),$$

where $f \in H(A, B)$, $\Phi_f \in H(B, H(A, B))$ in the category with which we are dealing. All remarks may be readily generalized to arbitrary (M, R) systems.
We put

$$S = H(A, B); \qquad \delta(a, f) = \Phi_f(f(a)), \quad a \in A; \qquad \lambda(a, f) = f(a).$$

Then the quintuple $(S, A, B, \delta, \lambda)$ comprises a sequential machine (see Section II.B). Hence, in general any (M, R) system may be regarded as a sequential machine (which need not, of course, be finite), in which the set of states of the machine corresponds to the set of possible "phenotypes" of the system, and the input and output alphabets correspond to the input and output sets of the (M, R) system, respectively.

The problem of *control* in this context is the following: Given a sequential machine in a particular initial state, is there a sequence of input alphabet symbols which will bring the system to a preassigned state?

---

†The results described in this section, and a variety of others, may be found in a series of papers by Rosen: [1964a, 1964b, 1966b]. Related results may be found in the paper by Arbib [1966b] and a series of papers by Demetrius [1966] in the same journal, the first of which is listed in the General References.

As is well known, the control problem must generally be answered in the negative for arbitrary sequential machines. The class of sequential machines for which this control problem may always be answered affirmatively is the class of *strongly connected* machines, and so we can say that we can answer the control problem in (M, R) systems affirmatively only if the machines corresponding to (M, R) systems are strongly connected.

In general, of course, these machines fail to be strongly connected; that is, there exist possible "phenotypes" of the system which may be unreachable from a given initial state by any sequence of environmental alterations. In the general theory of sequential machines, it is always possible to embed an arbitrary sequential machine into a larger one which is strongly connected, simply by enlarging the input alphabet, and appropriately extending the mappings δ and λ. In the case of (M, R) systems, this kind of embedding becomes a much more subtle problem for two reasons:

(a) The input set A and the state set $S = H(A, B)$ are functorially related in the (M, R) systems, and we cannot in general enlarge the input set without also enlarging the state set.

(b) By extending the mappings f and Φ_f in the (M, R) systems, we move these mappings from the sets $H(A, B)$ and $H(B, H(A, B))$ to new sets $H(A', B)$, $H(B, H(A', B))$, respectively.

But this last set must have the property described in the extended system for replication to be possible. This is not implied by replicability of the original system.

When thus cast into the formalism of sequential machines, the control problem for (M, R) systems becomes related to the general control problem for sequential machines. However, the question of whether, and how, to embed an (M, R) system into a larger system that is strongly connected raises a number of subtleties not found in the usual theory of sequential machines.

An exactly analogous development may be pursued for (M, R) systems realized as continuous input–output systems, especially in the linear case, using the notions sketched in Section II.B. The details, however, have not yet been published.

V. Structural and Relational Models Compared

Now that the reader has had some experience with a class of relational models, it is appropriate to seek to answer the question of how to construct relationships which will allow us to interpret our relational results in structural terms. As noted in the Section I, this is important in order to allow our relational models to make contact with the structural information which constitutes most of what we know about biological systems, and which, further, our experimental techniques are designed to obtain.

As we noted earlier, relational models are explicitly geared to capture those properties of organization shared by large classes of structurally diverse systems. Such classes are presented to us naturally by our intuitions, for example, the intuition which underlies our recognition of the fact that a particular system at which we are looking is animate or inanimate. There is, within the formalism of purely structural physics or biology, no way of comparing or classifying systems of different structure, beyond relatively trivial classifications into closed and open, conservative and nonconservative, and the like. Indeed, this seems to be why it has thus far proved impossible, despite great effort, to characterize living or animate systems in purely structural terms.

Relational biology is a systematic attempt to deal with such problems of functional similarity in a rigorous way, and as such, it deals with problems which fall outside the scope of purely structural studies. It does so by dealing from the outset with large classes of physically diverse systems which are defined precisely by the sharing of some element of functional similarity. It should be noted that, although structural considerations are initially excluded from the formulation of relational models, relational biology itself is as mechanistic a theory as any reductionist structural theory, but the aim is to organize our understanding of biological processes around functional rather than structural considerations. As such, it may be regarded as playing the same kind of unifying role in biology as, for example, the action principles do within physics. That is, it is possible to understand diverse branches of physics (such as mechanics, optics, and electrodynamics) in terms of the fact that they share analogous "principles of least action" rather than in terms of the fact that all systems are built out of the same kinds of elementary particles.

Like any sharp instrument, relational models cut several ways. On the one hand, we may pass to new levels of understanding when we exhibit basic properties of organisms as a consequence of overall system-theoretic functional properties, freed in large part from the relatively nonspecific physical structures which carry this organization. Thereby we gain enormously in insight and economy of thought, in a way which has all of the advantages of axiomatization within mathematics itself. For instance, it is obviously uneconomical to prove a general group-theoretic theorem individually for many different kinds of specific groups, using in each case the detailed properties of the group elements, when the theorem depends only on the group axioms and is independent of whether the group elements are matrices or numbers or differential operators. Relational biology does this for biology.

On the other hand, the very generality of relational arguments raises important questions about the way in which these arguments are to be related to the bulk of our knowledge of biological systems, and especially to the

ideas of molecular biology, which are essentially metric in character. Biologists ultimately want to answer specific questions about *specific organisms*,
and relational concepts, dealing as they do with *classes* of organisms, must be
considered incomplete if we cannot identify them in detail with specific
features of individual biological systems. At the moment, we can only indicate
an overall strategy whereby this binding together of metric and relational
ideas can be accomplished, and leave it to the future to determine whether
this strategy is sound or whether some new strategy will need to be developed.

The basic difficulty, as we have noted, is that relational ideas by their
very nature refer to large *classes* of systems, rather than to individual systems. And within these classes which share certain aspects of relational
structure, there are systems of utmost physical diversity, as was indicated by
our discussion of dynamical analogies earlier. Therefore it is obvious that
relational arguments *alone* cannot determine the physicochemical details
of structure of the systems belonging to such a class, although they do im-
pose some constraints on these details. We need some further general condition which, when imposed in addition to the relational structure defining
the class in question, will reduce the number of systems compatible with all
our relational constraints to a small number (ideally that number is one).
These few systems can then be characterized physicochemically to a far
greater degree than would otherwise be possible on the basis of relational
structure alone.

There is a natural candidate for such an additional constraint, and one
which has indeed received a good deal of study over the years. This is the
principle of optimal design, which asserts that any real biological system
must, in physicochemical terms, be optimal with respect to some overall
construction criterion or cost function. The rationale of this principle, and
its application in specific instances, has been investigated by numerous
authors.†

How can we apply this principle so that a relational argument, designed
to apply to a class of systems, actually bears on a single specific system in
the class? Let us illustrate by an example from engineering. Suppose we want
to specify an amplifier. Functionally, an amplifier is a device which multiplies any input $F(t)$ by a constant k. If we restrict attention to purely electrical
inputs, we have already placed one purely physical constraint on our system.
Nevertheless, the class of realizations of an electrical amplifier is diverse
in many ways. We can build an amplifier embodying a variety of different
physical principles, using vacuum tubes, or transistors, or resonating cavities,
or traveling wave tubes. Even if we specify the physical principle to be em-

†See Chapter 2A in Volume III.

ployed (one further constraint), we have an infinite number of possible circuit diagrams, and for each circuit diagram an infinite number of physically distinguishable wirings which realize the diagram.

If, however, we say something to the effect that (a) the system we are interested in shall contain a pair of triodes of a definite kind, and (b) the cost of construction of the amplifier shall be minimal, then we have sharply constrained those specific physical amplifying systems consistent with these constraints, and can make many assertions of a purely physical character about the specifics of such systems. The constraint (b), of course, is a typical kind of optimal design constraint, and illustrates the way in which optimal design can be used to supplement an initial physical *ansatz* to obtain more detailed physical information about those functional systems which are optimal. This kind of optimizing of a system around an initial positing of a physical property is much like a paleontologist's reconstruction of an entire skeleton once a single bone has been supplied.

To relate the (M, R) systems to actual individual physical cells, however, requires an identification of the components which comprise the (M, R) systems in terms of the structural characteristics of the cells themselves. This problem can be approached in a number of ways, all of them leading to the conclusion that there is no simple relationship between the functional notion of component and the structure which our methods of observation discern in cells; the functional organization seems to cut across physical structures and landmarks in a complicated way, while the same physical structure seems to be simultaneously and inextricably involved in a variety of functional activities. The relationship of a component to a discernible physical structure seems to be similar to the relationship discussed in Chapter 3, Volume I of this treatise between an active site and the molecules which cooperatively fold up to produce it; in physiological terms, what we have called a component seems most nearly related to what is ordinarily termed a compartment (see Chapter 5). At the present moment, however, it is impossible to be more specific.

Once it is possible to pass from classes of analogous systems sharing the same relational structure down to individual systems (or at least small classes of systems) in the class which share a particular aspect of physical structure (say as a result of minimizing some appropriate cost function) there are three distinct and exciting areas which suggest themselves for further exploration.

1. We may envisage the construction of artificial engineering systems manifesting organizational characteristics of biological organisms. Such attempts have been pursued for a long time, especially in the area of artificial intelligence, the design of artificial biological organs, and so forth, but never in a completely systematic manner.

2. We may envisage the specification, by the means described above, of the primitive eobiotic forms of life which have long since disappeared from the planet. This would bring many questions concerning the origin of life, about which only speculation is presently possible, under the possibility of rigorous scientific investigation.

3. We may envisage the specification of realizations of biological organization with a radically different physicochemical basis; it may be expected that such work would throw light on the possibility of extraterrestrial life on planets where terrestrial biochemistry is impossible.

The systematic pursuit of these three questions, in conjunction with the answering of questions about terrestrial organisms of great biomedical importance (see the discussion of control in (M, R) systems above) bespeaks the fact that relational biology, as yet in its earliest infancy, has a rich future before it.

General References

Arbib, M. A. [1966a]. "Brains, Machines, and Mathematics." McGraw-Hill, New York.
Arbib, M. A. [1966b]. Categories of (M, R)-systems, *Bull. Math. Biophys.* **28**, 511–517.
Courant, R., and Hilbert, D. [1953]. "Methods of Mathematical Physics," Vol. I. Wiley (Interscience), New York.
Demetrius, L. A. (1966). Abstract biological systems as sequential machines—Behavioral reversibility, *Bull. Math. Biophys.* **28**, 153–160.
Eilenberg, S., and MacLane, S. [1945]. General theory of natural equivalence, *Trans. Amer. Math. Soc.* **58**, 231–294.
Eilenberg, S., and Wright, J. B. [1967]. Automata in general algebras, *Amer. Math. Soc. Colloquium Lectures, 72nd Summer Meeting.*
Foster, B. L. [1966]. Re-establishability in abstract biology, *Bull. Math. Biophys.* **28**, 371–374.
Kan, D. M. [1958]. Adjoint functors, *Trans. Amer. Math. Soc.* **87**, 294–329.
Martinez, H. M. [1964]. Toward an optimal design principle in relational biology, *Bull. Math. Biophys.* **26**, 351–365.
Mesarović, M. D., ed. [1964]. "Views on General Systems Theory." Wiley, New York.
Mitchell, B. [1965]. "Theory of Categories." Academic Press, New York.
Rashevsky, N. [1954]. Topology and life: In search of general mathematical principles in biology and sociology, *Bull. Math. Biophys.* **16**, 317–348.
Rashevsky, N. [1955a]. Note on a combinatorial problem in topological biology, *Bull. Math. Biophys.* **17**, 45–50.
Rashevsky, N. [1955b]. Some theorems in topology and a possible biological implication, *Bull. Math. Biophys.* **17**, 111–26.
Rashevsky, N. [1955c]. Some remarks on topological biology, *Bull. Math. Biophys.* **17**, 207–218.
Rashevsky, N. [1955d]. Life, information theory, and topology, *Bull. Math. Biophys.* **17**, 229–235.
Rashevsky, N. [1956a]. The geometrization of biology, *Bull. Math. Biophys.* **18**, 31–56.
Rashevsky, N. [1956b]. Contributions to topological biology: Some considerations on the primordial graph and on some possible transformations, *Bull. Math. Biophys.* **18**, 113–128.

Rashevsky, N. [1956c]. What type of empirically verifiable predictions can topological biology make?, *Bull. Math. Biophys.* **18**, 173–188.

Rashevsky, N. [1957]. Remark on an interesting problem in topological biology, *Bull. Math. Biophys.* **19**, 205–208.

Rosen, R. [1958a]. A relational theory of biological systems I, *Bull. Math. Biophys.* **20**, 245–260.

Rosen, R. [1958b]. The representation of biological systems from the standpoint of the theory of categories, *Bull. Math. Biophys.* **20**, 317–341.

Rosen, R. [1959]. A relational theory of biological systems II, *Bull. Math. Biophys.* **21**, 109–128.

Rosen, R. [1961]. A relational theory of structural changes induced in biological systems by alterations in environment, *Bull. Math. Biophys.* **23**, 165–171.

Rosen, R. [1962]. A note on abstract relational biologies, *Bull. Math. Biophys.* **24**, 31–38.

Rosen, R. [1963a]. Some results in graph theory and their application to abstract relational biology, *Bull. Math. Biophys.* **25**, 231–241.

Rosen, R. [1963b] On the reversibility of environmentally induced alterations in abstract biological systems, *Bull. Math. Biophys.* **25**, 41–50.

Rosen, R. [1964a]. Abstract biological systems as sequential machines I, *Bull. Math. Biophys.* **26**, 103–111.

Rosen, R. [1964b]. Abstract biological systems as sequential machines II, *Bull. Math. Biophys.* **26**, 239–246.

Rosen, R. [1966a]. A note on replication in (M, R)-systems, *Bull. Math. Biophys.* **28**, 149–151.

Rosen, R. [1966b]. Abstract biological systems as sequential machines III, *Bull. Math. Biophys.* **28**, 141–148.

Rosen, R. [1967]. Further comments on replication in (M, R)-systems, *Bull. Math. Biophys.* **29**, 91–94.

Rosen, R. [1968]. Some comments on the physico-chemical description of biological activity, *J. Theor. Biol.* **18**, 380–386.

von Bertalanffy, L., and Rapoport, A., eds. [1959–1968]. "General Systems Theory Yearbook." Society for General Systems Research, Washington, D.C.

Woodger, J. H. [1952]. "Biology and Language." Cambridge Univ. Press, London and New York.

Zadeh, L. A., and Desoer, C. A. [1963]. "Linear Systems Theory." McGraw-Hill, New York.

Chapter 5

COMPARTMENTS

Aldo Rescigno

Department of Physiology
University of Minnesota
Minneapolis, Minnesota

James S. Beck

Division of Medical Biophysics
Faculty of Medicine
University of Calgary
Calgary, Alberta

I. The Concept of Compartment

A. Physiology

The concept of compartment has no fixed peculiarities for physiology as distinct from other fields. Yet circumstances which arise frequently in one field or another may serve the emphasize certain aspects of compartment theory. One of the least complicated instances of application of compartment theory occurs in the physiological studies of exchange of inert gases in mammals. The phenomenon of nitrogen absorption by and elimination from the various tissues via the lung and circulation was studied experimentally in dogs and man from the viewpoint of determining cardiac output, determining body composition, and prevention of decompression sickness by Behnke *et al.* [1935]. They represented their results (shown in Fig. 1), ob-

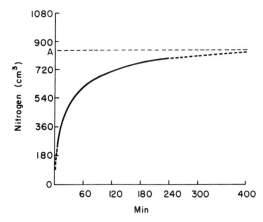

Fig. 1. Elimination of nitrogen from human subjects breathing pure oxygen, described by the equation $Y = A(1 - e^{-kt})$ according to Behnke *et al.* [1935].

tained by measuring accumulated nitrogen elimination from human subjects breathing pure oxygen, with the equation

$$Y = A(1 - e^{-kt}), \tag{1}$$

where Y is the amount of nitrogen eliminated up to time t, A is the total amount of nitrogen contained by the body at $t = 0$, the instant breathing of pure oxygen began.

We ignore their characterization of k and develop its meaning along more modern lines. This equation is a representation of the variation in the total amount of a substance lost to an infinite volume from a single finite volume. The assumptions implicit in this physical statement of the model are made clear by the differential equation from which the equation of Behnke *et al.* may be derived,

$$dY/dt = k(A - Y), \qquad Y(0) = 0.$$

The solution of this equation is the derived one if an only if k is a constant which represents the fraction of nitrogen in the body eliminated per unit time. As will be shown, this implies a closed system of two compartments with one-way transport.

Behnke *et al.* go on to a closer examination of their results, based on a longer observation time. They propose that the washout curve given by their data is a superposition of two curves, one representing the elimination of nitrogen from aqueous tissues and the other representing elimination from fat and other components of the body. These investigators did their experiments in such a way that the relatively rapid washout of nitrogen from the gas phase within the chest was not apparent. They expressed their results, then, as a sum of two exponential terms and a constant. In effect they were using a three-compartment closed system as a model, wherein one compartment represented the body fat plus lipoid tissues, another the aqueous spaces, and the third the expired breath and oxygen of the subject's closed environment. The effectively infinite volume of this last compartment makes the transport of nitrogen essentially one way.

The differential equations describing the dynamics of this system can be written in two ways, corresponding to two possible configurations of the system:

$$Z \xrightarrow{k_2} X \xrightarrow{k_1} Y$$

lipoid aqueous environ-
ment

Series model

$$Z \xrightarrow{k_3} Y$$
lipoid

$$X \xrightarrow{k_4} Y$$
aqueous enviroment

Parallel model

Series model		Parallel model	
$dY/dt = k_1 X,$	$X(0) = X_0,$	$dY/dt = k_3 Z + k_4 X,$	$X(0) = X_0,$
	$Z(0) = Z_0,$		$Z(0) = Z_0,$
$dX/dt = k_2 Z - k_1 X,$	$Y(0) = 0,$	$dX/dt = -k_4 X,$	$Y(0) = 0,$
	$X_0 + Z_0 = A,$		
$dZ/dt = -k_2 Z.$		$dZ/dt = -k_3 Z,$	$X_0 + Z_0 = A.$

The solutions for $Y(t)$, the observed variable in these sets of simultaneous differential equations, are both of the form

$$Y = A + B \exp(-k_1 t) + C \exp(-k_2 t),$$

and in both instances the constants k_i are fractional turnover rates per unit time for transport between two compartments. In both cases, also, the equilibrium value A is the same. The constants B and C are different for the two cases, however, and provide a way of determining the more appropriate model, assuming one result is reasonably close to the actual data and the other is not. It is easy to check that in the series model

$$B = k_2/(k_1 - k_2)Z_0 - X_0, \qquad C = k_1/(k_2 - k_1)Z_0$$

and in the parallel model

$$B = -X_0, \qquad C = -Z_0.$$

B. Pharmacology

In 1937 Teorell published a systematic study of the kinetics of drugs introduced into the mammalian body in various ways [1937a, b]. As in the analysis discussed above the assumption about the transport and the definition of the regions or compartments wherein measurements are to be made lead to a set of linear differential equations with constant coefficients. Beyond that, however, two other interesting considerations appear in this paper.

One is the idea of chemical transformation as a route between compartments where the latter term now has a more general meaning. Teorell's concern was the disappearance of a drug from blood or tissue. The activity of the drug, being dependent upon its chemical form, could decrease in kinetically identical ways by transport to another spatial region (elimination) or by transformation to another chemical form (inactivation). Thus, compartment is defined here as a state characterized by spatial localization and chemical nature. This is a useful generalization that will be discussed in the next section.

The other idea is the distinction between what one may call Fick kinetics and what one may call stochastic kinetics. We dealt with stochastic kinetics in the first example of transport of inert gases between compartments associated with pulmonary function. In this case the particles which collectively constitute a variable associated with a compartment each have a constant probability of transport from that compartment to any other. The instantaneous rate of loss thus is proportional to the number of particles (amount of the substance) present at that instant. A set of n such compartments then is represented by a set of n equations

$$dX_i/dt = \sum_{j \neq i}^{1 \cdots n} k_{ji} X_j - K_i X_i, \qquad i = 1, 2, \ldots, n, \tag{2}$$

where X_i is the amount of the substance (number of particles) in the ith compartment, the constant k_{ji} is the fraction of the substance in the jth compartment transported into the ith compartment per unit time, and the constant $K_i = \sum_{j \neq i}^{1 \cdots n} k_{ij}$ is the total fractional efflux from the ith compartment. This is the kinetic form attributed by Teorell to the resorption of a drug from a subcutaneous depot.

On the other hand, Teorell assumed for transport between blood and tissues what one may call Fick kinetics. This may be expressed by the equation

$$\phi = A(\psi_j - \psi_i),$$

where ϕ is the net flux from compartment j to compartment i, ψ_i is the activity in compartment i, and A is a constant. Here the driving force for transport is activity, a thermodynamic quantity, rather than an amount of substance. Then with the assumptions that the activity of a chemical entity is adequately approximated by its concentration and that the rate of change of concentration in a homogeneous constant volume is proportional to the net flux across its boundary, we have the relations

$$dC_i/dt = \sum_{j \neq i}^{1 \cdots n} h_{ij}(C_j - C_i), \qquad i = 1, 2, \ldots, n, \tag{3}$$

where h_{ij} is the *permeability constant* for the barrier of constant thickness

and area between the compartments i, j. These n equations represent the
kinetics of the system of compartments governed by Fick kinetics.

Equations (2) are more general than Eqs. (3). This is to be expected, for
the latter set follows from physical conditions that narrow the applicability.
We can show the relations between these sets of equations and the respective
parameters and variables as follows. Consider Eqs. (3), where the symbols
C and h have the meanings given above. Then, we define a variable X by the
equations

$$C_i = X_i/V_i, \tag{4}$$

where V is simply a parameter which is independent of time. Then, with
Eq. (4), we have from Eq. (3)

$$dX_i/dt = \sum_{j\neq i}^{1\cdots n} h_{ij}V_i(X_j/V_j - X_i/V_i). \tag{5}$$

Now we define new parameters,

$$k_{ji} = h_{ij}V_i/V_j, \qquad K_i = \sum_{j\neq i}^{1\cdots n} h_{ij}. \tag{6}$$

The definitions (6) substituted into (5) yield Eq. (2).

Formally, then, the Fick kinetics is a special case of stochastic kinetics
where the definitions (4) and (6) hold. Again formally, whatever X_i is, k_{ji}
is the instantaneous time rate of increase of X_i due to X_j expressed as a
fraction of X_j. Given the physical interpretations of C and h, one might
choose to regard V as a volume, which then leads to the interpretation of X
as an amount. Then k_{ji} becomes the *fractional turnover rate*, the fraction per
unit time of X_j contributed to X_i. Though Eqs. (3) are very restrictive, the
special case of Fick kinetics is an important one, having wide use as a model
for biological transport processes.

Returning to Teorell, we note some of his results. He wrote four differential equations representing (a) resorption, that is, passage between subcutaneous depot and blood; (b) elimination, the passage from blood to urine;
(c) tissue take-up, the exchange between blood and tissue; and (d) tissue
inactivation. Processes a and c were represented by Fick-type equations;
thus

$$\text{rate of resorption} = k_1(x/V_1 - y/V_2),$$
$$\text{rate of tissue take-up} = k_3(y/V_2 - z/V_3).$$

Here x is the amount of drug in the depot, y is the amount of drug in the
blood, z is the amount of drug in the tissue, V_1 is the volume of the depot,
V_2 is the volume of the blood, and V_3 is the volume of the tissue. However,
V_1 is very small compared with V_2, in man approximately 50 ml versus 5
liters. Therefore he simplified to

$$\text{rate of resorption} = (k_1/V_1)x.$$

Processes b and d were considered to be "monomolecular reactions," that is, reactions of order one, thus

$$\text{rate of elimination} = k_2 y / V_2,$$
$$\text{rate of inactivation} = k_4 z / V_3.$$

Combining these equations Teorell wrote

$$dx/dt = (-k_1/V_1)x,$$
$$dy/dt = (k_1/V_1)x - (k_2/V_2)y - [(k_3/V_2)y - (k_3/V_3)z],$$
$$dz/dt = (k_3/V_2)y - (k_3/V_3)z - (k_4/V_3)z,$$

and found that the amounts in blood and tissue as functions of time are sums of three exponential terms with constant coefficients. These are shown graphically in Fig. 2.

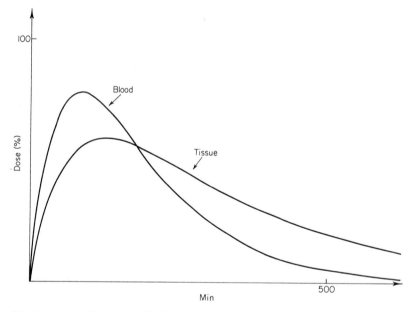

Fig. 2. Amount of a drug in the blood and in the tissue as functions of time, consisting of the sum of three exponential functions, according to Teorell. [Adapted from Teorell, 1937a].

C. Tracer Kinetics

The two examples discussed above represent the historical origins of the concept of compartment in physiological and pharmacological problems. The concept of compartment was introduced in the nitrogen problem as a geometric space defined by certain physical and physical chemical properties.

In Teorell's work the idea was extended to include chemical transformations. In both cases the substance followed was assumed to be chemically identifiable and the mathematics used was presented rather informally.

In a 1938 paper Artom *et al.* presented a radioactive tracer study of the formation of phospholipid as affected by dietary fat, in which they gave a more formal analysis than that of the two previous examples. They administered inorganic phosphate containing radioactive ^{32}P to rats and measured the radioactivity present in inorganic phosphate of blood, in the lipid of liver and in the skeleton at known times after administration. The physical correlate of compartment, then, is a state determined by the simultaneous existence of a particular location in space and a particular chemical state. For example, the variable representing the amount of ^{32}P in inorganic form in blood is a compartment and is distinct from the variable representing inorganic ^{32}P in the liver and distinct as well as from that representing lipid ^{32}P in blood.

As a basis for their analysis Artom *et al.* [1938] specify four assumptions:

(a) that the organism is incapable of distinguishing between ^{32}P and ^{31}P; (b) that the quantity of P fixed in any form whatever (for example, as lipid P) by a tissue per unit time is proportional to the amount of inorganic P in the blood; and, similarly, that the amount of inorganic P which, in the same time, is returned to the blood from the considered form is proportional to the amount of P present in that form in that tissue; (c) that the total amount of P in the tissues remains constant during the experiment; (d) that the quantity of P administered is sufficiently small that it does not modify the metabolism of the animal [p. 257].

They then define the following symbols:

1. N_s, N_f, N_ω represent the number of atoms of ^{31}P of the form of interest in blood, liver, and skeleton, respectively.

2. n_s, n_f, n_ω represent the analogous numbers of atoms of ^{32}P.

3. f/N_s represents the probability per unit time of fixation in the form of interest of a given atom of inorganic P by the liver.

4. ω/N_s represents the analogous probability for fixation by bone.

From these assumptions and definitions and the additional assumption that no other appreciable exchange of P occurs, three differential equations follow:

$$dn_s/dt = -(f + \omega)n_s/N_s + fn_f/N_f + \omega n_\omega/N_\omega,$$
$$dn_f/dt = fn_s/N_s - fn_f/N_f, \qquad (7)$$
$$dn_\omega/dt = \omega n_s/N_s - \omega n_\omega/N_\omega.$$

These three equations are analogous with Eqs. (2), where, say, $k_{fs} = f/N_f$, and so forth. The solutions as functions of time, Artom *et al.* go on to say, are in general sums of three exponentials. The constants of the exponents are characteristic of the system, that is, they depend upon f, ω, N_s, N_f, N_ω;

the coefficients on the other hand are constants dependent upon these parameters and the initial conditions of the experiment.

It is of interest to note here that the parameters f, ω play a two-way role in this case as does the permeability parameter h_{ij} in the case of Fick kinetics in Section I.B. The reason is quite different, however. In this case f, ω are numbers of atoms per unit time transported between compartments. Hence the number of atoms per unit time transported from blood inorganic P to liver lipid P, say, is f. The probability per unit time of transport for a single atom, then, is f/N_s and the number per unit time of radioactive atoms transported is fn_s/N_s. That the same parameter f appears in the term for transport from liver lipid P to blood inorganic P is required by the assumption (c) quoted above. It should be clear that the probabilities per unit time of transport between liver lipid and blood phosphate $(f/N_f, f/N_s)$ are not necessarily equal in the two directions. Furthermore, if there were a path for transport from liver to bone not including blood inorganic P, then this steady-state assumption would not imply the single parameter f for both directions.

The following sections will consist of a formal development of the concepts in the three examples examined and some generalizations which broaden the applicability of the analysis.

II. General System Properties

A. Definition of a System

A system is a set of interacting parts that are viewed as a whole; in all empirical sciences the system represents an abstraction that has to be used when investigating the properties of real objects. Indeed we cannot examine all attributes of the objects under study; we must confine ourselves to some of them. We shall consider those attributes which are quantities we can measure as functions of time; the set of these quantities defines the system.

If n is the number of the quantities defining a system, the set of n values of those quantities for a particular value of t defines a point P in an n-dimensional space called *phase space*. To describe the behavior of a system it is sufficient to describe the path of P in the phase space. If P does not move, the system is *static*; if P describes a continuous line, the system is *continuous*. The line described by P, seen as a parametric function of time, is called a *trajectory*.

We shall represent the quantities defining a system by the symbols $x_i(t)$, $x_j(t), \ldots$, and make the reasonable hypothesis that they are all functions defined and continuous for $t \geqslant 0$. A second, fundamental hypothesis is that the system is static in the absence of any external causes. This hypothesis seems at first to be rather restrictive, but it is easily met by redefining a system such that the defining quantities are the differences between the quantities of the original system and the corresponding quantities of the same system in the absence of external causes. The problem of defining the external causes need not be a difficult one; an external cause is any accident (in the Aristotelian sense) that was found experimentally to influence the system, that is, to cause a motion of P in the *phase space*, but that we choose not to include with the quantities defining the system.

A third hypothesis, not strictly necessary but very convenient for introducing a considerable simplification in the notation, is that all variables are zero when the system is at rest. When the system is redefined as above to make it static in the absence of external causes, this last condition is automatically met.

The functions $x_i(t)$, $x_j(t)$, . . . can measure either extensive properties or intensive properties of the system, and need not, for the present discussion, be dimensionally homogeneous quantities. More restrictive hypotheses will be necessary only later.

When an external cause is made to act on a system, point P leaves its position of rest and starts a trajectory. Suppose that the external cause is elementary enough to act directly on one quantity only of the system, say $x_i(t)$. Any other quantity, say $x_j(t)$, is not directly affected by that external cause, but may change as a consequence of the internal changes of the system. If it does so change, we will call $x_i(t)$ the *precursor* of $x_j(t)$, and $x_j(t)$ the *successor* of $x_i(t)$, inasmuch as the internal structure of the system is such that the variations of $x_i(t)$ are the causes of the variations of $x_j(t)$. Of course in general each function of a system may have many different precursors and many different successors.

B. Linear Systems

Let $x_i(t)$ be the precursor and $x_j(t)$ the successor in a given system; if there exists a function $g_{ij}(t, \tau)$ such that the equation

$$x_j(t) = \int_0^t x_i(\tau)g_{ij}(t, \tau)\, d\tau \qquad (8)$$

holds independently of the actual intensity and form of the cause acting on $x_i(t)$, then the system is said to be *linear* with respect to the variables $x_i(t)$ and $x_j(t)$. The linearity of a system may be restricted to a certain range of the variables $x_i(t)$ and $x_j(t)$.

If the two functions $y_i(t)$ and $y_j(t)$ represent the same variables as $x_i(t)$ and $x_j(t)$ in another experiment, and the system is linear, Eq. (8) becomes

$$y_j(t) = \int_0^t y_i(\tau)g_{ij}(t, \tau)\, d\tau; \qquad (9)$$

adding Eqs. (8) and (9) member by member yields

$$x_j(t) + y_j(t) = \int_0^t [x_i(\tau) + y_i(\tau)] g_{ij}(t, \tau) \, d\tau.$$

This equation shows that a linear system satisfies the *principle of superposition*; in words this principle may be stated, "To the precursor function $x_i(t)$ corresponds the successor function $x_j(t)$; to the precursor function $y_i(t)$ corresponds the successor function $y_j(t)$; then, if the system is linear, to the precursor function $x_i(t) + y_i(t)$ corresponds the successor function $x_j(t) + y_j(t)$."

The function $g_{ij}(t, \tau)$, whose existence is the defining property of a linear system, is a characteristic of a system and depends upon i and j. Note that a relationship of cause and effect between $x_i(t)$ and $x_j(t)$ is essential in the definition of linear system.

Sometimes $x_i(t)$ represents the amount of material present in a part of a system where it was initially introduced, and $x_j(t)$ the amount of the same material or of one of its products found in another part of the system; when this is the case, $x_i(t)$ may be considered the precursor and $x_j(t)$ the successor. However, if $x_i(t)$ and $x_j(t)$ are both the effects of a common cause, say the introduction of material into another part of the system, then a function $g_{ij}(t, \tau)$ may or may not exist, even if the system is linear.

Note that, according to Eq. (8), $x_j(0) = 0$; this is the third condition imposed in Section II.A. If $x_j(0) \neq 0$, Eq. (8) should have been written

$$x_j(t) = \int_0^t x_i(\tau) g_{ij}(t, \tau) \, d\tau + h_j(t),$$

and this equation could have been used to define a linear system; but in this case the principle of superposition could not have been shown to apply.

C. State-Determined Systems

The kernel in the integral of Eq. (8) may depend only on the difference $t - \tau$, and in this case that equation becomes

$$x_j(t) = \int_0^t x_i(\tau) g_{ij}(t - \tau) \, d\tau. \tag{10}$$

A system whose precursors and successors satisfy this equation is called *state determined*.

Equation (10) may be given a more evident physical meaning in this way. Define the function $x_i^*(t)$ as

$$\begin{aligned} x_i^*(t) &= 0 && \text{for} \quad t < t_0, \\ &= x_i(t - t_0) && \text{for} \quad t \geqslant t_0, \end{aligned} \tag{11}$$

and $x_j^*(t)$ as the function measuring the successor when the precursor is

$x_i^*(t)$. Because of Eq. (10) we can write

$$x_j^*(t) = \int_0^t x_i^*(\tau) g_{ij}(t - \tau) \, d\tau.$$

For $t < t_0$ the integral above is zero; therefore

$$x_j^*(t) = 0 \qquad \text{for} \quad t < t_0.$$

For $t \geqslant t_0$ the integral above may be written

$$\int_0^{t_0} x_i^*(\tau) g_{ij}(t - \tau) d\tau + \int_{t_0}^t x_i^*(\tau) g_{ij}(t - \tau) \, d\tau = \int_{t_0}^t x_i(\tau - t_0) g_{ij}(t - \tau) \, d\tau,$$

and with the change of variable $\tau = t_0 + \theta$,

$$\int_0^{t - t_0} x_i(\theta) g_{ij}(t - t_0 - \theta) \, d\theta;$$

therefore

$$x_j^*(t) = x_j(t - t_0) \qquad \text{for} \quad t \geqslant t_0.$$

In other words, by operating on $x_i(t)$ with the transformation (11), the function $x_j(t)$ has been transformed in the same way. The transformation (11) is a translation of the time axis, or a change in the origin of time. The relation between the variables of a state-determined system does not depend on the choice of the time origin.

The kernel of the integral in Eq. (10), that is, the function $g_{ij}(t)$, is called the *transfer function* between $x_i(t)$ and $x_j(t)$; as noted in the previous paragraph, the existence of a transfer function between two variables of a system necessarily implies a relationship of cause and effect between them.

Frequently $g_{ij}(t)$ as defined here is called the "weighting function," and the term "transfer function" is used for the Laplace transform of the weighting function. We regard this distinction as unnecessary and sometimes midleading, as will be shown.

D. Holonomic Systems

A system is called *holonomic* if there are l independent equations

$$\varphi_k(x_1, x_2, \ldots, x_n; t) = 0, \qquad k = 1, 2, \ldots, l, \qquad l \leqslant n. \tag{12}$$

The difference $n - l$ is called the *number of degrees of freedom* of the system.

These equations may be obtained by writing Eqs. (10) for all precursors and successors of the system. The total differentials of (12) are

$$\partial \varphi_k / \partial t \, dt + \partial \varphi_k / \partial x_1 \, dx_1 + \cdots + \partial \varphi_k / \partial x_n \, dx_n = 0$$

and we can therefore write the equations

$$\partial \varphi_k / \partial x_1 \, dx_1 / dt + \cdots + \partial \varphi_k / \partial x_n \, dx_n / dt = -\partial \varphi_k / \partial t; \qquad k = 1, 2, \cdots, l; \tag{13}$$

if the degree of freedom of the system is zero, there are n functions (12), and their Jacobian determinant

$$\begin{vmatrix} \partial\varphi_1/\partial x_1 & \cdots & \partial\varphi_1/\partial x_n \\ \cdot & & \cdot \\ \cdot & & \cdot \\ \cdot & & \cdot \\ \partial\varphi_n/\partial x_1 & \cdots & \partial\varphi_n/\partial x_n \end{vmatrix}$$

is not zero because they have been supposed independent; therefore Eqs. (13) have a unique solution for $dx_1/dt, \ldots, dx_n/dt$.

We thus can write

$$dx_i/dt = f_i(x_1, x_2, \ldots, x_n; t), \qquad i = 1, 2, \ldots, n, \qquad (14)$$

for a holonomic system with zero degrees of freedom, where the f's are functions of the x's and t through the partial derivatives of the φ's with respect to the x's and t.

If the degree of freedom $n - l > 0$, at least one determinant of order l formed with the $\partial\varphi_i/\partial x_j$ is not zero; suppose that determinant to be

$$\begin{vmatrix} \partial\varphi_1/\partial x_1 & \cdots & \partial\varphi_1/\partial x_l \\ \cdot & & \cdot \\ \cdot & & \cdot \\ \cdot & & \cdot \\ \partial\varphi_l/\partial x_1 & \cdots & \partial\varphi_l/\partial x_l \end{vmatrix};$$

if not, a relabeling of the x's will suffice. Now the $dx_1/dt, \ldots, dx_l/dt$ can be expressed by giving arbitrary values to the $dx_{l+1}/dt, \ldots, dx_n/dt$. If these last derivatives are all made equal to zero, then we can write

$$dx_i/dt = f_i(x_1, \ldots, x_l; c_1, \ldots, c_{n-l}; t), \qquad i = 1, 2, \ldots, l, \qquad (15)$$

where

$$c_j = x_{l+j} = \text{const}, \qquad j = 1, 2, \ldots, n - l.$$

We can conclude that the degree of freedom of a holonomic system can be reduced to zero by assigning some constant values to as many of its variables as the original degree of freedom.

The general integral of Eqs. (15) depends on l arbitrary constants, but for the hypotheses made in Section II.A the values of $x_1, \ldots, x_l$ are zero for $t = 0$, so that these l constants are determined. The integral thus can be written

$$x_i = h_i(t, c_1, c_2, \ldots, c_{n-l}), \qquad i = 1, 2, \ldots, l,$$

and depends on as many arbitrary constants as the degree of freedom of the system. The number l is the number of variables sufficient to describe the state of such a system at any time and its trajectory in phase space. If the system is linear with respect to all its variables, Eq. (8) can be written, using the Leibniz formula for differentiation of integrals,

$$dx_j/dt = \int_0^t x_i(\tau)\,(\partial/\partial t)g_{ij}(t, \tau)\,d\tau + x_i(t)g_{ij}(t, t) \qquad (16)$$

for any values of i and j $(i \neq j)$. This equation, compared with Eq. (14), shows that in a linear system all functions f_j are linear functions of their arguments except x_j and t. If in addition the system is state determined, Eq. (16) becomes

$$dx_j/dt = -\int_{\tau=0}^t x_i(\tau)\,dg_{ij}(t - \tau) + x_i(t)g_{ij}(0)$$
$$= -x_i(t)g_{ij}(0) + \int_{\tau=0}^t g_{ij}(t - \tau)\,dx_i(\tau) + x_i(t)g_{ij}(0)$$
$$= \int_0^t \dot{x}_i(\tau)g_{ij}(t - \tau)\,d\tau,$$

which shows that all functions f_j are linear functions of x_i and therefore of x_j, too, that they do not explicitly depend on t and that the coefficients of the x's are constant. Therefore, Eq. (14) can be written

$$dx_i/dt = a_{i1}x_1 + a_{i2}x_2 + \cdots + a_{in}x_n, \qquad i = 1, 2, \ldots, n, \qquad (17)$$

with the coefficients a_{ij} constant.

If the system had a degree of freedom $n - l$, we should have written only l equations (17) with $i = 1, 2, \ldots, l$ and the initial conditions

$$x_1 = x_2 = \cdots = x_l = 0 \qquad \text{for} \quad t = 0,$$

while the derivatives

$$dx_{l+1}/dt, \ldots, dx_n/dt$$

would have been undetermined, along with their corresponding functions $x_{l+1}, x_{l+2}, \ldots, x_n$. We therefore can pool these functions together and write the system equations

$$dx_i/dt = a_{i1}x_1 + \cdots + a_{il}x_l + \psi_i(t), \qquad i = 1, 2, \ldots, l, \qquad (18)$$

where $x_1 = x_2 = \cdots = x_l = 0$ for $t = 0$.

III. Systems of Compartments

A. The Operational Definition of a Compartment

A variable $x_i(t)$ of a system is called a *compartment* if it is governed by the differential equation

$$dx_i/dt = -K_i x_i + f_i(t) \qquad (19)$$

with K_i constant. All variables $x_i(t)$ of a linear state-determined holonomic system are compartments; in fact Eq. (18) is identical with Eq. (19) if we put

$$f_i(t) = a_{i1}x_1 + \cdots + a_{i,i-1}\,x_{i-1} + a_{i,i+1}\,x_{i+1} + \cdots + a_{il}x_l + \psi_i(t),$$
$$-K_i = a_{ii}.$$

A physical interpretation of a compartment may be the following: The variable $x_i(t)$ measures the quantity of a certain substance in a particular subdivision of a system, through which the concentration is uniform at any given time; that substance leaves that subdivision of the system at a rate proportional to its total amount there, that is, with a process of first order, with relative rate K_i; $f_i(t)$ measures the rate of entry of that substance into that subdivision of the system from other subdivisions or from outside the system. Equation (19) represents the behavior of $x_i(t)$ as determined by the behavior of $f_i(t)$; in this case $f_i(t)$ can be considered as the cause and $x_i(t)$ as the effect. This is so because if the function $f_i(t)$ is changed, the function $x_i(t)$ will change too, but Eq. (19) remains valid, that is, that equation is a sufficient description of the relationship between the precursor $f_i(t)$ and its successor $x_i(t)$ in all experimental conditions.

Integration of Eq. (19) is done with the substitution

$$x_i(t) = \exp(-K_i t)y(t)$$

and hence

$$dx_i/dt = -K_i \exp(-K_i t)y(t) + \exp(-K_i t)\,dy/dt$$
$$-K_i \exp(-K_i t)y(t) + \exp(-K_i t)\,dy/dt = -K_i \exp(-K_i t)y(t) + f_i(t)$$
$$dy/dt = \exp(K_i t)f_i(t)$$
$$y(t) = y(0) + \int_0^t \exp(K_i \tau)f_i(\tau)\,d\tau;$$

returning to the original function,

$$x_i(t) = x_i(0)\exp(-K_i t) + \int_0^t \exp[-K_i(t-\tau)]f_i(\tau)\,d\tau. \tag{20}$$

With the hypothesis $x_i(0) = 0$, we find the integral equation (10) with the transfer function $\exp(-K_i t)$ between the precursor $f_i(t)$ of a compartment and the compartment $x_i(t)$ itself.

The concept of compartment has been used extensively in the scientific literature in the last 30 years, though very few authors have tried to give it a rigorous definition. We refer the interested reader to two papers by Bergner [1961a, b] and a short note by Ličko [1965]. From an historical point of view, it is worth noting that Artom et al. [1938] (see Section I.C) were probably the first authors to use the concept of compartment in the sense of this section, though they did not use the term "compartment" at all; a footnote on p. 257 of their paper points to the limitations in the interpretation of the constants of the differential equations defining their compartments. The same problem is investigated carefully in the last chapter of a monograph by Bergner [1962].

Recently Brownell *et al.* [1968] have tried to unify the terminology used in tracer studies. Their first definition of compartment, naturally, applies only in the context of tracer experiments, but it coincides with the present one if the variables and constants of Eq. (19) are given the appropriate interpretation and if the mother substance ("tracee" in the authors' terminology) is in steady state; their second definition is more general and has no quantitative value. Earlier Berman *et al.* [1962] had given an operational definition of compartment substantially identical to the one we have adopted.

B. Pools versus Compartments

According to the previous definition, we cannot consider as a compartment the subdivision of a system whose state does not obey Eq. (19) or its equivalent (20). Here we list just a few examples of such cases.

Equation (1) given by Behnke *et al.* [1935] can be rewritten as

$$X(t) = X(0)e^{-kt},$$

where $X(t) = A - Y$ is the amount of nitrogen present in the body, and $X(0) = A$ is the total amount of nitrogen contained by the body at the instant pure oxygen breathing began. This equation corresponds to Eq. (20) if

$$X(t) = x_i(t), \qquad k = K_i, \qquad f_i(t) \equiv 0;$$

this last identity follows from the fact that no nitrogen is fed into the body after oxygen breathing begins. Equation (1) would thus indicate that the total nitrogen in the body would be a compartment, if k were a constant. If this were not the case, we would conclude that the total nitrogen in the body is not a compartment.

The equations given at the end of Section I.A, compared to Eq. (19), show that, if they are correct, both the nitrogen in the body fat plus lipoid tissues and the nitrogen in the aqueous spaces are compartments. Actually a more detailed analysis would show that even the separation of the nitrogen content of the body into two compartments is not realistic [see for instance, Gómez *et al.*, 1964].

Shemin and Rittenberg [1946] measured the turnover of erythrocytes in man using glycine labeled with ^{15}N. They found that the transfer function from hemin precursor to hemin cannot be described by a simple exponential function, but is an S-shaped curve, corresponding to the curve of survival of the erythrocytes [see Rescigno and Segre, 1966, pp. 167–170]. The hemin is therefore not a compartment.

The so-called "pulmonary transfer function," that is, the transfer function of the isotopically labeled blood from the pulmonary artery to the pulmonary vein, has been measured by Rescigno and Segre [1961a]; it is an asymmetric bell-shaped curve with a maximum between 4 and 5 sec. The blood in the lungs is therefore not a compartment [for more details see Rescigno and Segre, 1966, pp. 132 and 173; Segre *et al.*, 1965].

If two compartments x_1 and x_2 are so connected that x_1 is the precursor of x_2, then

$$dx_1/dt = -K_1 x_1 + f_1(t), \qquad dx_2/dt = -K_2 x_2 + x_1.$$

The integral of these differential equations is obtained by a simple substitution:

$$x_1 = \int_0^t \exp[-K_1(t - \tau)] f_1(\tau) \, d\tau,$$

$$x_2 = \int_0^t \exp[-K_2(t - \tau)] x_1(\tau) \, d\tau$$

$$= (K_2 - K_1)^{-1} \int_0^t \{\exp[-K_1(t - \tau)] - \exp(-K_2(t - \tau))\} f_1(\tau) \, d\tau$$

$$\text{if} \quad K_1 \neq K_2,$$

$$= \int_0^t (t - \tau) \exp[-K_1(t - \tau)] f_1(\tau) \, d\tau \qquad \text{if} \quad K_1 = K_2.$$

When the two compartments x_1 and x_2 are not observed separately, but their sum is measured instead, or more generally a linear combination $ax_1 + \beta x_2$, one has the expression

$$\alpha x_1 + \beta x_2$$

$$= (K_2 - K_1)^{-1} \int_0^t [\alpha(K_2 - K_1) + \beta] \exp[-K_1(t - \tau)] f_1(\tau) \, d\tau$$

$$- (K_2 - K_1)^{-1} \int_0^t \beta \exp[-K_2(t - \tau)] f_1(\tau) \, d\tau$$

$$\text{if} \quad K_1 \neq K_2$$

$$= \int_0^t [\alpha + \beta(t - \tau)] \exp[-K_1(t - \tau)] f_1(\tau) \, d\tau \qquad \text{if} \quad K_1 = K_2.$$

The transfer function between $f_1(t)$ and $ax_1 + \beta x_2$ is

$$(K_2 - K_1)^{-1}\{[\alpha(K_2 - K_1) + \beta] \exp(-K_1 t) - \beta \exp(-K_2 t)\}$$

$$\text{if} \quad K_1 \neq K_2,$$

$$(\alpha + \beta t) \exp(-K_1 t) \qquad \text{if} \quad K_1 = K_2;$$

in neither case is it the transfer function of a compartment, unless $K_1 = K_2 + \beta/\alpha$ or $\beta = 0$.

We thus have seen that, in general, the result of pooling together two or more compartments is not a compartment. Following Aubert and Bronner [1965], we shall call *pool* any subdivision of a system, whether it can be separated into a number of compartments or not.

Terms such as "nonhomogeneous compartment" or "discriminating compartment" have been used to imply that Eq. (19) does not apply; according to the present terminology, such an entity should be called a "pool." Of course any compartment can be considered a pool.

C. Classification of Systems

Thus far we have supposed that all variables of a system are equal to zero when the system is at rest, that is until external causes intervene. Though an external cause may in principle act on any number of compartments, we shall suppose that only one compartment, called *initial compartment*, is directly affected by an external cause. If there are n external causes, or if the same external cause acts on n compartments, then n initial compartments are present, and by the principle of superposition, the behavior of a system with n initial compartments is the sum of the behavior of n systems, each one of them with one initial compartment.

If the initial compartment is designated by x_1, then the equations of the system are, following the notation used in Eq. (18),

$$dx_1/dt = a_{11}x_1 + \cdots + a_{1l}x_l + \psi_1(t),$$
$$dx_j/dt = a_{j1}x_1 + \cdots + a_{jl}x_l, \qquad j = 2, 3, \ldots, l, \tag{21}$$

or, with a notation consistent with (19),

$$dx_1/dt = -K_1x_1 + \sum_{i=2}^{l} k_{i1}x_i + f(t),$$
$$dx_j/dt = -K_jx_j + \sum_{i \neq j}^{1 \cdots l} k_{ij}x_i, \qquad j = 2, 3, \ldots, l. \tag{22}$$

Two comments are made with respect to the change of notation from Eqs. (21) to Eqs. (22). First, we have switched from a_{ii} to $-K_i$; this change is justified by the interpretation of a compartment given in Section III.A; in the great majority of physical compartments the coefficient of x_i in Eq. (19) is negative, therefore it is convenient to call that coefficient $-K_i$. Second, the coefficients a_{ji} of Eqs. (18) or (21) have been substituted by the coefficients k_{ij}, with the order of the subscripts inverted; this inversion is not consistent with the usual algebraic notation, but has been used by several authors in compartment theory and offers some advantages in the manipulations that will be shown in later sections.

All compartments x_i in the above sums whose coefficients $k_{ij} \neq 0$, are called *precursors of order one* of x_j; x_j is their *successor of order one*. A precursor of order one of a precursor of order one of a particular compartment is a *precursor of order two* of that particular compartment. In this way a precursor of any order can be defined. A system is *connected* if its initial compartment is a precursor of any order of each of the other compartments. A system is *strongly connected* if all its compartments are precursors of all other compartments. A system is *mammillary* if it is connected and none of its noninitial compartments is a precursor of order one of any other non-initial compartment. A system is *catenary* if all its compartments can be ordered in such a way that each of them is the precursor of order one of the next, possibly of the preceding, and of none other.

When $K_i = 0$, compartment x_i is *closed*; otherwise it is *open*. Equation (21) of a closed compartment has one of the forms

$$dx_1/dt = \sum_{i=2}^{l} k_{i1}x_i + f(t),$$

$$dx_j/dt = \sum_{\substack{i \neq j}}^{1 \cdots l} k_{ij}x_i, \qquad j \neq 1,$$

which integrated gives the simple result

$$x_1 = \sum_{i=2}^{l} k_{i1} \int_0^t x_i(\tau)\, d\tau + \int_0^t f(\tau)\, d\tau,$$

$$x_j = \sum_{\substack{i \neq j}}^{1 \cdots l} k_{ij} \int_0^t x_i(\tau)\, d\tau, \qquad j \neq 1.$$

In other words, a closed compartment is a linear combination of the integrals of its precursors.

D. Topological Properties

The structure of a compartment system can be analyzed very efficiently using the methods of the theory of graphs. All properties of a compartment system can be seen as properties of its corresponding graph.

A *directed graph* is a set of *oriented arms* connecting a number of *nodes*. Each node represents a compartment and each oriented arm connects a precursor of order one to its successor. Equations (7) of Artom *et al.* [1938] represent a system of three compartments whose directed graph is illustrated in Fig. 3. As we shall consider only directed graphs and oriented arms, we can call them simply *graphs* and *arms*.

A *path* is formed by a succession of arms such that a node entered by an arm is the node at which the next arm starts. If one of the nodes is reentered, the path in which the starting node of the first arm and the entering node of the last arm coincide is called a *cycle*, or *closed path*. A path not containing

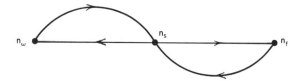

Fig. 3. Graph of the system described by Artom *et al.* [1938].

the same arm more than once is called *simple*. The *length* of a path is given by the number of its arms. A path is *elementary* if each node in it is entered only once. If a path does not contain a cycle it is called *open*.

A graph is *connected* if there is at least one node from which all other nodes can be reached. The minimum number of nodes (with their arms) that should be deleted from a connected graph in order to transform it into a nonconnected graph, is called the *connectivity* of the graph. A graph with connectivity p is also called p-connected. A *strongly connected graph* or *strong graph* is a graph where any node can be reached from any other node. A *subgraph* is a connected graph obtained by omitting some nodes from the given graph. A subgraph where each node occurs in one and only one cycle is a *linear subgraph*. Each group of cycles, in which each node of the subgraph occurs in one cycle only, is called a *strong component*.

To a graph with n nodes and m arms we can associate a square matrix A of order n (n rows and n columns); the element a_{ij} of the ith row and jth column is equal to 1 if an arm exists from node i to node j, or equal to 0 if such arm does not exist. Matrix A contains m elements equal to 1 and $n^2 - m$ equal to 0; it is called the *connectivity matrix* of the graph. All properties of a graph can be seen as properties of its connectivity matrix. For example, if there is a path formed by the nodes $i_1, i_2, \ldots, i_p$, then $a_{i_1 i_2} a_{i_2 i_3} \cdots a_{i_{p-1} i_p} = 1$.

Let us call A^2 the Boolean product $A \cdot A$, that is the matrix formed by the elements $a_{ij}^{(2)} = \sum_k a_{ik} a_{kj}$, where the sum is a Boolean sum (that is, it is equal to 0 when all terms are zero, and equal to 1 when at least one term is 1). The matrix A^2 shows the paths of length 2; its element $a_{ij}^{(2)}$ is 1 when there exists a path of length 2 from node i to node j. Similarly, the matrix A^r shows the paths of length r.

In the matrix A all elements of the principal diagonal are 0 by definition; in the matrix A^r an element $a_{ii}^{(r)}$ of the principal diagonal is 1 if there is a cycle of length r through node i.

If $A^r = 0$, that is, all elements of A^r are 0, but $A^{r-1} \neq 0$, then $r - 1$ is the length of the longest path, and of course there are no cycles. The original matrix A is called nilpotent. In a nilpotent matrix $r < n$ and no cycles are present.

The Boolean sum $A + A^2 + \cdots + A^k$ indicates the paths of any length up to k between any pair of nodes. If A is a nilpotent matrix, then all powers of A from the rth on are void, and the previous sum does not depend on k, as long as $k \geqslant r - 1$.

The matrix $R = \sum_{k=1}^{n} A^k$ is called the *reachability matrix*; any element r_{ij} of R is equal to 1 if there is a path of any length (including length 1) from node i to node j. If A is not nilpotent, the Boolean sum $A + A^2 + \cdots + A^k$ does not depend on k, as long as $k \geqslant m$, because any power of A higher than the mth shows only paths where at least one arm is contained more than once, and to each element equal to one in that matrix corresponds an element equal to 1 in a lower power of A. Therefore, when A is not nilpotent, the reachability matrix is given by $R = \sum_{k=1}^{m} A^k$.

The reachability matrix of a connected graph contains at least a row of elements equal to 1, with the possible exception of the diagonal element. The reachability matrix of a strong graph has all its elements equal to 1.

In the connectivity matrix A, the element a_{ij} shows whether i is a precursor of order one of j. In the matrix A^r, the element $a_{ij}^{(r)}$ shows whether i is a precursor of order r of j. In the reachability matrix R, the element r_{ij} shows whether i is a precursor of any order of j. The order of the precursor i with respect to the successor j is given by the minimum value of r for which $a_{ij}^{(r)} = 1$ in the matrix A^r.

The precursors of order one, or *direct precursors*, can be classified into the following types:

1. *Absolute precursor*: the only arm leaving i enters j, and the only arm entering j comes from i.

2. *Complete precursor*: the only arm leaving i enters j; if there is a cycle including j, no node of the cycle except i has an arm entering j.

3. *Complete precursor with recycling*: the only arm leaving i enters j; there is a cycle including j not passing through i.

4. *Unique precursor*: the only arm entering j comes from i.

5. *Total precursor*: there are no paths from i to j of length more than one; if there is a cycle including j, no node of the cycle except i has an arm entering j.

6. *Total precursor with recycling*: there are no paths from i to j of length more than one; there is a cycle including j not passing through i.

7. *Partial precursor*: there is at least one path of length more than one from i to j; if there is a cycle including j, it passes through i too.

8. *Partial precursor with recycling*: there is at least one path of length more than one from i to j; there is a cycle including j not passing through i.

For a more detailed analysis of the topological properties of the systems of compartments, see Rescigno and Segre [1964].

IV. Integration of the System Equations

The l equations (22) of the system are said to be written in *normal form* because they are resolved with respect to the derivatives of the functions $x_1, x_2, \ldots, x_l$.

If each of the l equations is differentiated successively $l - 1$ times, then we have a total of l^2 equations containing the functions $x_1, \ldots, x_l$ and their derivatives with respect to t up to the order l. With a process of elimination, the l^2 equations can be reduced to one containing only the function x_j and its derivatives up to the order l. In other words, the l differential equations of order one in l unknown functions $x_1, x_2, \ldots, x_l$ can be transformed into one differential equation of order l in one unknown function x_j. Of course this equation will contain the function $f(t)$ with its derivatives up to the order $l - 1$.

The general integral of such an equation is known to contain l arbitrary constants; those constants are to be determined by the initial values of the l functions $x_j, dx_j/dt, d^2x_j/dt^2, \ldots, d^{l-1}x_j/dt^{l-1}$. Remembering the condition we imposed in Section II.A, namely $x_1 = x_2 = \cdots = x_l = 0$ for $t = 0$, from the l Eqs. (22) we compute the initial values of $dx_1/dt, \ldots, dx_l/dt$; with these initial values and the derivatives of (22) we compute the initial values of $d^2x_1/dt^2, \ldots, d^2x_l/dt^2$; and so forth with the successive derivatives of (22) through the $(l - 1)$th ones. Thus, the initial values of $x_j, dx_j/dt, \ldots, d^{l-1}x_j/dt^{l-1}$ are formed by the coefficients k and K and by the initial values of $f(t)$ and of its derivatives up to the order $l - 1$.

A. The Operational Calculus of Mikusinski

The state equations of a system of compartments are easily manipulated with the use of Mikusinski operators. For a detailed exposition of these operators see Mikusinski [1959] or Erdélyi [1962]; this second book is more concise than the first, but requires a certain familiarity with the methods of abstract algebra. In this section we simply outline the main properties of the Mikusinski operators, without giving detailed proofs.

The integral $\int_0^t a(t - \tau)b(\tau)\,d\tau$, introduced in Section II.C, is called the *convolution* of the two functions $a(t)$ and $b(t)$. The functions usually associated with the systems of compartments are defined and continuous over the interval $0 \leqslant t < \infty$, therefore the convolution integral of those functions is defined and continuous over the same interval.

With the change of variable $t - \tau = \sigma$ in the previous integral, we have

$$\int_0^t a(t - \tau)b(\tau)\,d\tau = \int_0^t b(t - \sigma)a(\sigma)\,d\sigma;$$

therefore, the operation of convolution is commutative.

If we put

$$g(t) = \int_0^t a(t - \tau)b(\tau)\,d\tau, \qquad h(t) = \int_0^t b(t - \tau)c(\tau)\,d\tau,$$

then

$$\int_0^t g(t - \tau)c(\tau)\,d\tau = \int_0^t \left[\int_0^{t-\tau} a(t - \tau - \sigma)b(\sigma)\,d\sigma \right] c(\tau)\,d\tau$$

$$= \int_0^t \left[\int_\tau^t a(t - \omega)b(\omega - \tau)\,d\omega \right] c(\tau)\,d\tau$$

$$= \int_0^t a(t - \omega)\left[\int_0^\omega b(\omega - \tau)c(\tau)\, d\tau\right] d\omega$$

$$= \int_0^t a(t - \omega)h(\omega)\, d\omega;$$

therefore, the operation of convolution is associative. Moreover

$$\int_0^t a(t - \tau)[b(\tau) + c(\tau)]\, d\tau = \int_0^t a(t - \tau)b(\tau)\, d\tau + \int_0^t a(t - \tau)c(\tau)\, d\tau;$$

therefore, the operation of convolution is distributive with respect to addition.

These three properties of convolution are the same fundamental properties of the product as defined in arithmetic; therefore, the same formalism used for the traditional product may be used for the convolution. We shall call the above convolution integral "product of the functions $a(t)$ and $b(t)$," and represent it by the symbol $\{a(t)\}\{b(t)\}$. The brackets are used to distinguish between the product of functions and the ordinary arithmetic product. We shall represent the value of the function a at the point t by $a(t)$. Conversely, by the symbol, say 5, we mean the single value 5, and by the symbol $\{5\}$ we mean the function whose value is 5 for every value of the variable t. So

$$5 \cdot 6 = 30 \qquad \text{(product of numbers)},$$

$$\{5\} \cdot \{6\} = \left\{\int_0^t 5 \cdot 6 \cdot d\tau\right\} = \{30\, t\} \text{ (product of constant functions).}$$

We define the sum of two functions as the function whose values are the sums of the values of the two given functions:

$$\{a(t)\} + \{b(t)\} = \{a(t) + b(t)\}. \tag{23}$$

In the language of abstract algebra we thus have defined a *commutative ring* whose elements are functions, and whose two operations are (23) and

$$\{a(t)\} \cdot \{b(t)\} = \left\{\int_0^t a(t - \tau)b(\tau)\, d\tau\right\}.$$

We shall use the symbols

$$\{a(t)\}^2 = \{a(t)\} \cdot \{a(t)\},$$
$$\{a(t)\}^3 = \{a(t)\} \cdot \{a(t)\} \cdot \{a(t)\},$$

and so forth.

The constant function $\{1\}$ is of special importance because by definition of product

$$\{1\} \cdot \{a(t)\} = \left\{\int_0^t a(\tau)\, d\tau\right\},$$

that is, the product of a function by $\{1\}$ is equivalent to its integration between the limits 0 and t. It is easy to verify that

$$\{1\}^2 = \{t\},$$
$$\{1\}^3 = \{t^2/2\},$$

$$\cdot$$
$$\cdot$$
$$\cdot$$

$$\{1\}^n = \{t^{(n-1)}/(n-1)!\}. \tag{24}$$

An important property of the product of functions, known as the theorem of Titchmarsh, is that $\{a\} \cdot \{b\} = \{0\}$ if and only if either $\{a\} = \{0\}$ or $\{b\} = \{0\}$. The immediate consequence of this theorem is that the quotient of two functions $\{a\}$ and $\{b\}$, with $\{b\} \neq \{0\}$, can be defined as in arithmetic and is unique. By definition we write

$$\{a\}/\{b\} = \{c\}$$

when $\{b\} \neq \{0\}$ and a function $\{c\}$ exists such that

$$\{a\} = \{b\}\{c\};$$

the quotient $\{c\}$ is unique because if another function $\{c_1\}$ exists such that

$$\{a\} = \{b\}\{c_1\},$$

then

$$\{b\}(\{c\} - \{c_1\}) = \{0\},$$

and by the theorem of Titchmarsh either $\{b\} = \{0\}$ or $\{c\} = \{c_1\}$, both contrary to our hypotheses.

Although the product of two functions always exists, given two functions it is not always possible to find a third function equal to their quotient; for instance, there is no function $\{a\} = \{1\}/\{1\}$; it should be $\{1\} = \{\int_0^t a(\tau)\, d\tau\}$, which is clearly absurd.

To make the quotient of two functions always possible, with the only restriction that the function below the bar be not identically zero, we introduce the concept of *operator*. An operator is the quotient of two functions, and is sufficiently defined by the following properties:

1. Equality:

$$\{a\}/\{b\} = \{c\}/\{d\} \qquad \text{if and only if} \quad \{a\}\{d\} = \{b\}\{c\}.$$

2. Sum:

$$\frac{\{a\}}{\{b\}} + \frac{\{c\}}{\{d\}} = \frac{\{a\}\{d\} + \{b\}\{c\}}{\{b\}\{d\}}.$$

3. Product:

$$\frac{\{a\}}{\{b\}} \cdot \frac{\{c\}}{\{d\}} = \frac{\{a\}\{c\}}{\{b\}\{d\}}.$$

In the language of abstract algebra the operators form a *quotient field*.

Any function $\{f(t)\}$ can be considered an operator; in fact it can be written $\{f\}\{a\}/\{a\}$, and it is easy to verify that $\{f\} = \{g\}$ implies $\{f\}\{a\}/\{a\} = \{g\}\{b\}/\{b\}$. As already said, not all operators can be considered functions.

One of the most frequently used properties of the operators is

$$\frac{\{a\}}{\{b\}} = \frac{\{a\}\{c\}}{\{b\}\{c\}}$$

for any function $\{c\} \neq \{0\}$; it is an immediate consequence of the definition of equality of operators.

If $\{\alpha\}$ and $\{\beta\}$ are constant functions, the three defining properties of the operators become

$$\{\alpha\}/\{1\} = \{\beta\}/\{1\} \qquad \text{if and only if} \quad \{1\}\{\alpha\} = \{1\}\{\beta\}, \text{ that is}$$

$$\text{if and only if} \quad \alpha = \beta;$$

$$\{\alpha\}/\{1\} + \{\beta\}/\{1\} = (\{1\}\{\alpha\} + \{1\}\{\beta\})/\{1\}\{1\} = \{\alpha + \beta\}/\{1\};$$

$$\{\alpha\}/\{1\} \cdot \{\beta\}/\{1\} = \{\alpha\}\{\beta\}/\{1\}\{1\} = \left\{\int_0^t \alpha\beta \, d\tau\right\}/\{1\}\{1\}$$

$$= \{1\}\{\alpha\beta\}/\{1\}\{1\} = \{\alpha\beta\}/\{1\}.$$

The special operator $\{\alpha\}/\{1\}$ is called *numerical operator*; we can write for simplicity $\{\alpha\}/\{1\} = \alpha$, as the properties of this operator are shared by the ordinary numbers.

The product of a numerical operator and a function should be regarded as the product of two operators. Therefore

$$\alpha\{f\} = \{\alpha\}/\{1\} \cdot \{f\}\{a\}/\{a\} = \{\alpha\}\{f\}\{a\}/\{1\}\{a\} = \{\alpha\}\{f\}/\{1\}$$

$$= \left\{\int_0^t \alpha f(\tau) \, d\tau\right\}/\{1\} = \{1\}\{\alpha f\}/\{1\} = \{\alpha f\}.$$

For any operator $\{a\}/\{b\}$,

$$0 \cdot \{a\}/\{b\} = (\{0\}/\{1\}) \cdot \{a\}/\{b\} = \{0\}/\{1\}\{b\}$$

$$= \{0\}\{b\}/\{1\}\{b\} = \{0\}/\{1\} = 0.$$

The numerical operator 0 has the unique property of being equal to its corresponding function $\{0\}$; in fact,

$$0 = \{0\}/\{1\} = \{0\}\{1\}/\{1\} = \{0\}.$$

B. Differential Equations in Operational Notation

By definition, if $h = \{a\}/\{b\}$ is an operator, $\{b\}/\{a\}$ is the *inverse operator* of h. The inverse of the operator $\{1\}$ is called the *differential* operator and is represented by the letter s. As a consequence of this definition

$$s\{1\} = 1;$$

indeed, for any function $\{a\} \neq \{0\}$,

$$\{1\} = \{1\}\{a\}/\{a\} \qquad \text{and} \qquad s = \{a\}/\{1\}\{a\};$$

therefore

$$s\{1\} = \{a\}/\{1\}\{a\} \cdot \{1\}\{a\}/\{a\} = \{1\}/\{1\} = 1.$$

If $f(t)$ has a derivative $f'(t)$, then

$$\{f(t)\} = \left\{\int_0^t f'(\tau)\, d\tau\right\} + \{f(0)\}$$

or

$$\{f(t)\} = \{1\}\{f'(t)\} + \{1\}f(0);$$

multiplying both sides by s

$$s\{f(t)\} = \{f'(t)\} + f(0). \tag{25}$$

This justifies the name of the differential operator.

Equations (22) now can be written

$$s\{x_1\} = -K_1\{x_1\} + \sum_{i=2}^{l} k_{i1}\{x_i\} + \{f\},$$

$$s\{x_j\} = -K_j\{x_j\} + \sum_{\substack{i \neq j}}^{1\cdots l} k_{ij}\{x_i\}, \qquad j = 2, 3, \ldots, l,$$

or

$$(s + K_1)\{x_1\} - k_{21}\{x_2\} - \cdots - k_{l1}\{x_l\} = \{f\},$$
$$-k_{12}\{x_1\} + (s + K_2)\{x_2\} - \cdots - k_{l2}\{x_l\} = 0,$$

$$\vdots \tag{26}$$

$$-k_{1l}\{x_1\} - k_{2l}\{x_2\} - \cdots + (s + K_l)\{x_l\} = 0,$$

the solutions of which are given by

$$\{x_j\} = \frac{(-1)^{j+1}\Delta_{1;j}}{\Delta}\{f\}, \tag{27}$$

where Δ is the determinant

$$\begin{vmatrix} s + K_1 & -k_{21} & \cdots & -k_{l1} \\ -k_{12} & s + K_2 & \cdots & -k_{l2} \\ \cdot & & & \cdot \\ \cdot & & & \cdot \\ \cdot & & & \cdot \\ -k_{1l} & -k_{2l} & \cdots & s + K_l \end{vmatrix}$$

formed with the coefficients of the unknown functions in the system equations, and $\Delta_{i;j}$ is the determinant obtained with the suppression of the first row and the jth column from Δ. Observe that Δ can be written as a polynomial in s of degree equal to l, and that $\Delta_{i;j}$ can be written as a polynomial in s of degree equal to $l-1$ if $j=1$, equal to or less than $l-2$ if $j\neq 1$. This and other properties become more evident if we develop the determinant $\Delta_{1;j}$ according to the elements of its first column, and then each non-principal minor so obtained according to the elements of the column containing no binomial elements; the result, for $j\neq 1$, is

$$(-1)^{j+1}\Delta_{1;j} = k_{1j}{}^P\Delta_{1,j} + \sum_{i_1} k_{1i_1}k_{i_1 j}{}^P\Delta_{1,j,i_1}$$
$$+ \sum_{i_1 i_2} k_{1i_1}k_{i_1 i_2}k_{i_2 j}{}^P\Delta_{1,j,i_1,i_2} + \cdots,$$

where $^P\Delta_{a,b,\ldots}$ represents the principal minor of Δ obtained by suppressing from Δ the rows of order $a, b, \ldots$ and their corresponding columns. As all principal minors of Δ can be written as polynomials in s of degree equal to their order, with the coefficient of the term of highest degree equal to one, then the degree of the numerator in the (27), when $j\neq 1$, is $l-2$ if $k_{1j}\neq 0$, that is if x_1 is a precursor of order one of x_j; when $j=1$, the degree of the numerator is $l-1$ in any case.

The determinant Δ can be developed in a similar way to put in evidence some of its elements, for instance

$$\Delta = (s+K_i){}^P\Delta_i - \sum_{i_1} k_{1i_1}k_{i_1 i}{}^P\Delta_{i,i_1} - \sum_{i_1 i_2} k_{ii_1}k_{i_1 i_2}k_{i_2 i}{}^P\Delta_{i,i_1,i_2} - \cdots,$$

where all sums are extended to the values of $i_1, i_2, \ldots$, different from i and different from themselves [see Rescigno, 1956]. Some practical applications of the last two formulas will be shown later.

It is important to note that the right-hand member of the first Eq. (26) may be an operator instead of a function; in particular it may be a numerical operator x_0. This corresponds to the first Eq. (22), being

$$dx_1/dt = -K_1 x_1 + \sum_{i=2}^{l} k_{i1}x_i$$

with the initial condition

$$x_i = x_0 \qquad \text{for} \quad t=0.$$

In this way we have bypassed the condition imposed in Section II.A that all variables of a system are zero when the system is at rest. In fact we still consider that condition applicable to all systems we are dealing with, and we shall regard the external cause acting on the system as represented by the right hand member of the first Eq. (26), be it a function or an operator.

C. The Flow Graph

The set of equations (26) can be written

$$\{x_1\} = \sum_j \frac{k_{j1}}{s + K_1}\{x_j\} + \frac{1}{s + K_1}\{f\},$$

$$\vdots \tag{28}$$

$$\{x_i\} = \sum_j \frac{k_{ji}}{s + K_i}\{x_j\}, \qquad i \neq 1;$$

these equations can be represented by a graph, where the functions $\{x_1\}$, $\{x_j\}$ and $\{f\}$ are nodes, and the fractions $k_{j1}/(s + K_1)$, $k_{ji}/(s + K_i)$, and $1/(s + K_1)$ are arms. The arm $1/(s + K_1)$ connects node $\{f\}$ to node $\{x_1\}$, and the arm $k_{ji}/(s + K_i)$ connects node $\{x_j\}$ to node $\{x_i\}$, for any $i, j = 1$, $2, \ldots, l$.

Each Eq. (28) for a particular value of i introduces in the graph all arms, and only those arms, converging to node $\{x_i\}$; therefore, the graph does not contain more than an arm each way between any two nodes. From the way this graph has been constructed, this general rule can be formulated: *Each node is equal to the sum of the products of all arms entering it times their nodes of departure.* The node representing the function $\{f\}$ does not represent a real compartment; we shall call it the *source* of the graph. As shown in Section IV.B, $\{f\}$ may be substituted by an operator.

Most of the definitions given in Section III.D apply to this graph, with some additions. For instance, a path does have, in addition to a length (number of arms forming it), a *value*, given by the product of its arms.

Several properties of a system can be found by simple inspection of its flow graph. In particular, if a node p is chosen as *initial node* and a node q as *terminal node*, the transfer function between them can be written. The first rule to apply is that all nodes and arms not belonging to any path from p to q can be disregarded. This is a direct consequence of the general rule given above. Thus, the graph of Fig. 4 is transformed into the graph of Fig. 5. The second rule is that each path from p to q can be substituted by an arm equal to the value of the path, and then all those arms by a single arm equal to the sum of those values. Thus, from the graph of Fig. 5 we get the graph of Fig. 6 and then of Fig. 7. This rule is easy to apply when the number of paths from p to q is finite, that is, when there are no cycles, as in the graph just considered. However, the graph of Fig. 8 does not have a finite number of paths from p to q, due to the existence of the closed paths bfk and eik; indeed $abfj$, $abfk\,bfj$, $abfk\,bfk\,bfj$, $\ldots$ form an infinite succession of paths. The second rule should in this case be modified by first defining the

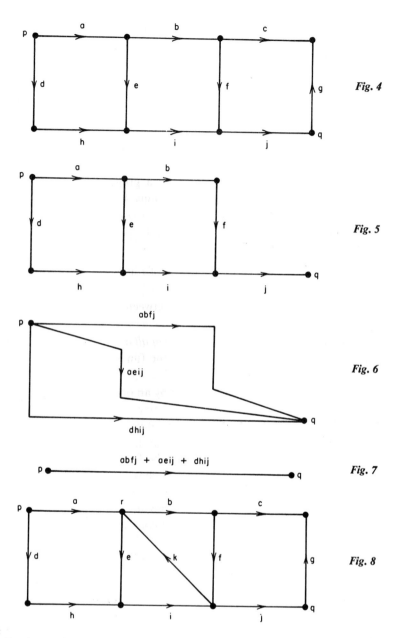

Fig. 4

Fig. 5

Fig. 6

Fig. 7

Fig. 8

Fig. 4. Graph containing no cycles; p is the initial and q the terminal node.

Fig. 5. The graph of Fig. 4 after removal of the arms not belonging to any path from p to q.

Fig. 6. The three paths from p to q of the graph of Fig. 5.

Fig. 7. The sum of the three paths of the graph of Fig. 6.

Fig. 8. A graph containing cycles; p is the initial and q the terminal node.

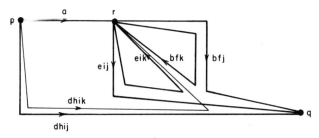

Fig. 9. The graph of Fig. 8 modified to show the paths between p, q and r; r is the chosen essential node.

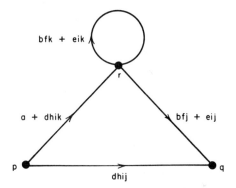

Fig. 10. Further simplification of the graph of Fig. 9.

essential nodes; they are the nodes that should be removed to interrupt all cycles. Their choice is not unique, but in any case it should be such that the essential nodes are minimum in number. Once the essential nodes have been chosen, the simplified graph contains:

(1) an arm from p to q,
(2) an arm from p to each essential node,
(3) an arm from each essential node to q,
(4) an arm from each essential node to each other essential node,
(5) an arm from each essential node to itself.

The value of each of these arms is the sum of the values of the simple paths between the nodes they connect, not passing through any of the other nodes of the simplified graph. Thus, from the graph of Fig. 8 we get the graph of Fig. 9 and then of Fig. 10. Of course, some of these possible arms may be missing, and q itself may be an essential node (see Figs. 11–13). The second rule, too, is a direct consequence of the general rule.

The third rule is that the closed arm of an essential node may be eliminated by dividing all arms entering that node by one minus the value of the closed

Fig. 11. A simplified graph where the terminal node q is an essential node.

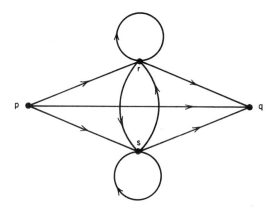

Fig. 12. A simplified graph with two essential nodes r and s, general case.

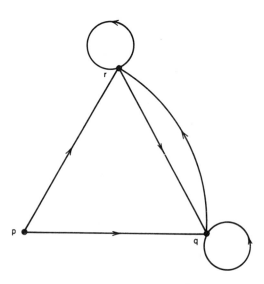

Fig. 13. A simplified graph with two essential nodes, one of them being the terminal node q.

arm. In fact, if x_i is an essential node, the general rule states

$$x_i = \sum_j a_{ji}x_j + a_{ii}x_i,$$

where a_{ji} is the value of the arm from x_j to x_i ($j = 1, 2, \ldots$) and a_{ii} is the value of the closed arm of x_i. Therefore

$$x_i = \sum_j a_{ji}x_j/(1 - a_{ii})$$

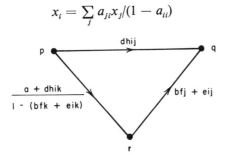

Fig. 14. The graph of Fig. 10 after removal of the closed arm.

(see Fig. 14). Once all closed arms have been eliminated, a further application of the second rule reduces the flow graph to a single arm from p to q; the value of that arm is the transfer function from p to q.

D. Partial Fraction Expansion

In the last two sections we have shown how to synthesize the transfer function of a system of compartments. The result so obtained is written in operational form; that is, it is a ratio of two polynomials in s, the numerator of lower degree than the denominator. That ratio can be written as a sum of simple fractions in the following way. The denominator of the transfer function can be written as the product of factors of the types

$$(s + \alpha)^\mu, \qquad [(s + \beta)^2 + \gamma^2]^\nu.$$

Then the transfer function can be written as the sum of fractions. To each factor of the first type correspond the fractions

$$\frac{a_0}{(s + \alpha)^\mu} + \frac{a_1}{(s + \alpha)^{\mu-1}} + \cdots + \frac{a_{\mu-1}}{s + \alpha},$$

and to each factor of the second type correspond the fractions

$$\frac{b_0 s + c_0}{[(s + \beta)^2 + \gamma^2]^\nu} + \frac{b_1 s + c_1}{[(s + \beta)^2 + \gamma^2]^{\nu-1}} + \cdots + \frac{b_{\nu-1} s + c_{\nu-1}}{(s + \beta)^2 + \gamma^2}.$$

The numerators of the partial fractions can be determined by transforming all partial fractions to the same denominator, then equating the numerator of the original total fraction with the sum of the numerators of the transformed partial fractions, and finally equating the coefficients of the corresponding powers of s. An example will show how simple the procedure is.

Given the operator

$$\frac{3s^7 + 5s^6 + 19s^5 - 26s^4 + 15s^3 - 57s^2 + 155s - 50}{s^8 + s^7 + 5s^6 - 11s^5 + 3s^4 - 29s^3 + 55s^2 - 25s},$$

we first try to decompose its denominator into factors. It is immediately evident that one factor is s and another factor is $s - 1$; by dividing the denominator by $s(s - 1)$, it is easy to check that the quotient contains the factor $s - 1$ two more times.

The denominator thus can be written

$$s(s - 1)^3(s^4 + 4s^3 + 14s^2 + 20s + 25);$$

this last factor is a perfect square and can be written $(s^2 + 2s + 5)^2$ or in the form $[(s + 1)^2 + 2^2]^2$. The given operator can be transformed therefore into the sum of partial fractions

$$\frac{a}{s} + \frac{b}{(s - 1)^3} + \frac{c}{(s - 1)^2} + \frac{d}{s - 1} + \frac{es + f}{[(s + 1)^2 + 2^2]^2} + \frac{gs + h}{(s + 1)^2 + 2^2},$$

where the coefficients a, b, c, d, e, f, g, and h are to be determined. The sum of these fractions is a fraction whose denominator is the same as the denominator of the given operator, and whose numerator is

$$a(s - 1)^3(s^2 + 2s + 5)^2 + bs(s^2 + 2s + 5)^2$$
$$+ cs(s - 1)(s^2 + 2s + 5)^2 + ds(s - 1)^2(s^2 + 2s + 5)^2$$
$$+ (es + f)s(s - 1)^3 + (gs + h)s(s - 1)^3(s^2 + 2s + 5),$$

obviously equal to the numerator of the given operator. We can therefore equate the coefficients of the corresponding powers of s of these two numerators, and we obtain thus the eight equations:

coefficient of $s^7 = a + d + g = 3$,
coefficient of $s^6 = a + c + 2d - g + h = 5$,
coefficient of $s^5 = 5a + b + 3c + 7d + e + 2g - h = 19$,
coefficient of $s^4 = -11a + 4b + 10c - 4d - 3e + f - 10g + 2h = -26$,
coefficient of $s^3 = 3a + 14b + 6c - d + 3e - 3f + 13g - 10h = 15$,
coefficient of $s^2 = -29a + 20b + 5c - 30d - e + 3f - 5g + 13h = -57$,
coefficient of $s \;\; = +55a + 25b - 25c + 25d - f - 5h = 155$,
coefficient of $s^0 = -25a = -50$.

Their solution is

$$a = 2, \quad b = 1, \quad c = 0, \quad d = 1,$$
$$e = 2, \quad f = 0, \quad g = 0, \quad h = 1.$$

Therefore, the given operator can be written as the sum of the partial fractions

$$\frac{2}{s} + \frac{1}{(s - 1)^3} + \frac{1}{s - 1} + \frac{2s}{[(s + 1)^2 + 2^2]^2} + \frac{1}{(s + 1)^2 + 2^2}.$$

E. The Operational Form of Some Simple Functions

Using the formula (25) we can write

$$s\{e^{-\alpha t}\} = \{-\alpha e^{-\alpha t}\} + 1,$$

thence

$$\{e^{-\alpha t}\} = 1/(s + \alpha). \tag{29}$$

In the same way

$$s\{\sin \gamma t\} = \{\gamma \cos \gamma t\}, \qquad s\{\cos \gamma t\} = \{-\gamma \sin \gamma t\} + 1,$$

thence

$$\{\sin \gamma t\} = \gamma/(s^2 + \gamma^2), \tag{30}$$

$$\{\cos \gamma t\} = s/(s^2 + \gamma^2). \tag{31}$$

Formulas (29)–(31) show to what functions some of the operators found in Section IV.D correspond. To find the correspondence of all other operators of that section, one could use repeatedly Eq. (25), but it is simpler to introduce the two operations T^β and D.

The operation T^β on a function is defined by the formula

$$T^\beta\{f(t)\} = \{e^{-\beta t} f(t)\}; \tag{32}$$

this operation has the three properties

$$T^{\beta_1} T^{\beta_2}\{f\} = T^{\beta_1 + \beta_2}\{f\},$$

$$T^\beta(\{f\} + \{g\}) = T^\beta\{f\} + T^\beta\{g\},$$

$$T^\beta(\{f\} \cdot \{g\}) = T^\beta\{f\} \cdot T^\beta\{g\};$$

the first two properties are easily proved by a simple substitution; to prove the third property we write

$$T^\beta(\{f\} \cdot \{g\}) = T^\beta\left\{\int_0^t f(t - \tau)g(\tau)\, d\tau\right\}$$

$$= \left\{e^{-\beta t}\int_0^t f(t - \tau)g(\tau)\, d\tau\right\}$$

$$= \left\{\int_0^t e^{-\beta(t-\tau)} f(t - \tau)e^{-\beta\tau}g(\tau)\, d\tau\right\}$$

$$= \{e^{-\beta t} f(t)\}\{e^{-\beta t}g(t)\} = T^\beta\{f\} \cdot T^\beta\{g\}. \qquad \text{Q.E.D.}$$

On an operator $p = \{f\}/\{g\}$ the operation T^β is defined by

$$T^\beta p = T^\beta\{f\}/T^\beta\{g\}; \tag{33}$$

it has the properties, analogous to the previous ones,

$$T^{\beta_1} T^{\beta_2} p = T^{\beta_1 + \beta_2} p, \tag{34}$$

$$T^\beta(p + q) = T^\beta p + T^\beta q, \tag{35}$$

$$T^\beta(p \cdot q) = T^\beta p \cdot T^\beta q, \tag{36}$$

as it is easy to verify.

For a numerical operator

$$T^\beta \alpha = \alpha;$$

in fact

$$T^\beta \alpha = T^\beta \frac{\{\alpha\}}{\{1\}} = \frac{T^\beta\{\alpha\}}{T^\beta\{1\}} = \frac{\{\alpha e^{-\beta t}\}}{\{e^{-\beta t}\}} = \frac{\alpha\{e^{-\beta t}\}}{\{e^{-\beta t}\}} = \alpha. \qquad \text{Q.E.D.}$$

It is also easy to prove that

$$T^\beta\{\gamma f\} = \gamma T^\beta\{f\}$$

and

$$T^\beta\{f\}^{-1} = (T^\beta\{f\})^{-1}.$$

For the differential operator

$$T^\beta s = s + \beta; \qquad (37)$$

in fact

$$T^\beta s = T^\beta\{1\}^{-1} = \{e^{-\beta t}\}^{-1},$$

and using Eq. (29), Eq. (37) follows.

Using repeatedly property (36),

$$T^\beta s^2 = (s + \beta)^2, \quad T^\beta s^3 = (s + \beta)^3, \ldots, \quad T^\beta s^n = (s + \beta)^n.$$

If $R(s)$ is a rational expression of s, then

$$T^\beta R(s) = R(s + \beta), \qquad (38)$$

where $R(s + \beta)$ is the same expression with s substituted with $s + \beta$.

Equation (24) can be written

$$s^{-n} = \{t^{n-1}/(n - 1)!\}, \qquad (39)$$

and using Eqs. (38) and (39),

$$(s + \alpha)^{-n} = \{t^{n-1}e^{-\alpha t}/(n - 1)!\}; \qquad (40)$$

this formula is a generalization of Eq. (29), to which it coincides for $n = 1$.

From Eqs. (30) and (31) we get

$$\{e^{-\beta t} \sin \gamma t\} = \frac{\gamma}{(s + \beta)^2 + \gamma^2}, \qquad (41)$$

$$\{e^{-\beta t} \cos \gamma t\} = \frac{s + \beta}{(s + \beta)^2 + \gamma^2}. \qquad (42)$$

F. The Algebraic Derivative of an Operator

To find more correspondences between functions and operators, we introduce the operation D defined by the formula

$$D\{f(t)\} = \{-t f(t)\}; \qquad (43)$$

this operation has the two properties

$$D(\{f\} + \{g\}) = D\{f\} + D\{g\},$$
$$D(\{f\}\cdot\{g\}) = D\{f\}\cdot\{g\} + \{f\}\cdot D\{g\}.$$

The first property is easily proved by simple substitution; to prove the second property we write

$$D(\{f\} \cdot \{g\}) = D\left\{\int_0^t f(t - \tau)g(\tau)\,d\tau\right\} = \left\{-t\int_0^t f(t - \tau)g(\tau)\,d\tau\right\}$$
$$= \left\{\int_0^t (-t + \tau)f(t - \tau)g(\tau)\,d\tau\right\}$$
$$+ \left\{\int_0^t f(t - \tau)(-\tau)g(\tau)\,d\tau\right\}$$
$$= \{-tf(t)\}\{g(t)\} + \{f(t)\}\{-tg(t)\}$$
$$= D\{f\}\cdot\{g\} + \{f\}\cdot D\{g\}. \qquad \text{Q.E.D.}$$

The operation D on an operator $p = \{f\}/\{g\}$ is defined by the formula

$$Dp = \frac{D\{f\}\cdot\{g\} - \{f\}\cdot D\{g\}}{\{g\}^2}; \tag{44}$$

it has the properties, analogous to the previous ones,

$$D(p + q) = Dp + Dq,$$
$$D(p\cdot q) = Dp\cdot q + p\cdot Dq. \tag{45}$$

It is easy to verify by simple substitution the additional property

$$D\frac{p}{q} = \frac{Dp\cdot q - p\cdot Dq}{q^2}. \tag{46}$$

For a numerical operator

$$D\alpha = 0;$$

in fact

$$D\alpha = D\frac{\{\alpha\}}{\{1\}} = \frac{D\{\alpha\}\cdot\{1\} - \{\alpha\}D\{1\}}{\{1\}^2} = \frac{\{-\alpha t\}\{1\} - \{\alpha\}\{-t\}}{\{1\}^2}$$
$$= \frac{\{-\tfrac12\alpha t^2\} - \{-\tfrac12\alpha t^2\}}{\{1\}^2} = 0. \qquad \text{Q.E.D.}$$

For the differential operator

$$Ds = 1;$$

in fact

$$Ds = D\frac{1}{\{1\}} = \frac{D1\cdot\{1\} - 1\cdot D\{1\}}{\{1\}^2} = \frac{-\{-t\}}{\{t\}} = 1. \qquad \text{Q.E.D.}$$

Using repeatedly property (45),

$$Ds^2 = 2s, \qquad Ds^3 = 3s^2, \ldots, \qquad Ds^n = ns^{n-1}.$$

If $R(s)$ is a rational expression of s, then because of these last formulas and Eq. (46), the operator $DR(s)$ can be computed by formal differentiation of $R(s)$ with respect to s, as if s were a variable. This fact is expressed by the formula

$$DR(s) = dR(s)/ds. \qquad (47)$$

Equation (47) applied to Eq. (29) gives

$$-1/(s + \alpha)^2 = \{-te^{-\alpha t}\},$$

and repeating the operation we obtain the general formula

$$1/(s + \alpha)^\mu = \{[1/(\mu - 1)!]t^{\mu-1}e^{-\alpha t}\}, \qquad (48)$$

identical with Eq. (40) obtained in a different way.

As $DR(s)$ is a rational expression of s, the same rules of formal differentiation applied to $R(s)$ can be applied to $DR(s)$. Therefore we can use the formula

$$DDR(s) = d^2 R(s)/ds^2.$$

G. The Operational Form of Other Trigonometric Functions

For any natural number $n > 1$ we can write

$$\frac{d}{ds} \frac{s}{(s^2 + \gamma^2)^{n-1}}$$

$$= \frac{1}{(s^2 + \gamma^2)^{n-1}} - \frac{2(n-1)s^2}{(s^2 + \gamma^2)^n}$$

$$= -2(n-1)\left[\frac{s^2 + \gamma^2}{(s^2 + \gamma^2)^n} - \frac{\gamma^2}{(s^2 + \gamma^2)^n}\right] + \frac{1}{(s^2 + \gamma^2)^{n-1}}$$

$$= 2(n-1)\gamma^2(s^2 + \gamma^2)^{-n} - [2(n-1) - 1](s^2 + \gamma^2)^{-(n-1)},$$

thence

$$\frac{1}{(s^2 + \gamma^2)^n} = \frac{1}{2\gamma^2}\left[\frac{1}{n-1}\frac{d}{ds}\frac{s}{(s^2 + \gamma^2)^{n-1}} + \frac{2n-3}{n-1}\frac{1}{(s^2 + \gamma^2)^{n-1}}\right]. \qquad (49)$$

For any natural number $n > 2$ we can write

$$\frac{d}{ds}\frac{1}{(s^2 + \gamma^2)^{n-2}} = \frac{-2(n-2)s}{(s^2 + \gamma^2)^{n-1}},$$

thence

$$\frac{s}{(s^2 + \gamma^2)^{n-1}} = -\frac{1}{2(n-2)}\frac{d}{ds}\frac{1}{(s^2 + \gamma^2)^{n-2}}.$$

Substitution of this expression into Eq. (49) gives

$$\frac{1}{(s^2 + \gamma^2)^n}$$

$$= \frac{1}{2\gamma^2}\left[-\frac{1}{2(n-1)(n-2)}\frac{d^2}{ds^2}\frac{1}{(s^2 + \gamma^2)^{n-2}} + \frac{2n-3}{n-1}\frac{1}{(s^2 + \gamma^2)^{n-1}}\right], \qquad (50)$$

valid for any natural number $n > 2$.

For $n = 1$ we have the formula (30)

$$1/(s^2 + \gamma^2) = \{\gamma^{-1} \sin \gamma t\}; \tag{51}$$

for $n = 2$ we can use Eq. (49) in conjunction with Eqs. (30) and (31),

$$\frac{1}{(s^2 + \gamma^2)^2} = \frac{1}{2\gamma^2}\left[\frac{d}{ds}\{\cos \gamma t\} - \{\gamma^{-1} \sin \gamma t\}\right],$$

$$\frac{1}{(s^2 + \gamma^2)^2} = \frac{1}{2\gamma^2}[\{-t \cos \gamma t\} + \{\gamma^{-1} \sin \gamma t\}]. \tag{52}$$

The recursive formula (50), keeping in mind Eqs. (51) and (52), and the fact that

$$(d^2/ds^2)\{f\} = \{t^2 f\},$$

can be written in the general form

$$\frac{1}{(s^2 + \gamma^2)^n} = \frac{1}{(2\gamma^2)^{n-1}}[\{A_n(\gamma^2 t^2) \gamma^{-1} \sin \gamma t\} - \{B_n(\gamma^2 t^2) t \cos \gamma t\}], \tag{53}$$

valid for any natural number $n > 0$, where

$$A_1(x) = 1, \qquad B_1(x) = 0, \qquad A_2(x) = 1, \qquad B_2(x) = 1,$$

because of Eqs. (51) and (52). The functions $A_n(x)$ and $B_n(x)$ can be computed in the following way. Calculate

$$\frac{d}{ds}\frac{1}{(s^2 + \gamma^2)^{n-1}} = \frac{-2(n-1)s}{(s^2 + \gamma^2)^n}; \tag{54}$$

thence

$$\frac{d^2}{ds^2}\frac{1}{(s^2 + \gamma^2)^{n-1}} = -2(n-1)\left[\frac{1}{(s^2 + \gamma^2)^n} - \frac{2ns^2}{(s^2 + \gamma^2)^{n+1}}\right]$$

$$= 4n(n-1)\left[\frac{s^2 + \gamma^2}{(s^2 + \gamma^2)^{n+1}} - \frac{\gamma^2}{(s^2 + \gamma^2)^{n+1}}\right] - \frac{2(n-1)}{(s^2 + \gamma^2)^n}$$

$$= \frac{2(2n-1)(n-1)}{(s^2 + \gamma^2)^n} - \frac{4\gamma^2 n(n-1)}{(s^2 + \gamma^2)^{n+1}},$$

and by rearranging,

$$\frac{1}{(s^2 + \gamma^2)^{n+1}} = \frac{1}{4\gamma^2 n(n-1)}\left[\frac{2(2n-1)(n-1)}{(s^2 + \gamma^2)^n} - \frac{d^2}{ds^2}\frac{1}{(s^2 + \gamma^2)^{n-1}}\right];$$

now using (53)

$$\frac{1}{(s^2 + \gamma^2)^{n+1}} = \frac{2n-1}{2\gamma^2 n}\frac{1}{(2\gamma^2)^{n-1}}[\{A_n(\gamma^2 t^2) \gamma^{-1} \sin \gamma t\} - \{B_n(\gamma^2 t^2) t \cos \gamma t\}]$$

$$- \frac{1}{4\gamma^2 n(n-1)}\frac{1}{(2\gamma^2)^{n-2}}[\{t^2 A_{n-1}(\gamma^2 t^2) \gamma^{-1} \sin \gamma t\}$$

$$- \{t^2 B_{n-1}(\gamma^2 t^2) t \cos \gamma t\}];$$

comparing now this expression with Eq. (53) written for $n + 1$ instead of n, we have

$$\frac{1}{(2\gamma^2)^n} A_{n+1}(\gamma^2 t^2) = \frac{2n-1}{n(2\gamma^2)^n} A_n(\gamma^2 t^2) - \frac{t^2\gamma^2}{n(n-1)(2\gamma^2)^n} A_{n-1}(\gamma^2 t^2),$$

$$\frac{1}{(2\gamma^2)^n} B_{n+1}(\gamma^2 t^2) = \frac{2n-1}{n(2\gamma^2)^n} B_n(\gamma^2 t^2) - \frac{t^2\gamma^2}{n(n-1)(2\gamma^2)^n} B_{n-1}(\gamma^2 t^2)$$

or

$$A_{n+1}(x) = [(2n-1)/n]A_n(x) - [x/n(n-1)]A_{n-1}(x),$$

$$B_{n+1}(x) = [(2n-1)/n]B_n(x) - [x/n(n-1)]B_{n-1}(x).$$

With the functions A_1, B_1, A_2, and B_2 already calculated, we can find recursively the functions A_n and B_n for any value of n. Table I gives the functions $A_n(x)$ and $B_n(x)$ for n up to 5.

TABLE I

n	$A_n(x)$	$B_n(x)$
1	1	0
2	1	1
3	$\frac{3}{2} - \frac{1}{2}x$	$\frac{3}{2}$
4	$\frac{5}{2} - x$	$\frac{5}{2} - \frac{1}{6}x$
5	$35/8 - (15/8)x + (1/24)x^2$	$35/8 - (5/12)x$

Using Eq. (54), we can write, provided $n > 1$, the additional formula

$$s/(s^2 + \gamma^2)^n = \frac{1}{2(n-1)(2\gamma^2)^{n-2}}[\{A_{n-1}(\gamma^2 t^2)\,(t/\gamma)\sin\gamma t\}$$
$$-\{B_{n-1}(\gamma^2 t^2)\,t^2\cos\gamma t\}]. \qquad (55)$$

Equations (53) and (55) can be made more general by using the operation T^β on them. Thus

$$\frac{1}{[(s+\beta)^2 + \gamma^2]^n} = \frac{1}{(2\gamma^2)^{n-1}}[\{A_n(\gamma^2 t^2)\,\gamma^{-1}\,e^{-\beta t}\sin\gamma t\}$$
$$-\{B_n(\gamma^2 t^2)te^{-\beta t}\cos\gamma t\}], \qquad (56)$$

$$\frac{s+\beta}{[(s+\beta)^2 + \gamma^2]^n} = \frac{1}{2(n-1)(2\gamma^2)^{n-2}}[\{A_{n-1}(\gamma^2 t^2)(t/\gamma)e^{-\beta t}\sin\gamma t\}$$
$$-\{B_{n-1}(\gamma^2 t^2)t^2\,e^{-\beta t}\cos\gamma t\}]. \qquad (57)$$

These last two formulas, together with Eqs. (39) and (40), allow us to write as functions all operators we dealt with in Section IV.D.

H. The Transfer Function

As seen in Section IV.C, the transfer function between two compartments of a system, or between the source and a compartment, can be obtained in operational form by a number of elementary operations on the operators of the types $1/(s + K_1)$ and $k_{ji}/(s + K_i)$. In this section we shall study the genesis of the transfer function from such operators.

The first step in the simplification of the flow graph is to determine the essential nodes and reduce the graph to one containing only the initial node, the terminal node, and the essential nodes, if any. Each arm in this reduced graph is the sum of the values of the simple paths between the nodes it connects. The value of a simple path from node a to node b is a product

$$\prod k_{ji}/(s + K_i)$$

which is equal to a fraction whose numerator is formed by a string of constants k_{ji}, such that the first subscript of the first k is a, the second subscript of each k is the same as the first subscript of the following k, and the second subscript of the last k is b; the denominator of this fraction is a product of factors $s + K_i$, where the subscripts of the K's included in it are those of all nodes along the considered path, excluding a but including b. In summing the values of the simple paths between the same two nodes, the above fractions should first be transformed into fractions with the same denominator; then the numerators can be added. The common denominator of those fractions is a product of factors $s + K_i$, where the subscripts of the K's included in it are those of all nodes, each of them taken once, belonging to any path from a to b, excluding a but including b. The numerator is the sum of the numerators of the above fractions, each one multiplied by the factors $s + K_i$ that did not appear in the corresponding denominator; that is, the numerator is a sum of as many terms as there are simple paths from a to b, each term being a product of the constants k_{ji} along a particular path, and of factors $s + K_i$, where the K's included are those corresponding to arms not belonging to that particular path but to a path from a to b; this implies that each product is formed by the same number of factors, some of the form k_{ji} and some of the form $s + K_i$. If in any path the arm $1/(s + K_1)$ is included, the above rules can still be used if we put that arm equal to $k_{01}/(s + K_1)$, with $k_{01} = 1$.

Suppose now that the graph has no essential nodes; the reduced graph is a single arm. Its denominator is the product of all factors $s + K_i$ belonging to the paths from the initial to the terminal node. Its numerator is the sum of a number of strings of k_{ji}, each string multiplied by a product of factors $s + K_i$. By expanding these last products, we find that the numerator of the transfer function is a polynomial in s of degree equal to the degree of the denominator minus the length of the shortest path from the initial to the terminal node; the coefficient of the highest power of s in the numerator is the product of the k's along the shortest path from the initial to the terminal node, or the sum of such products if there are more than one path of minimal length.

For the case when the initial node is the source, the same result was obtained in Section IV.B. With the present hypothesis that there are no essential
nodes, all products of the type $k_{ii_1}k_{i_1i}$, $k_{ii_1}k_{i_1i_2}k_{i_2i}$, ..., are void; therefore

$$\Delta = (s + K_i)^p \Delta_i;$$

but i was any number from 1 to l; therefore

$$\Delta = \prod_{i=1}^{l} (s + K_i).$$

This last product includes the terms $s + K_i$ not belonging to any path from
the initial to the terminal node, if any; such factors did not appear in the
denominator of the transfer function as computed from the flow graph. The
seeming discrepancy is resolved by observing that those same factors appear
in the numerator $(-1)^{j+1}\Delta_{1;j}$, and therefore they cancel each other. As for
this numerator, its development as shown in Section IV.B puts in evidence
the products of the k's along all possible paths from the source to the terminal
nodes; of course, no products containing fewer k's than the length of the
shortest path appear.

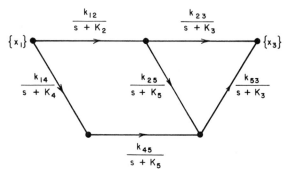

Fig. 15. The graph of a simple compartment system. The transfer function from $\{x_1\}$
to $\{x_3\}$ is shown in the text.

A simple example is given in Fig. 15; the corresponding transfer function
from $\{x_1\}$ to $\{x_3\}$ is

$$\frac{k_{12}k_{23}}{(s + K_2)(s + K_3)} + \frac{k_{12}k_{25}k_{53}}{(s + K_2)(s + K_5)(s + K_3)} + \frac{k_{14}k_{45}k_{53}}{(s + K_4)(s + K_5)(s + K_3)}$$

$$= \frac{k_{12}k_{23}(s + K_4)(s + K_5) + k_{12}k_{25}k_{53}(s + K_4) + k_{14}k_{45}k_{53}(s + K_2)}{(s + K_2)(s + K_3)(s + K_4)(s + K_5)}.$$

The degree in s of the denominator is 4; the degree of the numerator is 2;
the difference is the length of the shortest path from $\{x_1\}$ to $\{x_2\}$; the coefficient of s^2 in the numerator is $k_{12}k_{23}$, formed by the constants k along the
shortest path.

1. Effects of Closed Arms on the Transfer Function

If the reduced graph contains essential nodes and closed arms, the transfer function will contain in its denominator other factors besides the factors $s + K_i$. We shall discuss only a simple case which avoids many analytical complications but is sufficiently representative of the general problem.

Fig. 16. Simplified graph where the terminal node b is an essential node. The transfer function from a to b is shown in the text.

If there is only one essential node and it coincides with the terminal node, the reduced graph has the form of Fig. 16; its two arms have the form indicated in Section IV.H. By the third rule given in Section IV.C, the transfer function from a to b is

$$\frac{p_{ab}/q_{ab}}{1 - p_{bb}/q_{bb}} = \frac{p_{ab}q_{bb}}{q_{ab}(q_{bb} - p_{bb})}.$$

This last fraction can sometimes be simplified; q_{ab} and q_{bb} are products of the factors $s + K_i$ belonging, respectively, to all elementary paths from a to b, and to all cycles from b to b not going through a; now some arms may belong to both classes of paths, therefore q_{ab} and q_{bb} can be divided by the factors $s + K_i$ they have in common and the transfer function becomes

$$\frac{p_{ab}q'_{bb}}{q'_{ab}(q_{bb} - p_{bb})},$$

where q'_{ab} represents the product of the factors $s + K_i$ belonging to all elementary paths from a to b but not any cycle from b to b, and q'_{bb} represents the product of the factors $s + K_i$ belonging to all cycles from b to b not going through a, excluding all terms belonging to any path from a to b. The denominator of this transfer function, in addition to the said factors $s + K_i$, contains a polynominal in s of degree equal to the number of different nodes along any path from b to b not going through a, and with its coefficients formed by the constants K and k associated with such paths.

Here again the development of Δ given in Section IV.B shows how the products $k_{ii_1}k_{i_1i}$, $k_{ii_1}k_{i_1i_2}k_{i_2i}$, ..., corresponding to all cycles build up in the formation of the factor $q_{bb} - p_{bb}$ of the denominator.

The numerator of the transfer function, as in the case with no essential nodes, is a polynomial in s, and the coefficient of the highest power of s is the product of the k's along the shortest path from a to b, or the sum of such products if there are more than one path of minimal length.

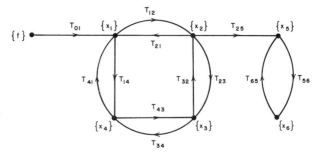

Fig. 17. Graph of a system; $\{f\}$ is the initial node; either $\{x_1\}$ or $\{x_2\}$ is regarded as the terminal node.

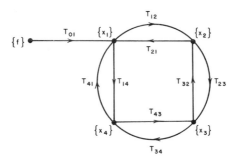

Fig. 18. The graph of Fig. 17 after removing the arms not belonging to any path from $\{f\}$ to either $\{x_1\}$ or $\{x_2\}$.

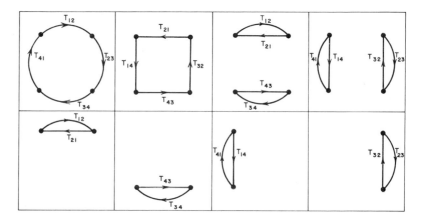

Fig. 19. The eight different strong components of the subgraphs of the graph of Fig. 18.

A simpler way to see the formation of the transfer function from the flow graph, irrespective of the number of essential nodes, is by using the concept of strong components, see Section III.D, as defined by Rescigno and Segre [1965], and we refer the reader to that paper for all proofs.

The transfer function from the source $\{f\}$ to any node $\{x_i\}$ is equal to the ratio $-M_i/M$, and the transfer function from any node $\{x_j\}$ to any node $\{x_i\}$ connected with it is equal to the ratio M_j/M_i.

To compute the functions M, $-M_i$, and $-M_j$ the first rule of Section IV.C is applied to the graph, thus eliminating all arms not belonging to any path from the initial to the terminal node chosen. M is equal to 1 plus the sum of the products formed with the values of the arms forming a strong component of a subgraph of the simplified graph, the sum being extended to all possible strong components of all possible subgraphs, and each product has the sign $+$ or $-$ according to whether in the corresponding strong component there is an even or an odd number of cycles; $-M_i$ is equal to the sum of the products of the arms forming an elementary path from the source to $\{x_i\}$, times a function obtained as is M, but excluding from the graph all nodes entered by the said paths. Given the graph of Fig. 17, if we choose as initial node $\{f\}$, and as terminal node either $\{x_1\}$ or $\{x_2\}$, using the first rule we can disregard the nodes $\{x_5\}$ and $\{x_6\}$ and the arms T_{25}, T_{56}, and T_{65}, obtaining the graph of Fig. 18. Its subgraphs with their corresponding strong components are illustrated in Fig. 19. Therefore

$$M = 1 - T_{12}T_{23}T_{34}T_{41} - T_{14}T_{43}T_{32}T_{21} + T_{12}T_{21}T_{34}T_{43}$$
$$+ T_{14}T_{41}T_{23}T_{32} - T_{12}T_{21} - T_{34}T_{43} - T_{14}T_{41} - T_{23}T_{32},$$
$$-M_1 = T_{01}(1 - T_{34}T_{43} - T_{23}T_{32}),$$
$$-M_2 = T_{01}T_{12} + T_{01}T_{14}T_{43}T_{32}.$$

Note that the values T_{ji} of the arms need not be of the form $k_{ji}/(s + K_i)$; therefore this method can be applied either before or after the graph has been simplified according to the second rule of Section IV.C.

2. Analysis of the Transfer Function

As seen in the two previous sections, some of the properties of a system can be analyzed immediately from its transfer function. Other properties can be analyzed only if additional hypotheses are made on the system; for instance, if some of the constants k are set equal to zero. As an example we analyze here the results of an experiment by Matthews [1957], where albumin labeled with ^{131}I was injected into the plasma of a rabbit and the concentration of ^{131}I was subsequently measured in the plasma and in the urine. The transfer function from the site of injection to the plasma was found to be

$$0.33e^{-0.080t} + 0.53e^{-1.00t} + 0.14e^{-14.4t}$$

and from the plasma to the urine, 0.20. Calling the source $\{f\}$, ^{131}I in plasma $\{x_1\}$ and ^{131}I in urine $\{x_2\}$, we have in operational form

$$\{x_1\}/\{f\} = \frac{0.33}{s + 0.080} + \frac{0.53}{s + 1.00} + \frac{0.14}{s + 14.4}$$

$$= \frac{s^2 + 12.91s + 5.37}{s^3 + 15.48s^2 + 15.63s + 1.15} \tag{58}$$

$$\{x_2\}/\{x_1\} = 0.20/s. \tag{59}$$

From Eq. (59) we deduce that the flow graph of this system contains only a single path from $\{x_1\}$ to $\{x_2\}$, and that this path is formed by a single arm of value $0.20/s$, that is

$$k_{12} = 0.20, \qquad K_2 = 0.$$

The transfer function (58) has a denominator of degree 3 and a numerator of degree 2; therefore, there are three nodes forming all possible paths from $\{f\}$ to $\{x_1\}$, and the length of the shortest of these paths is $3 - 2 = 1$. This shortest path is unique, and has the value $1/(s + K_1)$. The other two nodes besides $\{x_1\}$ should form one or two cycles. We can therefore write

$$\frac{\{x_1\}}{\{f\}} = \frac{1/(s + K_1)}{1 - p_{11}/q_{11}} = \frac{q_{11}}{(s + K_1)(q_{11} - p_{11})},$$

where p_{11}/q_{11} is the value of the closed arm in the reduced graph and where $\{x_1\}$ is the essential node. The operator q_{11} is a polynomial in s of degree 3 and contains the factor $s + K_1$, as all possible paths forming it end in $\{x_1\}$; p_{11} is a polynomial in s of degree 1 or 0, as all possible paths forming it have length 1 or 2. Therefore, the difference $q_{11} - p_{11}$ is a polynomial in s of degree 3 and in which the coefficient of s^2 is the same as in the polynomial q_{11}. We can write now

$$\{x_1\}/\{f\} = q'_{11}/(q_{11} - p_{11}),$$

where $q'_{11} = q_{11}/(s + K_1)$, and comparing with Eq. (58)

$$q_{11} = (s + K_1)(s^2 + 12.91s + 5.37),$$

$$q_{11} - p_{11} = s^3 + 15.48s^2 + 15.63s + 1.15.$$

From these two equations we have

$$K_1 + 12.91 = 15.48,$$

and therefore

$$K_1 = 2.57.$$

Substituting this value into the first equation

$$q_{11} = s^3 + 15.48s^2 + 38.55s + 13.80.$$

Subtracting the second equation, we have

$$p_{11} = 22.92s + 12.65.$$

Because p_{11} is found to be a polynomial of degree 1 in s, the shortest cycle has length 2. A detailed analysis of the cycles of this graph would lead us to the formula

$$\frac{p_{11}}{q_{11}} = \frac{k_{13}k_{31}(s + K_4) + k_{14}k_{41}(s + K_3) + k_{13}k_{34}k_{41} + k_{14}k_{43}k_{31}}{(s + K_1)[(s + K_3)(s + K_4) - k_{34}k_{43}]}.$$

It follows that

$$k_{13}k_{31} + k_{14}k_{41} = 22.92,$$
$$k_{13}k_{31}K_4 + k_{14}k_{41}K_3 + k_{13}k_{34}k_{41} + k_{14}k_{43}k_{31} = 12.65,$$
$$K_3 + K_4 = 12.91,$$
$$K_3 K_4 - k_{34}k_{43} = 5.37,$$

and we have four equations with eight unknowns. The problem is highly indeterminate. The author made the supplementary hypotheses

$$k_{34} = k_{43} = 0, \qquad k_{31} = K_3, \qquad k_{41} = K_4,$$

so the problem becomes determinate, yielding

$$k_{13}K_3 + k_{14}K_4 = 22.92, \qquad k_{13}K_3K_4 + k_{14}K_3K_4 = 12.65,$$
$$K_3 + K_4 = 12.91, \qquad\qquad K_3K_4 = 5.37.$$

From the last two of the above equations,

$$K_3 = 12.48, \qquad K_4 = 0.43,$$

and from the first two

$$k_{13} = 1.82, \qquad k_{14} = 0.54.$$

The equations of this system can now be written

$$\begin{aligned}
dx_1/dt &= -2.57x_1 &+ 12.48x_3 &+ 0.43x_4 \\
dx_2/dt &= 0.20x_1 \\
dx_3/dt &= 1.82x_1 &- 12.48x_3 \\
dx_4/dt &= 0.54x_1 && - 0.43x_4.
\end{aligned}$$

For a different approach to this problem, see Berman and Schoenfeld [1956].

Many computer programs are available for the analysis of the transfer function, the discussion of which transcends the limits of this chapter [see, for instance, Berman, 1965].

V. Analysis of the Precursor–Successor Relationship

From Eq. (10) we can write

$$x_j^{(p)}(t) = \int_0^t x_i(\tau) g_{ij}^{(p)}(t - \tau)\, d\tau + \sum_{a=0}^{p-1} x_i^{(a)}(t) g_{ij}^{(p-1-a)}(0),$$

where $x^{(p)}(t)$ indicates the pth derivative of $x(t)$ with respect to t including the function $x(t)$, itself considered as its oth derivative. If m is the order of the lowest derivative of $x_i(t)$ not equal to zero at $t = 0$, that is

$$x_i^{(a)}(0) = 0 \qquad \text{for any} \quad a < m, \qquad x_i^{(m)}(0) \neq 0,$$

and if the first n derivatives of $g_{ij}(t)$ are zero at $t = 0$, that is

$$g_{ij}^{(a)}(0) = 0 \qquad \text{for any} \quad a < n, \tag{60}$$

then

$$\lim_{t \to 0} x_j^{(m+n+1)}(t) = x_i^{(m)}(0) g_{ij}^{(n)}(0). \tag{61}$$

We can now prove that

$$\lim_{t \to 0} \frac{d^{m+n}}{dt^{m+n}}[t^n x_i(t)] = \frac{(m+n)!}{m!} x_i^{(m)}(0). \tag{62}$$

This is obviously true for $n = 0$. Assume it is valid for a given n; we can write

$$\frac{d^{m+n+1}}{dt^{m+n+1}}[t^{n+1} x_i(t)] = \frac{d^{m+n}}{dt^{m+n}}[(n+1)t^n x_i(t) + t^{n+1} x_i^{(1)}(t)]$$

$$= \frac{d^{m+n}}{dt^{m+n}}[t^n y(t)],$$

where

$$y(t) = (n+1)x_i(t) + t x_i^{(1)}(t).$$

Now using Eq. (62) for this last function

$$\lim_{t \to 0} \frac{d^{m+n+1}}{dt^{m+n+1}}[t^{n+1} x_i(t)] = \lim_{t \to 0} \frac{d^{m+n}}{dt^{m+n}}[t^n y(t)] = \frac{(m+n)!}{m!} y^{(m)}(0);$$

but

$$y^{(m)}(t) = (n+1)x_i^{(m)}(t) + t\, x_i^{(m+1)}(t) + m\, x_i^{(m)}(t),$$

thence

$$y^{(m)}(0) = (m+n+1)x_i^{(m)}(0)$$

and

$$\lim_{t \to 0} \frac{d^{m+n+1}}{dt^{m+n+1}}[t^{n+1} x_i(t)] = \frac{(m+n+1)!}{m!} x_i^{(m)}(0), \tag{63}$$

which is exactly Eq. (62) with $n + 1$ instead of n. Therefore, we have proved that Eq. (62) is valid for any integer $n > 0$.

Now using l'Hospital's rule and Eqs. (61) and (63),

$$\lim_{t \to 0} \frac{x_j(t)}{t^{n+1} x_i(t)} = \lim_{t \to 0} \frac{x_j^{(m+n+1)}(t)}{\dfrac{d^{m+n+1}}{dt^{m+n+1}} [t^{n+1} x_i(t)]}$$

$$= \frac{x_i^{(m)}(0) g_{ij}^{(n)}(0)}{[(m+n+1)!/m!] x_i^{(m)}(0)}$$

$$= \frac{m!}{(m+n+1)!} g_{ij}^{(n)}(0);$$

therefore

$$g_{ij}^{(n)}(0) = \frac{(m+n+1)!}{m!} \lim_{t \to 0} \frac{x_j(t)}{t^{n+1} x_i(t)}. \qquad (64)$$

Equation (64) can be used only if all derivatives of $g_{ij}(t)$ up to the $(n-1)$th are zero for $t = 0$. The correct use of this formula is therefore the following: With $n = 0$, Eq. (64) becomes

$$g_{ij}(0) = (m+1) \lim_{t \to 0} x_j(t)/t x_i(t);$$

by plotting on graph paper the fraction $x_j/t x_i$ as a function of t, then extrapolating for $t \to 0$, one finds the value of $g_{ij}(0)$. If this value is zero, then Eq. (64) can be used for $n = 1$. Thus

$$g_{ij}^{(1)}(0) = (m+2)(m+1) \lim_{t \to 0} x_j(t)/t^2 x_i(t).$$

The same graphical procedure as before is used to find the value of $g_{ij}^{(1)}(0)$. If this value, too, is zero, then Eq. (64) can be used for $n = 2$. Thus

$$g_{ij}^{(2)}(0) = (m+3)(m+2)(m+1) \lim_{t \to 0} x_j(t)/t^3 x_i(t).$$

These operations are carried on until the maximum value of n satisfying condition (60) is found; this number plus one is the precursor order of $\{x_i\}$ with respect to $\{x_j\}$. For more details on this method see Beck and Rescigno [1964]. The value of $g_{ij}^{(n)}(0)$ is called the *precursor's principal term* [see Rescigno and Segre, 1961b].

As shown in Sections IV.H and IV.H.1, if $n+1$ is the precursor order of $\{x_i\}$ with respect to $\{x_j\}$, the operator $\{g_{ij}^{(n)}(t)\} = s^n \{g_{ij}(t)\}$ is a ratio of two polynomials in s, the numerator having a degree one less than the degree of the denominator; we can therefore write

$$\{g_{ij}^{(n)}(t)\} = \frac{a s^{l-1} + b s^{l-2} + \cdots}{s^l + c s^{l-1} + \cdots} = \frac{a}{s} + \frac{1}{s} \frac{(b - ac) s^{l-1} + \cdots}{s^l + c s^{l-1} + \cdots}$$

or

$$g_{ij}^{(n)}(t) = a + \int_0^t f(\tau) \, d\tau, \qquad (65)$$

where

$$\{f(t)\} = \frac{(b - ac)s^{l-1} + \cdots}{s^l + cs^{l-1} + \cdots}$$

is a function, and a, according to Sections IV.H and IV.H.1, is the product of the k's along the shortest path from $\{x_i\}$ to $\{x_j\}$, or the sum of such products if there are more than one path of minimal length. Taking the limit of Eq. (65) for $t \rightarrow 0$,

$$g_{ij}^{(n)}(0) = a,$$

that is, the precursor's principal term is equal to a.

Some additional information about the function $\{g_{ij}(t)\}$ can be obtained using Eq. (64). If we define a function $\{\bar{g}_{ij}\}$ whose values are given by

$$\bar{g}_{ij}(t) = g_{ij}(t) - g_{ij}(0) - tg_{ij}^{(1)}(0) - (t^2/2!)g_{ij}^{(2)}(0) - \cdots$$
$$- [t^{\nu-1}/(\nu - 1)!]g_{ij}^{(\nu-1)}(0)$$

for any integer $\nu \geqslant 1$ that we choose, then this function $\{\bar{g}_{ij}\}$ coincides with $\{g_{ij}\}$ if $\nu \leqslant n$; for $\nu > n$

$$\bar{g}_{ij}^{(n)}(0) = g_{ij}^{(n)}(0)$$

and also condition (60) is satisfied for the new function $\{\bar{g}_{ij}(t)\}$. The output $\{\bar{x}_j(t)\}$ of a system the input of which is $\{x_i(t)\}$ and the transfer function of which is $\{\bar{g}_{ij}(t)\}$, is given by

$$\{\bar{x}_j\} = \{x_i\}\{\bar{g}_{ij}\} = \{x_i\}\{g_{ij}\} - \{x_i\}\{g_{ij}(0)\} - \{x_i\}\{tg_{ij}^{(1)}(0)\} - \cdots$$
$$- \{x_i\}\{[t^{\nu-1}/(\nu - 1)!]g_{ij}^{(\nu-1)}(0)\}$$

or, using Eq. (24) for all terms after the first in the right-hand side,

$$\{\bar{x}_j\} = \{x_j\} - g_{ij}(0)\{1\}\{x_i\} - g_{ij}^{(1)}(0)\{1\}^2\{x_i\} - \cdots - g_{ij}^{(\nu-1)}(0)\{1\}^\nu\{x_i\}$$

or

$$\bar{x}_j = x_j - g_{ij}(0)\int_0^t x_i - g_{ij}^{(1)}(0)\int_0^t\int_0^t x_i - \cdots - g_{ij}^{(\nu-1)}(0)\int_0^t\int_0^t \cdots \int_0^t x_i,$$
$$(66)$$

where the abbreviated integral symbols at the right-hand side mean integration of the function $x_i(t)$ one, two, . . . , ν times between the limits 0, t.

For the values of n not satisfying condition (60) for $g_{ij}(t)$, Eq. (64) can be modified by substituting $x_j(t)$ with $\bar{x}_j(t)$ as given by Eq. (66). Thus the initial values of the transfer function and of all its derivatives of any order can be obtained if the input and the output of a system are known.

VI. Material Transport through Compartments

When $\{x_i\}$, for any i, is a concentration or a mass, considerations of conservation of mass require that

$$k_{ij} \geqslant 0 \qquad \text{for any} \quad i, j, \qquad (67)$$

and that

$$K_i \geqslant \sum_j k_{ij} \qquad \text{for any} \quad i. \tag{68}$$

Inequalities (67) follow from the fact that if, for some value of t and a given value of i, it is

$$f(t) = 0, \qquad x_j(t) = 0 \qquad \text{for any} \quad j \neq i,$$

then Eqs. (22) become

$$dx_j/dt = k_{ij}x_i \qquad \text{for any} \quad j \neq i,$$

and if inequalities (67) are not true, x_j will become negative.

Inequalities (68) follow from the fact that, adding all equations (22) together, we get

$$(d/dt) \sum_j x_j = \sum_i (\sum_j k_{ij} - K_i)x_i + f(t),$$

and if the coefficients of x_i are not negative, the sum $\sum_j x_j$ will increase without limit, even when $f(t) = 0$.

From here on, we shall consider Eqs. (67) and (68) to be always valid, unless different hypotheses are explicitly made.

When all Eqs. (68) are strict equalities, the system is said to be *closed*; if at least one of Eqs. (68) is a proper inequality, the system is said to be *open*.

In Section III.C a compartment x_i was called closed if

$$K_i = 0;$$

with the present hypotheses it should be in addition

$$k_{ij} = 0 \qquad \text{for any} \quad j.$$

The definitions of closed system and closed compartment can be extended to a subsystem. The subsystem formed by the compartments $x_{i_1}, x_{i_2}, \ldots, x_{i_m}$ is called closed when

$$K_i = \sum_j^{i_1 \cdots i_m} k_{ij}, \qquad i = i_1, i_2, \ldots, i_m, \tag{69}$$

and consequently

$$k_{ij} = 0, \qquad i = i_1, i_2, \ldots, i_m; \qquad j \neq i_1, i_2, \ldots, i_m. \tag{70}$$

A number of properties can be proved for a system of compartments when the two sets of inequalities (67) and (68) hold. A paper by Hearon [1963] reviews many of these properties. We shall prove here the fundamental property, known as the theorem of Hearon [1953], in a weaker form, but sufficient for our purposes.

Theorem. All factors of Δ (see Section IV.B) have one of the forms

$$s, \quad (s+\alpha)^\mu, \quad [(s+\beta)^2 + \gamma^2]^\nu,$$

with $\alpha > 0, \beta > 0, \gamma \neq 0; \mu, \nu = 1, 2, \ldots$.

PROOF: All factors of Δ can be obtained by looking for the roots of the equation $\Delta = 0$, where the operator s is substituted by a complex variable s^*. The roots are not altered if we change the columns with the rows in Δ; then this equation becomes

$$\begin{vmatrix} s^* + K_1 & -k_{12} & \cdots & -k_{1l} \\ -k_{21} & s^* + K_2 & \cdots & -k_{2l} \\ \cdot & \cdot & & \cdot \\ \cdot & \cdot & & \cdot \\ \cdot & \cdot & & \cdot \\ -k_{l1} & -k_{l2} & \cdots & s^* + K_l \end{vmatrix} = 0,$$

and it represents the condition for having a set of complex numbers y_1, $y_2, \ldots, y_l$, not all void, such that

$$(s^* + K_1)y_1 - k_{12}y_2 - \cdots - k_{1l}y_l = 0$$
$$-k_{21}y_1 + (s^* + K_2)y_2 - \cdots - k_{2l}y_l = 0$$
$$\cdot$$
$$\cdot$$
$$\cdot$$
$$-k_{l1}y_1 - k_{l2}y_2 - \cdots + (s + K_l)y_l = 0.$$

If y_i has the largest modulus among $y_1, y_2, \ldots, y_l$, we can write

$$s^* + K_i = \sum_j k_{ij}(y_j/y_i),$$

$$|s^* + K_i| \leqslant \sum_j k_{ij} |y_j|/|y_i| \leqslant \sum_j k_{ij},$$

and for Eq. (68),

$$|s^* + K_i| \leqslant K_i.$$

If the root sought is real, then we can put

$$s^* = -\alpha,$$

and the above inequality becomes

$$-\alpha + K_i \leqslant K_i$$

or

$$\alpha \geqslant 0.$$

If the root is complex, then we can put

$$s^* = -\beta + \gamma\sqrt{-1}$$

with $\gamma \neq 0$, and the above inequality becomes

$$|-\beta + \gamma\sqrt{-1} + K_i| \leqslant K_i,$$
$$\beta^2 - 2K_i\beta + \gamma^2 \leqslant 0,$$

which requires that

$$\beta \geqslant K_i - (K_i^2 - \gamma^2)^{1/2} > 0.$$

The case $s^* = 0$ happens when and only when the determinant

$$\begin{vmatrix} K_1 & -k_{21} & \cdots & -k_{l1} \\ -k_{12} & K_2 & \cdots & -k_{l2} \\ \cdot & \cdot & & \cdot \\ \cdot & \cdot & & \cdot \\ \cdot & \cdot & & \cdot \\ -k_{1l} & -k_{2l} & \cdots & K_l \end{vmatrix} \tag{71}$$

is void. This is true when the system or one of its subsystems is closed. In fact if the system is closed, Eq. (68) can be written

$$K_i - \sum_j k_{ij} = 0, \qquad i = 1, 2, \ldots, l,$$

and they represent the same linear relationship among the elements of each column in the determinant (71). If a subsystem is closed, that subsystem either includes the initial compartment (see Section III.C) or not; in the first case, the initial compartment is connected only to the compartments of the closed subsystem, and all other compartments can be neglected; therefore the whole system can be considered closed. If the closed subsystem does not include the initial compartment, all compartments can be relabeled in such a way that the closed subsystem is formed by the compartments x_{m+1}, x_{m+2}, $\ldots$, x_l; then with Eq. (70), Δ becomes

$$\begin{vmatrix} s + K_1 & \cdots & -k_{m1} & 0 & \cdots & 0 \\ \cdot & & \cdot & \cdot & & \cdot \\ \cdot & & \cdot & \cdot & & \cdot \\ \cdot & & \cdot & \cdot & & \cdot \\ -k_{1m} & \cdots & s + K_m & 0 & \cdots & 0 \\ -k_{1,m+1} & \cdots & -k_{m,m+1} & s + K_{m+1} & \cdots & -k_{1,m+1} \\ \cdot & & \cdot & \cdot & & \cdot \\ \cdot & & \cdot & \cdot & & \cdot \\ -k_{1l} & \cdots & -k_{ml} & -k_{m+1,l} & \cdots & s + K_l \end{vmatrix}$$

$$= \begin{vmatrix} s + K_1 & \cdots & -k_{m1} \\ \cdot & & \cdot \\ \cdot & & \cdot \\ \cdot & & \cdot \\ -k_{1m} & \cdots & s + K_m \end{vmatrix} \cdot \begin{vmatrix} s + K_{m+1} & \cdots & -k_{1,m+1} \\ \cdot & & \cdot \\ \cdot & & \cdot \\ \cdot & & \cdot \\ -k_{m+1,l} & \cdots & s + K_l \end{vmatrix},$$

and the second determinant of the product for Eq. (69) is the determinant of a closed system; therefore, it includes the factor s.

The factor s cannot be a multiple factor of Δ. In fact this would require that the determinant (71) be void together with the sum of all its principal minors of order $l - 1$; but all these minors are nonnegative, therefore their sum is void only when they are all void. This requires that all subsystems formed by $l - 1$ compartments be closed, evidently absurd. Q.E.D.

The immediate consequence of this theorem is that the transfer functions from the source to any variable of a system of compartments in which (67) and (68) hold, are a linear combination of operators of the form

$$1/s, \quad 1/(s + \alpha)^\mu, \quad 1/[(s + \beta)^2 + \gamma^2]^\nu, \quad (s + \beta)/[(s + \beta)^2 + \gamma^2]^\nu,$$

the first one being present only in a closed system; to these operators correspond the functions given by formulas (39), (40), (56), and (57), with the following limitations: in Eq. (39), $n = 1$ only; in Eq. (40), $\alpha > 0$; in Eqs. (56), and (57), $\beta > 0$ and $\gamma \neq 0$.

It follows that, if $g_j(t)$ is the transfer function from the source to the variable $x_j(t)$, then

$$\lim_{t \to \infty} g_j(t) = a_j,$$

where

$$a_j = 0$$

when the system is open and it does not include any closed subsystem. That the system or a subsystem be closed is a necessary, but not sufficient, condition for $a_j \neq 0$. In fact $a_j \neq 0$ when s is a factor of Δ, but at the same time s is not a factor of $\Delta_{1;j}$ (see Section IV.B).

The actual value of a_j may be computed using Eq. (77).

A. Turnover

The term *turnover* is borrowed from biochemistry, where it generally refers to the synthesis (anabolism) by an organism of new molecules of a particular type that is balanced by the degradation (catabolism) of pre-existing molecules of the same type. The interested reader may refer to a series of letters published by *Nature* on the most appropriate definition of turnover in biochemistry [Kleiber, 1955; Zilversmit, 1955; Mawson, 1955].

In the context of the theory of compartments, we prefer to define axiomatically the *turnover time* of a particular variable x_i as the inverse of the constant K_i introduced in Eq. (19). When x_i is the amount of a tracer and the corresponding mother substance or tracee is at steady state, its turnover time $1/K_i$ is the time necessary for the renewal of an amount of tracee equal to the amount present. In fact the equation for the tracer is

$$dx_i/dt = -K_i x_i + f_i(t), \tag{72}$$

and if X_i is the amount of tracee, the hypothesis of steady state requires that

$$dX_i/dt = 0 \quad \text{or} \quad X_i = \text{const.}$$

The tracer is eliminated from the compartment at the fractional rate K_i, and the tracee is eliminated and introduced at exactly the same rate; therefore

$K_i X_i$ is the amount of tracee renewed in a unit time, and the amount renewed during the time $1/K_i$ is

$$K_i X_i K_i^{-1} = X_i,$$

the amount of tracee present. This fact justifies the name "turnover time" given to the constant $1/K_i$.

When the tracee is not at steady state, or when x_i is not the amount of a tracer, the turnover time as defined here does not have the physical meaning just shown, though it still is dimensionally a time, as can be seen in Eq. (72). The term *time constant* is used sometimes to indicate the constant $1/K_i$ in Eq. (72). A constant strictly related to it is the *half-value time*, defined as the time necessary for the variable x_i to reduce its value to one-half in the absence of any input; from Eq. (72) with $f_i(t) \equiv 0$ we obtain

$$dt = -dx_i/K_i x_i$$

and the half-value time $t_{1/2}$ is therefore

$$t_{1/2} = \int_{2x_i}^{x_i} - (dx_i/K_i x_i) = -(1/K_i)(\ln x_i - \ln 2x_i) = (1/K_i) \ln 2$$

or approximately

$$t_{1/2} = 0.693(1/K_i)$$

In Section III.C we called "closed" a compartment whose K_i is void; we can now redefine a closed compartment as one whose turnover time is infinite.

The inverse of the turnover time is called *turnover rate* by some authors [for instance, Brownell *et al.*, 1968], or *fractional turnover rate* by others [for instance, Zilversmit, 1960]; we think the introduction of a special term for this quantity is not necessary in this chapter; nevertheless, consistent usage of these terms is important. "Turnover rate," we think, should always be modified as "mass turnover rate," "concentration turnover rate," or the like; the fractional turnover rate as defined by Zilversmit is of a more general nature and deserves this special name.

In the example given in Section IV.H.2, though the general problem was highly indeterminate, the turnover times of two of the compartments could be computed immediately.

A general method of computing the turnover time of a compartment x_i is to compute the integral of x_i from 0 to ∞, if this integral converges. In fact, Eq. (72) in operational form is

$$s\{x_i\} = -K_i\{x_i\} + \{f_i\},$$

and with a few simple transformations we have

$$\{x_i\} = \{f_i\}/(s + K_i), \tag{73}$$

$$\left\{ \int_0^t x_i(\tau)\,d\tau \right\} = \{f_i\}/s(s + K_i) = (1/K_i)\{f_i\}[1/s - 1/(s + K_i)]$$

$$= (1/K_i)(\{f_i\}\{1\} - \{f_i\}\{\exp(-K_i t)\}).$$

The last product may be written

$$\{f_i\}\{\exp(-K_i t)\} = \left\{\int_0^t f_i(\tau) \exp[-K_i(t-\tau)]\, d\tau\right\}$$

$$= \left\{\exp(-K_i t)\int_0^t f_i(\tau) \exp(K_i \tau)\, d\tau\right\};$$

its value for $t \to \infty$ is given by

$$\lim_{t\to\infty}\int_0^t f_i(\tau) \exp(K_i \tau)\, d\tau\Big/\exp(K_i t),$$

and if

$$\lim_{t\to\infty} f_i(t) = 0$$

it vanishes. Therefore

$$\int_0^\infty x_i(\tau)\, d\tau = (1/K_i)\int_0^\infty f_i(\tau)\, d\tau;$$

that is, the turnover time is given by the ratio of the integrals

$$1/K_i = \int_0^\infty x_i(\tau)\, d\tau\Big/\int_0^\infty f_i(\tau)\, d\tau. \tag{74}$$

The use of this formula requires the knowledge of the input function $f_i(t)$, or at least of its integral between 0 and ∞. If the input is not a function but the operator

$$x_i(0) = \{x_i(0)\}/\{1\},$$

in the formulation above $\{f_i\}\{1\}$ will be substituted by $\{x_i(0)\}$; the value of this function is constantly $x_i(0)$, therefore, Eq. (74) becomes

$$1/K_i = \int_0^\infty x_i(\tau)\, d\tau\Big/x_i(0).$$

Equation (74) may be made more general; from Eq. (73), using Eq. (44),

$$D\{x_i\} = \frac{D\{f_i\}(s+K_i) - \{f_i\}D(s+K_i)}{(s+K_i)^2};$$

therefore

$$\{t x_i\} = \frac{\{t f_i\}}{s+K_i} + \frac{\{f_i\}}{(s+K_i)^2},$$

and

$$\left\{\int_0^t \tau x_i(\tau)\, d\tau\right\}$$

$$= \frac{\{t f_i\}}{s(s+K_i)} + \frac{\{f_i\}}{s(s+K_i)^2}$$

$$= \frac{1}{K_i}\{t f_i\}\left(\frac{1}{s} - \frac{1}{s+K_i}\right) + \frac{1}{K_i^2}\{f_i\}\left(\frac{1}{s} - \frac{K_i}{(s+K_i)^2} - \frac{1}{s+K_i}\right)$$

$$= \frac{1}{K_i}\{t f_i\}\{1 - \exp(-K_i t)\} + \frac{1}{K_i^2}\{f_i\}\{1 - K_i t \exp(-K_i t) - \exp(-K_i t)\}.$$

With the same hypothesis on $f_i(t)$ as before, for $t \to \infty$ we obtain

$$\int_0^\infty \tau x_i(\tau)\,d\tau = \frac{1}{K_i}\int_0^\infty \tau f_i(\tau)\,d\tau + \frac{1}{K_i^2}\int_0^\infty f_i(\tau)\,d\tau;$$

thence, using Eq. (74),

$$\frac{1}{K_i} = \frac{\int_0^\infty \tau x_i(\tau)\,d\tau}{\int_0^\infty x_i(\tau)\,d\tau} - \frac{\int_0^\infty \tau f_i(\tau)\,d\tau}{\int_0^\infty f_i(\tau)\,d\tau}. \tag{75}$$

The two fractions on the right-hand side of Eq. (75) have the dimensions of time; we shall call them *times of exit* from the compartment x_i and from its precursor f_i, respectively; this name is justified by the fact that, when x_i or f_i measure the quantity or the concentration of a substance in a specified site, those fractions measure the average time at which the particles of that substance leave that particular site.

An obvious generalization of Eq. (75) is to write

$$t_{ij} = \frac{\int_0^\infty \tau x_j(\tau)\,d\tau}{\int_0^\infty x_j(\tau)\,d\tau} - \frac{\int_0^\infty \tau x_i(\tau)\,d\tau}{\int_0^\infty x_i(\tau)\,d\tau}, \tag{76}$$

where between x_i and x_j there is a relationship as described in Section II.C, that is, x_i is a precursor of x_j of any order; t_{ij} is called *transfer time* from x_i to x_j. When x_i and x_j measure the quantity or the concentration of a substance, the transfer time t_{ij} measures the average interval of time elapsing from the moment a particle of that substance leaves x_i to the moment it leaves x_j. If x_i is the precursor of order one of x_j, then Eq. (76) coincides with Eq. (75) and t_{ij} is the turnover time $1/K_j$ of x_j.

To be able to write Eq. (76) in a more concise form we are going to prove that, if $f(t)$ is any function such that

$$\{f\} = p(s)/q(s),$$

then

$$\lim_{t \to \infty} f(t) = \frac{sp(s)}{q(s)}\bigg|_{s=0} \tag{77}$$

where the symbol on the right-hand side represents the value obtained when the operator $sp(s)/q(s)$ is written with 0 instead of s.

Two cases are possible. If $q(s)$ does not contain the factor s, we can write

$$q(s) = a + bs + cs^2 + \cdots$$

with $a \neq 0$, and the function $f(t)$ contains only exponential and trigonometric terms whose limit for $t \to \infty$ is zero for the hypotheses made in Section VI; in this case $sp(s)/q(s)$ does contain the factor s; therefore

$$\frac{sp(s)}{q(s)}\bigg|_{s=0} = 0.$$

The other case is that $q(s)$ contains the factor s, so we can write

$$p(s) = \alpha + \beta s + \cdots, \qquad q(s) = bs + cs^2 + \cdots$$

with $b \neq 0$, then

$$\{f\} = \frac{\alpha + \beta s + \cdots}{s(b + cs + \cdots)} = \frac{\alpha/b}{s} + \frac{(\beta - \alpha c)/b + \cdots}{b + cs + \cdots},$$

and therefore

$$\lim_{t \to \infty} f(t) = \alpha/b;$$

on the other hand

$$\frac{sp(s)}{q(s)} = \frac{\alpha + \beta s + \cdots}{b + cs + \cdots}$$

and

$$\left.\frac{sp(s)}{q(s)}\right|_{s=0} = \frac{\alpha}{b}. \qquad \text{Q.E.D.}$$

Using Eq. (77) we can write

$$\int_0^\infty x_i(\tau)\, d\tau = s\{1\}\{x_i\}\big|_{s=0} = \{x_i\}\big|_{s=0}$$

$$\int_0^\infty \tau x_i(\tau)\, d\tau = s\{1\}(-D\{x_i\})\big|_{s=0} = -D\{x_i\}\big|_{s=0}$$

and Eq. (76) becomes

$$t_{ij} = \left.\frac{D\{x_i\}}{\{x_i\}}\right|_{s=0} - \left.\frac{D\{x_j\}}{\{x_j\}}\right|_{s=0}. \tag{78}$$

For the properties of the operation D shown in Section IV.F, we can make the convention

$$D\{f\}/\{f\} = D \ln\{f\}$$

and use the formula

$$\ln(\{f\}/\{g\}) = \ln\{f\} - \ln\{g\},$$

so Eq. (78) becomes

$$t_{ij} = D \ln\{x_i\}/\{x_j\}\big|_{s=0},$$

$$t_{ij} = -D \ln\{g_{ij}\}\big|_{s=0}. \tag{79}$$

In particular, if x_i is the absolute precursor of order one of x_j,

$$t_{ij} = -D \ln[k_{ij}/(s + K_j)]\big|_{s=0} = 1/K_j,$$

that is, the transfer time from x_i to x_j is the turnover time of x_j.

B. Graphs and Turnover

Equation (78) is of immediate application to all operations made on the flow graphs. The three kinds of operations are: (a) two arms in series are substituted by an arm equal to their product, (b) two arms in parallel are substituted by an arm equal to their sum, (c) the closed arm of an essential node can be eliminated by dividing all arms entering it by one minus the value of the closed arm.

In terms of transfer times the first operation becomes

$$t_{ij} = -D \ln(g_{ii_1} g_{i_1 j})\Big|_{s=0}$$
$$= -D \ln g_{ii_1}\Big|_{s=0} - D \ln g_{i_1 j}\Big|_{s=0} = t_{ii_1} + t_{i_1 j} \tag{80}$$

that is, to the product of transfer functions corresponds the sum of transfer times.

The second operation becomes

$$t_{ij} = -D \ln(g'_{ij} + g''_{ij})\Big|_{s=0}$$
$$= -\frac{dg'_{ij}/ds + dg''_{ij}/ds}{g'_{ij} + g''_{ij}}\Big|_{s=0}$$
$$= \frac{-\dfrac{dg'_{ij}/ds}{g'_{ij}} g'_{ij} - \dfrac{dg''_{ij}/ds}{g''_{ij}} g''_{ij}}{g'_{ij} + g''_{ij}}\Bigg|_{s=0},$$
$$t_{ij} = \frac{g'_{ij}\Big|_{s=0} t'_{ij} + g''_{ij}\Big|_{s=0} t''_{ij}}{g'_{ij}\Big|_{s=0} + g''_{ij}\Big|_{s=0}}, \tag{81}$$

that is, to the sum of transfer functions corresponds the weighted mean of transfer times, the *weights* being the transfer functions themselves with s substituted by 0.

The third operation becomes

$$t_{ij} = -D \ln[g'_{ij}/(1 - g_{jj})]\Big|_{s=0},$$

where g'_{ij} is the transfer function from x_i to x_j in the absence of the cycle; calling t'_{ij} the corresponding transfer time,

$$t_{ij} = -D \ln g'_{ij}\Big|_{s=0} + D \ln(1 - g_{jj})\Big|_{s=0} = t'_{ij} - \frac{dg_{jj}/ds}{1 - g_{jj}}\Big|_{s=0},$$
$$t_{ij} = t'_{ij} + \frac{g_{jj}}{1 - g_{jj}}\Big|_{s=0} t_{jj}; \tag{82}$$

that is, to division of a given transfer function by $(1 - g_{jj})$ corresponds addition to the given transfer time of the transfer time corresponding to g_{jj} multiplied by the factor $[g_{jj}/(1 - g_{jj})]|_{s=0}$.

If the transfer time is all the information wanted from a flow graph, then each arm of the graph should indicate its transfer time and its weight. For the complete graph these are $1/K_j$ and k_{ij}/K_j, respectively, on the arm from $\{x_i\}$ to $\{x_j\}$. Then one performs the operations indicated in this section on the transfer times and the operations indicated in Section IV.C on the weights. When the graph is reduced to a single arm from $\{x_i\}$ to $\{x_j\}$, its transfer time is the quantity defined by Eq. (76) and its weight is $g_{ij}|_{s=0}$. The transfer function can be written

$$g_{ij} = \{x_j\}/\{x_i\} = s\left\{\int_0^t x_j(\tau)\,d\tau\right\}\bigg/s\left\{\int_0^t x_i(\tau)\,d\tau\right\}$$

and, using Eq. (77),

$$g_{ij}\bigg|_{s=0} = \int_0^\infty x_j(\tau)\,d\tau\bigg/\int_0^\infty x_i(\tau)\,d\tau. \tag{83}$$

When the two functions $x_i(t)$ and $x_j(t)$ represent the amount of a substance in a particular site as a function of time and the turnover time of x_j is not infinite, the ratio

$$\int_0^\infty K_j x_j(\tau)\,d\tau\bigg/\int_0^\infty K_i x_i(\tau)\,d\tau = r_{ij}$$

measures the fraction of the substance leaving x_i that will eventually leave x_j, that is the *yield* of the transport from x_i to x_j. Equation (83) then becomes

$$g_{ij}\bigg|_{s=0} = r_{ij}\,K_i/K_j.$$

The yield of a single arm $k_{ij}/(s + K_j)$ is

$$[k_{ij}/(s + K_j)]\bigg|_{s=0} K_j/K_i = k_{ij}/K_i,$$

the ratio of the transfer constant from x_i to x_j to the sum of the transfer constants from x_i to all other compartments. The yield between any two compartments is the weight as calculated above times the turnover time of the initial compartment divided by the turnover time of the terminal compartment.

VII. Noncompartmented Systems

In a system that cannot be described in terms of compartments, the constants k_{ij} and K_i as defined in Sections III.A and IV lose their original meaning.

In a pool (see Section III.B) the ratio

$$\int_0^\infty \tau x_i(\tau)\,d\tau\bigg/\int_0^\infty x_i(\tau)\,d\tau \tag{84}$$

is the average time at which the particles whose quantity is measured by the function $x_i(t)$ leave the pool. This average time, in the case of a compartment where the particles enter all at once at time $t = 0$ and where they are not recycled into it, is the turnover time of the compartment, that is, $1/K_i$.

We can now define a constant, $1/K_i$ of a pool, and call it the turnover time of that pool, as the ratio (84) computed when the particles whose quantity is measured by $x_i(t)$ enter the pool all at once at time $t = 0$ and they are not recycled into it.

The experimental circumstances required for these measurements might not always be realizable, but this definition is still valid and could lead to an indirect measurement of K_i, as shown in the following. The usefulness of this definition is limited by the constancy of the ratio (84) under different experimental conditions; in fact the average time might vary according to the site of entrance and to the total dose entering the pool (that is, the integral in the denominator).

If $x_i(t)$ and $x_j(t)$ measure the quantity of particles in two compartments, and if $\{x_i\}$ is an absolute or unique precursor of $\{x_j\}$ of order one (that is, all particles entering $\{x_j\}$ come from $\{x_i\}$), then

$$\int_0^\infty k_{ij} x_i(\tau)\, d\tau = \int_0^\infty K_j x_j(\tau)\, d\tau, \tag{85}$$

which means that all particles leaving $\{x_i\}$ for $\{x_j\}$ will at some time leave $\{x_j\}$. The integral on the right-hand side of Eq. (85) is the total amount of particles that can be collected in all compartments following $\{x_j\}$, and can be called Q_j; k_{ij} is constant, therefore

$$k_{ij} = Q_j \bigg/ \int_0^\infty x_i(\tau)\, d\tau. \tag{86}$$

Formula (86) can be used as a definition of the constant k_{ij} of two pools connected in such a way that the second is fed only by the first. This definition is useful under experimental conditions that guarantee the proportionality of $\int_0^\infty x_i(\tau)\, d\tau$ and Q_j, even if this ratio cannot be measured directly.

Now that the two fundamental constants k_{ij} and K_i have been defined for the pools of a system, its flow graph can be constructed in the same way as described in Section VI.B, under the single condition that the system be linear. In such a flow graph the nodes represent an extensive property of the pools, but the arms do not have the same meaning as in Section IV.C, because Eqs. (26) do not hold. Each arm instead is given two values, namely, its transfer time and its weight. In the complete graph these two constants are $1/K_j$ and k_{ij}/K_j, respectively, for the arm connecting pool $\{x_i\}$ to pool $\{x_j\}$.

We shall prove now that the three operations described in Section VI.B for a compartment graph apply to this graph too. If two pools $\{x_i\}$ and $\{x_j\}$

are connected by an elementary path, the transfer time between them is the sum of the transfer times between all pools of the given path. Writing equation (85) for each arm of the path, then eliminating all functions except $\{x_i\}$ and $\{x_j\}$, we obtain

$$\int_0^\infty k_{ii_1} k_{i_1 i_2} \cdots k_{i_n j} x_i(\tau) \, d\tau = \int_0^\infty K_{i_1} K_{i_2} \cdots K_j x_j(\tau) \, d\tau.$$

Therefore, this elementary path can be substituted by a single arm whose transfer time is the sum of the transfer times and whose weight is the product of the weights, of the single arm forming it.

If two pools $\{x_i\}$ and $\{x_j\}$ are connected by a number of elementary paths, the transfer time between them is the weighted mean of the transfer times along the different paths, the weights being proportional to the fraction of material flowing through each path; from definition (86) it follows that the fraction of material flowing through each path is k_{ij}/K_j, that is, the weight of the respective path. Also the fraction of material flowing through the combined paths is the sum of the weights of the paths. Therefore a number of arms connecting the same pools can be substituted by a single arm whose transfer time is the weighted mean of the transfer times, and whose weight is the sum of the weights, of the single arms.

When a closed arm starts and ends in pool $\{x_j\}$, to the transfer time $1/K_j$ represented by the open arm from $\{x_i\}$ to $\{x_j\}$ we should add the *recycling time*, that is, a term to account for the time the material spends for returning in $\{x_j\}$ when it is recycled one, two, ... times. If the time for one such cycle is $1/K_c$, the recycling time is $1/K_c$ multiplied by the factor

$$r_c + r_c^2 + r_c^3 + \cdots = r_c/(1 - r_c),$$

where r_c is the ratio between the rate of arrival in $\{x_j\}$ from the cycle and the rate of disappearance from $\{x_j\}$, that is, the weight of the cycle. Also the total amount of material flowing through the open and closed paths combined is equal to the amount flowing to the open path plus the amount flowing through the closed path multiplied by the factor

$$1 + r_c + r_c^2 + \cdots = 1/(1 - r_c),$$

to account for the repeated return of material to $\{x_j\}$. Therefore the cycle in $\{x_j\}$ can be eliminated by adding to the transfer time of the open arm entering $\{x_j\}$ the term $(1/K_c)r_c/(1 - r_c)$, and by multiplying the weight of that open arm by the factor $1/(1 - r_c)$.

All these transformations are identical with those defined in Section VI.B for compartments. Therefore to a system of pools can be associated a graph whose nodes represent an extensive property of the pools, and whose arms are given the values $k_{ij}/(s + K_j)$, where s is not the differential operator, but a letter used for convenience. All operations on graphs described in Section IV.C are applicable to this graph.

VIII. Dimensional Analysis of the System Equations

The mathematical representation of systems has been cast in terms of unspecified variables and their variations in time. The system equations themselves may be taken to express relations among variables and constants at points in time (or more abstractly at points in the domain of the independent variable t). Consideration of what physical realities these variables and parameters may represent leads to the necessity of a dimensional analysis. If, for example, the physical entities (such as concentrations, amounts, turnover times) represented by the mathematical entities (compartments, pools, coefficients) are assigned values which are multiples of some unit (such as moles per liter, gram, second) then the values are clearly dependent upon the units chosen. If this dependence is homogeneous—that is, a single constant multiplies all terms in both members of the equation to convert from one set of units to another—then any consistent choice of units does not disturb the relations implicit in the equations. However, this need not be the case.

The dimensions of the variables and parameters is determined by the nature of the system the equations are intended to describe and by choice of the one who writes the equations. This choice is usually a matter of past experience and cognizance of what measurements are to be made when testing the model. Generally, in using equations to describe physical processes, variables and parameters are chosen such that the equations are *dimensionally homogeneous;* all the terms have the same dimensions (are dimensional products of the same powers of the fundamental quantities). Dimensional homogeneity does not necessarily render the magnitude of a member of an equation independent of the units chosen; it does, however, assure that consistent changes in the units will result in equally consistent representations of the process which the equations describe. The system equations in differential form (22) used in this chapter have two types of terms: (a) derivatives and (b) variables multiplied by constant coefficients. Dimensional homogeneity, then, requires that all terms have the dimensions of the derivative, which will be the product of the first power of the dimensions of the dependent variable x_i with the inverse power of the dimensions of the independent variable t. Indicating dimensionality by [ ],

$$[dx_i/dt] = [x_i]/[t] = [x_i][t]^{-1},$$

where the square brackets [ ] are read "the dimensions of" followed by the name of the entity enclosed. The other terms will be products whose dimensions must also be $[x_i][t]^{-1}$. That is,

$$[k_{ji}][x_j] = [x_i][t]^{-1}.$$

Clearly, if all the variables x_k have the same dimensions, then

$$[k_{ji}] = [t]^{-1} \qquad \text{for all} \quad i, j.$$

In this special case

$$[K_i] = [\textstyle\sum k_{ij}] = [t]^{-1}.$$

The integral form of the system equations is subject to the same sort of analysis.

The possibility of any choices for $[x_j]$, subject only to the restriction that $[x_j][k_{ji}]$ must be the same for all terms and the same as $[dx_i/dt]$, extends even to use of both extensive and intensive variables in the same equation. For example, the intensive variable concentration represented by x_m, say, and the extensive variable mass represented by x_n, say, may both appear in the right-hand member of Eq. (22), as long as

$$[k_{mj}x_m] = [k_{nj}x_n], \qquad j = 1, 2, \ldots.$$

Such a condition might be satisfied if, for example,

$$[x_m] = [\text{mole} \cdot \text{liter}^{-1}]; \qquad [x_n] = [\text{gram}];$$
$$[k_{mj}] = [\text{mole}^{-1} \cdot \text{liter} \cdot \text{gram} \cdot \text{second}^{-1}];$$
$$[k_{nj}] = [\text{second}^{-1}]; \qquad j = 1, 2, \ldots.$$

The physical meaning of such parameters k_{ij} may seem obscure compared to that of fractional turnover rates multiplying dimensionless numbers such as fractions or probabilities. But the meaning may be clarified by considering these impure parameters as products of fractional turnover rates and particular conversion factors. One might have

$$k_{mj} = k'_{mj}k''_m k'''_m,$$

where $[k'_{mj}] = [\text{second}^{-1}]$; $[k''_m] = [\text{liter}]$; $[k'''_m] = [\text{gram} \cdot \text{mole}^{-1}]$. Then it is comforting to consider k''_m to be simply the volume of the compartment, k'''_m to be some conversion factor, in this case the molecular weight of the mother substance (tracee) in the mth compartment of a tracer system, thus making the product $x_m k''_m k'''_m$ dimensionally homogeneous with x_n, and k'_{mj} with k_{nj}. In this way one may see that the generality of the system equations can be preserved while the dimensions of the factors in the individual terms may be adjusted for convenience.

However, one must recognize that in solving the equations, certain manipulations require certain dimensional restrictions. In particular, when one writes

$$\sum_{j \neq i} k_{ij} = K_i,$$

it is necessary not only that

$$[k_{ij}] = [k_{il}] \qquad \text{for all} \quad i, l$$

but also that

$$[k_{ij}] = [x_i]^{-1}[x_j][t]^{-1}.$$

This is so because Eq. (22) requires that

$$[dx_j/dt] = [k_{ij}][x_i].$$

As a formalism for following the dimensionality of the terms and equations in operational form, one may consider the brackets { } and the differential operator s to have dimensions defined in terms of the independent variable t. If t is time, then we may write

$$[t] = T,$$

where T simply represents time as a dimension, unit unspecified. Then the dimensional analysis of the definition of product

$$\{f(t)\}\{g(t)\} = \left\{ \int_0^t f(t - \tau)g(\tau)\, d\tau \right\}$$

goes as follows (where we mean by "[[ ]]" the "dimension generated by the operation of enclosure in { }"):

$$[[f(t)\}\{g(t)\}]] = [[\{[f(t)][g(t)][t]\}]],$$
$$[[\ ]][f(t)][[\ ]][g(t)] = [[\ ]][f(t)][g(t)][t],$$
$$[[\ ]] = [t] = T,$$

regardless of the dimensions of f and g.

Now the dimension of s may be determined:

$$s\{f(t)\} = \{f'(t)\} + f(0),$$
$$[s][[\ ]][f(t)] = [[\ ]][f'(t)] = [f(0)],$$
$$[s][[\ ]][f(t)] = [f(t)] = [f(0)];$$

this requires that the dimension of s be the inverse of the dimension of { }, that is, the inverse of the dimension of the independent variable, in this case T^{-1}. This is consistent with defining s as the inverse of the function {1} where the 1 is a pure number.

IX. The Use and Abuse of Compartment Analysis

Coming full circle, we return now to the paper of Behnke *et al.* [1935] with a consideration of what criteria exist for choosing and evaluating the models by which one may describe the system under investigation. The authors of that paper represented their data with Eq. (1):

$$Y/A = 1 - e^{-kt},$$

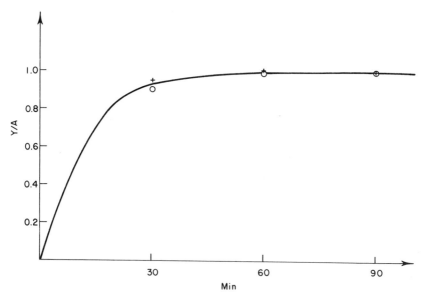

Fig. 20. The data obtained by Behnke *et al.* [1935] (see Fig. 1) can be approximated with two different equations, whose values are plotted at intervals of 30 min: (+) $Y/A = 1 - e^{-0.10t}$; (○) $Y/A = 1 - 2e^{-0.10t} + e^{-0.20t}$.

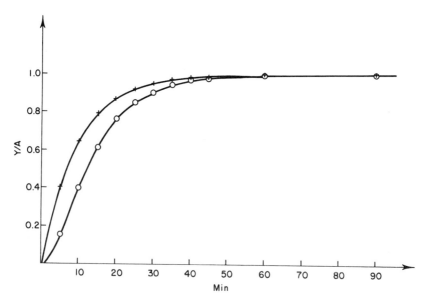

Fig. 21. The two equations given in the previous figure are plotted at intervals of 5 min.

which is consistent with a two-compartment model with one-way transport, Y being the amount accumulated in the closed (see Section III.C) compartment where $1/k$ is the turnover time and A is the total amount of material in the system, all in the open compartment at $t = 0$. In Fig. 20 we show the points plotted for two equations at intervals of 30 min beginning at $t = 0$. The second equation represents a model in which another compartment with a slower turnover time appears. These compartments may be representative of two classes of tissue with respect to nitrogen washout. The greatest measured difference between Y_1/A and Y_2/A is 5 % of the observed values. Without being overly conservative one might regard such a difference as insignificant for this kind of experiment. Thus there is no evidence for the added compartment; that is, the two models are equally consistent with the behavior of the system in this experiment.

Figure 21 displays data points computed from the same equations at intervals of 5 min. The largest difference between measured values occurs at $t = 5$ min and the differences at the first four or five sampling times are likely to be detectable. Thus the same two models now are made distinguishable by earlier sampling.

Given the situation described above, let us suppose that the actual data were close to the values computed from the equation for Y_2 and inconsistent with those computed from the equation for Y_1. By "close" we mean only that the data are not significantly different from values computed at corresponding times, given the precision of the measurements. The experimenter, then, is obliged to discard the model which generates Y_1. But this does not make the other model true. It can only be said, from this experiment, that it is consistent with the data. However, there are an infinity of compartment models which are consistent with the same data to the same or better closeness— but none with fewer than three compartments (provided k's have been optimally chosen).

A good discussion of the utilization of data in the formulation of compartment models was given by Berman [1963a]. Section V of this chapter gives a method of getting a relational parameter, precursor order, from data subject to the usual restrictions. Berman [1963b] has also proposed a method of determining parameters of a model from sets of data produced by the same experiment on a system in unperturbed and perturbed states.

Beyond these limitations, consistency depends upon choices more fundamental than details such as number of compartments or values of parameters, for there are classes of functions other than that appropriate to describe a compartment model which are consistent with the same data. And, in fact, there are an infinity of sets of parameters within each of these classes which are consistent with the data. Indeed, any continuous function can be approximated to an arbitrary closeness (quantitatively defined in some way, such as sum of squares of differences) by a sum of exponentials, by a sum of sine and cosine terms, by a polynomial, or by any number of other functions. Thus, not only is consistency arbitrarily defined and dependent upon precision of measurement and times of sampling, but the criterion of consistency itself, often referred to as "fitting the data," is not sufficient, is in fact often weak, in choosing models.

There are, of course, other criteria outside the immediate context of a particular experiment. The choice of classical mechanics as a model for a game of billiards seems appropriate before the game begins. Hertz [1900] placed three requirements upon acceptable theories. They must be (a) logically permissible, (b) correct, and (c) appropriate. Correctness is, in terms of models, what we have spoken of as consistency with observation. This includes two sorts of requirement upon a model: (a) it must, to a reasonable accuracy, reflect the observations from which it was formulated or by which it was selected, and (b) it must predict system responses to new situations to a reasonable accuracy. For a sufficiently limited system a set of tables might fulfill these requirements quite well. And this might be a satisfactory model for applications. However, it is not likely to represent any dynamic relations or mechanisms, or to contribute to any conceptual understanding or to promote any new insights, or to suggest the design of new experiments. These sorts of contributions of a model fit, in part at least, into Hertz' third attribute, appropriateness. It is here that models have their scientific, as opposed to, say, technological, uses. It is here that the insights and instincts of the scientist visualizing mechanisms and unquantified dependencies in the realm of what Nooney [1965] calls imagery come into play in formulating and using models. This is where discovery occurs.

Using the paper of Behnke *et al.* [1935], we have seen how judgements about consistency are dependent upon precision and accuracy of measurement and sampling times. But what of the realm of appropriateness and imagery and how might these considerations produce results? The choice of a compartment model in the case of inert gas exchange in humans could well be motivated by analogy with inanimate systems wherein such stochastic transport seems conceptually satisfying as well as reliably predictive. The choice of two compartments (one within the body) within this class of models may result from the conjecture that the relaxation from a nonequilibrium state among tissue types occurs over very much shorter times than the relaxation from nonequilibrium between body as a whole and atmosphere. But then observation, as we have seen, may reveal inconsistency; perhaps a single compartment added to the model restores consistency and the experimenters know something new, in the context of their class of model. Similar new knowledge might appear with respect to a turnover time. One might know from other experiments that interstitial pulmonary fibrosis does not affect the transport of nitrogen from blood to tissues. (We claim no knowledge here; this is a hypothetical example.) Then using data from a diseased patient and the three-compartment (two within the body) model, one might find the turnover time from atmosphere to blood to be half the normal value. This result, even though its meaning derives from a compartment model, may suggest that the experimenter make histological observations of the alveolar wall or measure lung compliance, which observations may in turn increase understanding of the diseased and normal, states, independent of the compartment model.

The careful use of models holds as much promise as the careless use holds danger. The difference between realizing the promise and encountering the danger lies in genuine acknowledgement of limitations and productive interplay of knowledge and imagination, of discipline and adventure, of seriousness and humor.

General References

The number of papers about compartment theory or its applications is very large. We do not intend to present here a complete bibliography, which would be useless if not accompanied by summaries and detailed indexes. We list only those papers and books that were mentioned explicitly in this chapter, either as sources of examples or as suggested readings.

Artom, C., Sarzana G., and Segre, G. [1938]. Influence des grasses alimentaires sur la formation des phospholipides dans les tissues animaux (nouvelles recherches), *Arch. Internat. Physiol.* **47**, 245–276.

Aubert, J.-P., and Bronner F. [1965]. A symbolic model for the regulation by bone metabolism of the blood calcium level in rats, *Biophys. J.* **5**, 349–358.

Beck, J. S., and Rescigno, A. [1964]. Determination of precursor order and particular weighting functions from kinetic data, *J. Theor. Biol.* **6**, 1–12.

Behnke, A. R., Thomson, R. M., and Shaw, L. A. [1935]. The rate of elimination of dissolved nitrogen in man in relation to the fat and water content of the body, *Amer. J. Physiol.* **114**, 137–146.

Bergner, P.-E. E. [1961a]. Tracer dynamics: I. A tentative approach and definition of fundamental concepts, *J. Theor. Biol.* **1**, 120–140.

Bergner, P.-E. E. [1961b]. Tracer dynamics: II. The limiting properties of the tracer system, *J. Theor. Biol.* **1**, 359–381.

Bergner, P.-E. E. [1962]. The significance of certain tracer kinetical methods, especially with respect to the tracer dynamic definition of metabolic turnover, *Acta Radiol. Suppl.* 210.

Berman, M. [1963a]. The formulation and testind of models, *Ann. N.Y. Acad. Sci.* **108**, 182–194.

Berman, M. [1963b]. A postulate to aid in model building, *J. Theor. Biol.* **4**, 229–236.

Berman, M. [1965]. Compartmental analysis in kinetics, *in* "Computers in Biomedical

Research" (R. W. Stacy and B. D. Waxman, eds.), Vol. II, pp. 173–201. Academic Press, New York.

Berman, M., and Schoenfeld, R., Jr. [1956]. Invariants in experimental data on linear kinetics and the formulation of models, *J. Appl. Phys.* 27, 1361–1370.

Berman, M., Weiss, M. F., and Shahn, E. [1962]. Some formal approaches to the analysis of kinetic data in terms of linear compartmental systems, *Biophys. J.* 2, 289–316.

Brownell, G. L., Berman, M., and Robertson, J. S. [1968]. Nomenclature for tracer kinetics, *Internat. J. Appl. Radiation Isotopes* 19, 249–262.

Erdélyi, A. [1962] "Operational Calculus and Generalized Functions." Holt, New York.

Gómez, D. M., Briscoe, W. A., and Comming, G. [1964]. Continuous distribution of specific tidal volume throughout the lung, *J. Appl. Physiol.* 19, 683–692.

Hearon, J. Z. [1953]. The kinetics of linear systems with special reference to periodic reactions, *Bull. Math. Biophys.* 15, 121–141.

Hearon, J. Z. [1963]. Theorems on linear systems, *Ann. N. Y. Acad. Sci.* 108, 36–68.

Hertz, H. [1900]. "Principles of Mechanics," pp. 1–2. Reprinted by Dover, New York.

Kleiber, M. [1955]. Meaning of "turnover" in biochemistry, *Nature* 175, 342.

Ličko, V. [1965]. On compartmentalization, *Bull. Math. Biophys.* 27 (Special Issue), 15–19.

Matthews, C. M. E. [1957]. The theory of tracer experiments with ^{131}I-labelled plasma proteins, *Phys. Med. Biol.* 2, 36–53.

Mawson, C. A. [1955]. Meaning of "turnover" in biochemistry, *Nature* 176, 317.

Mikusiński, J. [1959]. "Operational Calculus." Pergamon, Oxford.

Nooney, G. C. [1965]. Mathematical models, reality and results, *J. Theor. Biol.* 9, 239–252.

Rescigno, A. [1956]. A contribution to the theory of tracer methods, Part II; *Biochim. Biophys. Acta* 21, 111–116.

Rescigno, A., and Segre, G. [1961a]. Calcolo della funzione di trasferimento polmonare da dati radiocardiografici, *Minerva Nucl.* 5, 296–298.

Rescigno, A., and Segre, G. [1961b]. The precursor–product relationship, *J. Theor. Biol.* 1, 498–513.

Rescigno, A., and Segre, G. [1964]. On some topological properties of the systems of compartments, *Bull. Math. Biophys.* 26, 31–38.

Rescigno, A., and Segre, G. [1965]. On some metric properties of the systems of compartments, *Bull. Math. Biophys.* 27, 315–323.

Rescigno, A., and Segre, G. [1966]. "Drug and Tracer Kinetics." Ginn (Blaisdell), Boston, Massachusetts.

Segre, G., Turco, G. L. and Ghemi, F. [1965]. Determination of the pulmonary weighting function, of the mean pulmonary transit time and of the pulmonary blood volume in man by means of radiocardiograms, *Cardiologia* 46, 295–311.

Shemin, D., and Rittenberg, D. [1946]. The biological utilization of glycine for the synthesis of the protoporphyrin of hemoglobin, *J. Biol. Chem.* 166, 621–625.

Teorell, T. [1937a]. Kinetics of distribution of substances administered to the body. I. The extravascular modes of administration, *Arch. Internat. Pharmacodynamie Thérapie* 57, 205–225.

Teorell, T. [1937b]. Kinetics of distribution of substances administered to the body. II. The intravascular modes of administration, *Arch. Internat. Pharmacodynamie Thérapie* 57, 226–240.

Zilversmit, D. B. [1955]. Meaning of turnover in biochemistry, *Nature* 175, 863.

Zilversmit, D. B. [1960]. The design and analysis of isotopes experiments, *Amer. J. Med.* 29, 832–848.

AUTHOR INDEX

Numbers in italics refer to the pages on which the complete references are listed.

A

Anfinsen, C. B., Jr., 6, *76*
Apter, M. J., 139, *139,* 178, 203, *212*
Arbib, M. A., 168, 169, 170, 174, 180, 188, 193, 194, *212, 213,* 223, 247, *252*
Artom, C., 261, 268, 272, *321*
Aubert, J.-P., 270, *321*

B

Balzer, R., 160, *213*
Bar-Hillel, Y., 185, *213*
Barzdin, Y. M., 174, *213*
Beck, J. S., 301, *321*
Behnke, A. R., 255, 256, 269, 317, 318, 320, *321*
Bensam, A., 137, *140*
Bergner, P.-E. E., 268, 269, *321*
Berman, M., 269, 299, 319, *321, 322*
Berrill, N. J., 30, *76*
Blum, M., 180, *213*
Bonner, J. T., 70, *76*

Boyer, P. D., 138, *139*
Briscoe, W. A., 269, *322*
Bronner, F., 270, *321*
Brownell, G. L., 269, 307, *322*
Bullough, S., 155, *213*
Burton, A. C., 138, *139*

C

Cairns-Smith, A. G., 205, *213*
Campbell, R. D., 55, *76*
Caspar, D. L. D., 6, 7, 15, 22, *76*
Chance, B., 117, 137, *139*
Changeaux, J. P., 15, *76,* 138, *140*
Child, C. M., 75, *76*
Chomsky, N., 182, *213*
Codd, E. F., 193, 200, 204, *213*
Coffin, R. W., 179, *214*
Cohen, M. H., 152, 155, *213*
Comming, G., 269, *322*
Courant, R., 225, *252*
Crane, H. R., 8, 9, *76*
Crick, F. H. C., 26, *76,* 206, 208, *213*

323

SUBJECT INDEX

74041